The Globalization of Energy

International Comparative Social Studies

VOLUME 21

The Globalization of Energy

China and the European Union

Edited by

M. Parvizi Amineh and Yang Guang

BRILL

LEIDEN • BOSTON
2010

Cover illustration design: © Sandra van Merode

This book is printed on acid-free paper.

Library of Congress Cataloging-in-Publication Data

The globalization of energy : China and the European Union / edited by Mehdi Parvizi Amineh and Yang Guang.
p. cm. — (International comparative social studies ; 21)
Includes bibliographical references and index.
ISBN 978-90-04-18112-0 (hardback : alk. paper)
1. Power resources—China. 2. Power resources—European Union countries. 3. Globalization—China. 4. Globalization—European Union countries. I. Amineh, Mehdi Parvizi. II. Guang, Yang. III. Title: Globalisation of energy. IV. Series.

HD9502.C62G56 2010
333.79—dc22

2009043279

ISSN 1568-4474
ISBN 978 90 04 18112 0

PRINTED IN THE NETHERLANDS

CONTENTS

PART TWO

RENEWABLE ENERGY & SUSTAINABLE DEVELOPMENT

ACKNOWLEDGMENTS

This volume is the result of a joint research program between the Energy Program Asia of the International Institute for Asian Studies, Leiden, the Netherlands, and the Institute of West-Asian and African Studies of the Chinese Academy of Social Sciences (CASS), Beijing, China, in cooperation with the Institute of Industrial Economy, CASS.

The successful completion of the first phase of this research project depended on the cooperation and input of researchers with diverse disciplinary and intellectual backgrounds. Therefore, first and foremost, we thank them as contributors to this volume.

This project has received financial support from both the Dutch Royal Academy of Social Sciences (KNAW) and the Chinese Academy of Social Sciences. We would like to thank Ms. Liu Yingxiang (International Cooperation Bureau-CASS) and Mr. A. J. de Wit and Marise Bantjes (China Exchange Program-KNAW) for their organizational efforts.

We owe special thanks to the Director of IIAS, Prof. Max Sparrenboom, who from the beginning was an enthusiastic supporter of the joint program between IIAS and IWAAS.

We wish to thank the following IIAS staff members for otherwise assisting in the finalization of this project: Manon Osseweijer, Heleen van der Minne, Anna Yeadell, Berry Bock, Manuel Haneveld, Thomas Voorter, and Amparo de Vogel.

Last but not least we are thankful to Kurt Radtke, Patrizia Arena, Sander de Rijke and Pieter Smit for their cheerful assistance and useful advisement.

Yang Guang, Beijing
M. Parvizi Amineh, Amsterdam

LIST OF CONTRIBUTORS

M. Parvizi Amineh Ph.D. (1998, Political Sciences, University of Amsterdam), is Senior Research Fellow at the International Institute for Asian Studies (IIAS), University of Leiden, The Netherlands and Programme Director of the Energy Programme Asia (EPA) at the same institute. He is also Adjunct Professor of International Relations at Webster University, Leiden, Senior Lecturer at the International School for Humanities and Social Sciences (ISHSS), and Affiliated Fellow at the Amsterdam School for Social Science Research (ASSR), University of Amsterdam. His publications include among others: *The Greater Middle East in Global Politics: Social Science Perspective on the Changing Geography of the World Politics* (2007), *Central Eurasia in Global Politics: Conflict, Security and Development* (2004/2005) with Henk Houweling, *Globalization, Geopolitics and Energy Security in Central Eurasia and the Caspian Region* (2003), *Towards the Control of Oil Resources in the Caspian Region* (1999/2000), and *Die Globale Kapitalistische Expansion und Iran: Eine Studie der Iranischen Politischen Ökonomie* (1999).

Robert M. Cutler is Senior Research Fellow at the Institute of European, Russian and Eurasian Studies, Carleton University, Canada. He was educated at MIT and the University of Michigan, where he earned a Ph.D. in political science, and has specialized and consulted in the international affairs of Europe, Russia, and Eurasia since the late 1970s. He has held research and teaching positions at major universities in the United States, Canada, France, Switzerland, and Russia, and is a prolific contributor in three languages to leading academic journals and policy reviews, as well as the print and electronic mass media.

Chen Mo was born on July 28, 1962 in China. She graduated from Beijing Normal University in 1985. She then joined the Middle East Studies Program at the Institute of West-Asian and African Studies, Chinese Academy of Social Sciences (IWAAS CASS) as a Research Apprentice. From 1990 to 2000 she was Assistant Research Fellow and from 2001 to 2009 Associate Professor in IWAAS CASS.

Her research interests include energy economy, economic development, and Middle Eastern and African economies. Some of her publications are *African Market Organization*, China Encyclopedia Publishing House, 1995, "Oil and Economy of Saudi Arabia" in *The New theory on the Gulf Oil*, 2001, China Social Science Documentation Publishing House, "The EU's new Mediterranean policy and its impact on the South-Mediterranean countries" in *West-Asia and Africa*, No. 11, 2007, and "The origin and future of Mediterranean Alliance" in *West-Asia and Africa*, No. 10, 2008.

Eva Patricia Rakel is lecturer in International Relations at the University of Amsterdam and Affiliated Fellow at the International Institute for Asian Studies, the Netherlands. Her research interests include history, politics, economy, and foreign policy of the Middle East and Central Eurasia, foreign policy of the European Union, European Union, Middle East and Asia relations, and elite studies. Her most recent publications are: "The Political Elite in the Islamic Republic of Iran: From Khomeini to Ahmadinejad", *Comparative Studies of South Asia, Africa, and the Middle East*, 29(1), 105–125 (forthcoming 2009), *Power, Islam, and Political Elite in Iran—A Study on the Iranian Political Elite from Khomeini to Ahmadinejad*, International Comparative Social Studies, vol. 18, (Leiden, Boston, and London: Brill Academic Publishers, 2009).

Daniel Scholten (MSc Political Science and International Relations) is a Ph.D. candidate in the Economics of Infrastructures Department at Delft University of Technology, the Netherlands. His research focuses on the coevolution of institutions and technology in energy transitions. More specifically, he asks how governments can align institutions (as modes of industry organization) to the technical changes inherent in a transition to hydrogen. He is also assistant editor of the international journal *Competition and Regulation in Network Industries* (CRNI). Scholten is also an Affiliated (Research) Fellow of the Energy Programme Asia (EPA) at the International Institute for Asian Studies (IIAS). His current efforts involve contributing a chapter to the upcoming book entitled *Domestic and Geopolitical Challenges to Energy Security for China and the European Union*. He also gives guest lectures at the EPA lecture series. Finally, he is the lecturer of the third year BA course "International Political Economy" at the Webster University in Leiden.

S. Philip Sen is formerly a journalist and reported on military operations and technology for Jane's Information Group and *The Engineer* magazine. Between 2003 and 2005 he lived and worked in Asia, particularly in India and China, where he taught in the Joint Undergraduate Program at Shanghai University of Finance and Economics. Educated at the Universities of Oxford and Cardiff in the UK, Philip Sen took his Masters in International Relations at the University of Amsterdam. He now works in the communications unit of United Nations Volunteers, a UNDP agency based in Bonn, Germany.

Raquel Shaoul received her Ph.D. from London University in 2002 and has since been teaching in the East Asian Studies Department at Tel-Aviv University. She is a Research Fellow at the Energy Programme Asia (EPA) at the International Institute for Asian Studies (IIAS), and at the Center for Iranian Studies (CIS), Tel-Aviv University. Dr. Shaoul's fields of interest are Japanese foreign policy, defense policy, and energy policies with a focus on Japan's energy supply security from the Persian Gulf and its impact on the Japanese-Chinese relationship and defense policy issues.

Frank Umbach, born in 1963, is head of the "Security Policies in Asia-Pacific" and "International Energy Security" programs at the Research Institute of the German Council on Foreign Relations (DGAP). He is also Co-Chair of the European Committee of the Council for Security Cooperation in Asia-Pacific (ESCSCAP or CSCAP-Europe), the most important regional 'track two-diplomacy' institution for security policies in the Asia-Pacific. He worked as a research fellow from 1991 to 1994 at the Federal Institute for East European and International Affairs (BIOst) in Cologne, and from 1995 to 1996 at the Research Institute of the Japanese Foreign Ministry (Japan Institute for International Affairs/JIIA) in Tokyo. In 1992 he worked for one year as a Research Assistant in the Office of the "Special Advisor for Central and East European Affairs", office of the General Secretary of NATO in Brussels and was then a consultant there until the summer of 2003 and a presenter at official NATO conferences (such as for NATO ambassadors). Dr. Umbach also conducted extensive research in US (Washington, Fort Leavenworth, and Santa Monica) and Moscow over several stays in the 1990s. He has received research grants from NATO, the Kennan Institute for Advanced Russian Studies (Washington), and the Japan Foundation. He studied political science, East European history, and international law at the universities of Marburg and Bonn, and received his Ph.D. in 1996 (Bonn). He is a consultant of the German Foreign and Defense Ministries, and has published reports on behalf of the European Commission in Brussels, the European Parliament in Strasbourg, the House of Lords in London, and NATO in Brussels. He is also, inter alia, a member of the International Institute for Strategic Studies (IISS), a lecturer for the next generation of diplomats in Germany's Foreign Ministry, and author of some 200 publications in more than 20 countries, including the books: *Cooperation or Conflict in Asia-Pacific? China's Tying into Regional Security Structures and the Implications for Europe* (Munich: Oldenbourg-Verlag, 2002, 395 pp. in German); and *Global Energy Security: Strategic Challenges for the European and German Foreign Policy* (Munich: Oldenbourg-Verlag, 2003, in German).

Eduard B. Vermeer holds doctoral degrees in Sinology and in History (Leiden University) and wrote his Ph.D. on Water Conservancy and Irrigation in China He was professor of Contemporary Chinese History at the University of Turku and President of the Graduate School of Asian Studies in Finland, and is co-editor of the quarterly journal China Information. He is author of a dozen books on China, *Ooggetuigen van de Chinese Geschiedenis* (*Eyewitnesses of Chinese History*), Bert Bakker 2009. Eduard Vermeer's research focuses on China's contemporary economy and society, environmental and energy issues, and cooperative organizations. He has conducted large-scale socioeconomic surveys in China, and participated in cooperative projects with administrative, commercial, educational and research organizations at national and local levels. He has been mission leader or member in several EU programs in China, including its 20-Cities Dairy Development Programme, the EU-China Centre for Agrotechnical Exchange, the EU-China Management School, and the rural governance project. (with MoCA).

Shi Dan is Senior Research Fellow and Director at the Energy Economics Research Center of the Institute of Industrial Economics at the Chinese Academy of Social Sciences (CASS). Shi Dan received her Ph.D. degree from Huazhong Science and Technology University and masters degrees in economics separately from Australian National University and Renmin University of China. Her bachelors degree is in engineering from Changchun Industrial Engineering University.

Her research field is the energy economy, policy, price, security and market reform in China in the last decade. Shi has been granted 13 research awards from CASS and the Development Research Center of State Council and Ministry of National Science and Technology for her outstanding work. Shi has published more than 80 research reports and papers including "The Chinese Energy Industry Marketization Reform Research Report", Economic Management Publish House, 2006, "Regional Differences in China's Energy Efficiency and Conservation Potentials", *China & World Economy*, Jan.–Feb., 2007.

Yang Guang studied at the Beijing Foreign Language Institute and the Graduate School of the Chinese Academy of Social Sciences (CASS). He has also been an intern at the Institute of Political Studies of Paris, France and a visiting scholar at the University of Wisconsin in the US. He has conducted research on the Middle East and Africa as well as on international energy security over the past 30 years, and currently serves as Director-General of the Institute of West-Asian and African Studies (IWAAS) of CASS, President of the Chinese Associations of Middle East Studies, Executive President of the Chinese Society of African Studies, and editor-in-chief of the academic journal *West Asia and Africa*. He is the co-author of a number of books, including *The Middle East Markets* (1993), *Organizations of Petroleum Exporting Countries* (1995), *Developing Economies' Challenges towards the 21st Century* (1999), and *Annual Reports on Development in the Middle East and Africa* (1997–2009).

MAPS, TABLES, AND FIGURES

Maps

Tables

Figures

LIST OF ABBREVIATIONS

ABWR	Advanced Boiling Water Reactors
ACD	Asian Cooperation Dialogue
AEC	Atomic Energy Commission
AIOC	Anglo-Iranian Oil Company
ANRE	Agency for Natural Resources and Energy
APEC	Asia-Pacific Economic Cooperation
APOC	Anglo-Persian Oil Company
APWR	Advanced Pressurized Water Reactor
ART	Advanced Thermal Reactor
ASEAN	Association of Southeast Asian Nation
Bbl/d	barrels per day
Bcf/d	billion cubic feet pet day
Bcm/y	billion cubic meters per year
BNFL	British Nuclear Fuels
BP	British Petroleum
bpd	barrels per day
BTC	Baku-Tbilis-Ceyhan (pipeline)
CCS	Carbon Capture and Storage
CEA	Central Eurasia
CO2	Carbon dioxide
CGNPC	China Guangdong Nuclear Power Company
CNNC	China National Nuclear Corp
CPC	Caspian Pipeline Consortium
CUTE	Clean Urban Transport for Europe
DOE	U.S. Department of Energy
ECC	Energy Cooperation Center
ECN	Energy Centre of the Netherlands
ECS	Energy Charter Secretariat
EIHP	European Integrated Hydrogen Project
EOS	National Energy Research Strategy
EU	European Union
EZ	Dutch Ministry of Economic Affairs
EU	European Union
FBR	Fast Breeder Reactor
FC(V)	Fuel Cell (Vehicle)
FEPC	Federation of Electric Power Companies of Japan

Fin	Dutch Ministry of Finance
Puf	Fissile plutonium
GNEP	Global Nuclear Energy Partnership
GWe	Gigawatt electric
H_2	Hydrogen
HVDC	High Voltage Direct Current
IAEA	International Atomic Energy Agency
IEA	International Energy Agency
IEF	International Energy Forum
ILSA	Iran-Libya Sanctions Act
IPE	International Political Economy
IPI	Iran-Pakistan-India pipeline
IPR	Intellectual property rights
IR	International Relation
IRGC	Islamic Revolutionary Guards Corps
IRI	Islamic Republic of Iran
ISA	Iran Sanctions Act
JAEA	Japan Atomic Energy Agency
JAEC	Japan Atomic Energy Commission
kWh	kilowatt hours
LDP	Liberal Democratic Party
LWR	Light water reactors
LNG	Liquefied Natural Gas
LNV	Dutch Ministry of Agriculture, Nature and Food Quality
LPG	Liquefied Petroleum Gas
LTVE	Long-Term Energy Supply Strategy (the Netherlands 2000)
Mcf/d	million cubic feet per day
METI	Ministry of Economy, Trade and Industry
MEXT	Ministry of Education, Culture, Sports, Science and Technology
MHI	Mitsubishi Heavy Industries Ltd
MITI	Ministry of International Trade and Industry
MOFA	Ministry of Foreign Affairs
MOU	Memorandum of Understanding
MOX	Mixed-oxide fuel
Mtoe	Million tons of oil equivalent
MWe	Megawatt electric
NDRC	National Development and Reform Commission
NEDO	New Energy and Industrial Technology Development Organization

NGO	Non-Governmental Organizations
NIOC	National Iranian Oil Company
NMP4	Fourth National Environmental Policy Plan (the Netherlands 2000)
NOx	Nitrogen oxide
NPT	Nuclear Non-Proliferation Treaty
NPT	Non-Proliferation Treaty
NWO/ACTS	Netherlands Organisation for Scientific Research/ Advanced Chemical Technologies for Sustainability
OBP	Odessa-Brody Pipeline
OECD	Organisation for Economic Co-operation and Development
OS	Dutch Ministry of Development Cooperation (part of Foreign Affairs)
PEM	Proton Exchange Membrane (fuel cell type)
PIT	Project Implementation Transition management
PV	Photovoltaic
R&D	Research and Development
RE	Renewable Energy
SAUA	Sino-Australia Uranium Agreement
SCP	South Caucasus Pipeline
SCO	Shanghai Cooperation Organization
SMR	Steam Methane Reforming
SO^2	Sulfur dioxide
SO	Solid oxide (fuel cell type)
SOE	State of enterprise
TAR	Trans-Asia Railway Network
TCGP	Trans-Caspian Gas Pipeline
Tcm	thousand cubic meters
Tcf	trillion cubic feet
TEPCO	Tokyo Electric Power Co.
UAE	United Arab Emirates
UKR	Unique Chances Subsidies Scheme
UN	United Nations
US	United States
V&W	Dutch Ministry of Transport, Public Works and Water Management
VROM	Dutch Ministry of Housing, Spatial Planning and the Environment
WNA	World Nuclear Association

Base 802714AI (R00152) 3-01

Map 1 Peoples Republic of China.

Map 2 The European Union.

PART ONE

ENERGY & GEOPOLITICS

CHAPTER ONE

INTRODUCTION. TAPPING GLOBAL ENERGY STOCKS: ENERGY SECURITY CHALLENGES FOR THE EUROPEAN UNION AND CHINA

M. Parvizi Amineh and Yang Guang

1. Introduction

Energy security challenges top the policy agendas of the European Union (EU) and China today. Consequently, policy makers of both import-dependent polities continue to look for new responses. These include not only the diversification of source and origin to encourage supply security, but also support for substitutes of fossil energy and improving efficiency in energy use. In these endeavors, the urgency of geopolitical concerns seems to draw more attention than long-term plans of transitions to renewable energy systems. As a result, the perception that the EU and China might be competitors in the geopolitical arena for access to foreign markets seems to overshadow their common interest in developing renewable energy and sharing efficiency-improving technology. However, it is often neglected that the EU and China at the same time are well positioned to cooperate vis-à-vis producer countries and compete in the development of high-end renewable technologies. This volume[1] challenges such dominant perceptions and aims to provide the wider public with a more balanced account of EU—Chinese energy relations. To this end, this volume not only focuses on geopolitical realities that affect energy relations among both polities, but also on energy efficiency and the development of alternatives and renewables.

Global primary energy demand, according to the EIA International Energy Outlook 2008, is projected to increase by 50% between 2005 and 2030, although this is greatly dependent on the level of economic growth until then. Most of this growth is expected to come from newly

[1] The result of our joint research project will be published in several volumes.

industrializing Asia, mainly China and India, while the growth of demand in industrialized or Organization for Economic Cooperation and Development (OECD) countries, like the EU and US, will be more modest, given that these already have high levels of per capita use. As a result, it is predicted that by 2030 more than 25% of world energy demand will come from developing Asia, particularly China and India (18% in 2005). By contrast, the share of the US in world consumption is expected to diminish between 2005 and 2030 from 22% to 17%. According to EIA International Energy Outlook 2008 estimates, world oil consumption will rise from 84 million barrels per day in 2005 to 113 million in 2030. The EIA also anticipates a substantial growth in the global consumption of natural gas for the period 2005–2030 from 104 trillion cubic feet (tcf) in 2005 to 158 tcf in 2030. Gas import dependence will grow substantially in all major consumer markets, except in East and Southeast Asia, where import dependence is already very high. The import dependency of the EU and China will grow to about 70% of domestic consumption by 2030 (EU Green paper 2000 and EIA IEO 2008).

In the period to 2030, global oil and gas supplies are predicted to originate in fewer countries than today. This is due to the fact that proven oil and gas reserves are unevenly distributed in the world and that only a few countries are surplus producers. The total global oil stock as of January 1, 2008 was estimated at 1,332 billion barrels (bbl) proven oil reserves, around 70% of which is located in OPEC countries, and 30% of which in non-OPEC countries. Just five countries (Saudi Arabia, Iraq, the UAE, Kuwait, and Iran) hold about 55% of global proven oil reserves. So far, the Gulf has been a critical factor in meeting global demand, followed by the states of the former Soviet Union. It is expected that world oil supply will need to be 28 million barrels per day more in 2030 than in 2005. To meet this demand OPEC and non-OPEC countries combined are expected to produce 49 and 63 million barrels per day, respectively, according to the EIA IEO 2008. This results in a market share of OPEC of 44% in 2030 and continues a longstanding growth trend. Moreover, non-OPEC supplies are maturing, resulting in increased demands on OPEC oil in the long-term future.

Proven gas reserves are slightly less concentrated than oil reserves. Russia (including the Caspian Sea region) and the Middle East each represent about 1/3 and 2/5 of proven global reserves, respectively. Moreover, Russia and Iran hold about 42.5% of the global gas reserves as of 2008 (January 1, 2008, EIA International Energy Outlook 2008). The Middle East has substantial gas potential, but it remains largely untapped. This is due to the difficulty or cost at which these gas reserves can be developed and brought to market. Compared to the international oil market, the international gas market is still very much a regional one, divided into Asia's LNG market, the Russian-European market, and the North American market. Non-OECD Europe and Eurasia and the Middle East account for around 40% of global production in 2005, and are expected to account for 45% of the increase in production between 2005 and 2030. OECD countries will decline their share of global production from 39% to 27%. Hence, it is estimated that by 2030 supplies of gas for the world market will originate in fewer countries than today because some of the existing sources will dry up.

Global oil and gas markets look bleak as the result of ever-growing energy consumption, an increasing exhaustion of reserves, and an increasing geographical concentration of production. Against this background, it is likely that state and non-state actors will assign more significance to economic and resource concerns and energy relations will increasingly politicize. On the one hand, the growing energy imports of countries such as China and India adds to that of the European Union and the United States. In addition, the anticipation of future supply disturbances is reflected in generally rising oil and gas prices, especially their increasing volatility and the inelastic demand by major consumers. On the other hand, based on the location and increasing scarcity of world oil and gas reserves, a geographical concentration of energy supplies is expected to materialize in the politically unstable producer countries of the Gulf, Russia, and the Caspian Sea region. Moreover, internal conflicts may arise in countries where oil and gas are the main source of income, especially when accompanied by ethnic hostility, terrorism, religious fundamentalism, economic injustice, corruption, and political competition. Hence, both competition and cooperation for energy supplies among consumer countries and between consumer and producer countries are likely to intensify in the coming decades, the more so because policy responses of consumer countries cannot be understood in isolation of one another.

The global demand for oil and gas, rising political instability in many producer countries, and the approach of the "peak-oil" situation (2010–2020) are beginning to change the overall balance of power in the relationship between energy producer and consumer states in a way that strengthens the latter. This dilemma is further worsened by the renationalization wave experienced by the energy industries of producer states. Whereas in the early nineties energy sectors spearheaded market reforms, giving TNCs considerable power over energy resources, they are now slowly but surely transforming into government-controlled sectors.[2] With oil and gas companies already state-owned in most Middle Eastern OPEC states, recently nationalized in Russia and Venezuela, and subject to government control in China and India, trust in the market seems critically misplaced. Hence a reliance on the market and transnational energy corporations for providing adequate supplies no longer seems sufficient in a world in which oil and gas are both a source of energy and power, and bilateral contracts increasingly bypass the market.

To summarize, the combination of increasing oil and gas consumption, diminishing reserves, and geopolitical rivalry creates a setting for both the EU and China that can be characterized as one of demand-induced, supply-induced, and structural scarcity, or a combination thereof.[3] Demand-induced scarcity refers to a situation in which population growth, a rising per capita income resulting in higher levels of consumption, and technological change that renders fossil fuels more essential for the production of wealth and power, increase domestic demand for fossil fuels. Supply-induced scarcity refers to a situation in which a decrease of stock (or market-efficient access to it), inefficient use of supplies, and a lack of adequate productive capacity and pipeline infrastructure, decrease the supply of energy resources. Structural scarcity refers to a situation in which there is a supply-induced scarcity caused by the deliberate action of a major power or non-state actors, like transnational oil companies and producer cartels like OPEC. For example, in the current unipolar military order, the US can opt to induce scarcity for allies, competitors, and enemies alike by interdicting the maritime transport of oil and gas. That option, however, is available only after oil and gas have been brought to ports and ships from the territory of extraction. America, by extending the country's defense perimeter into the heartland of energy supply, is equipping itself with the capacity to induce structural scarcity for contenders by diverting flows on land.[4]

---

[2] The Economist 2005 'A survey of oil', vol. 375, no. 8424.
[3] Amineh and Houweling 2005: 80–81.
[4] Amineh and Houweling 2004/2005: Ch. 1.

Environmental constraints and advances in technology also affect energy security. According to the IEA World Energy Outlook, the world is facing a twin energy threat: that of energy supply security (at affordable prices) and that of environmental harm. Global CO2 emissions are expected to increase by 55% between 2004 and 2030 (1.7% annually). Emissions thus grow faster than energy demand. This is alarming, because it contrasts with the 25-year-long opposite trend towards cleaner energy sources, and testifies that our future energy use will be more 'dirty.' A likely cause for this is the switch back to coal that occurs in response to the oil and gas scarcity in many countries. Coal resources are more abundant and geographically less concentrated, though they have higher levels of carbon than oil and gas. This shift coincides not accidentally with the fact that developing countries will overtake OECD countries as the biggest emitters of CO2 shortly after 2010, and will account for more than 50% of global emissions by 2030. China alone is expected to represent 39% of the rise in emissions until then.

While pollution creates cross-border tensions, innovations in alternative and renewable resources, alongside efficiency measures, can reduce energy import dependence and contribute to reducing greenhouse gas emissions. However, deploying new technologies involves more than a simple replacement of oil and gas by other energy sources. It requires building new production facilities, new storage and distribution means, and even new end-use applications. Apart from time and money, such energy infrastructure transitions also necessitate continued government and popular support, especially when it is likely to be driven by social, political, and environmental benefits, and the technical and economic sides cannot yet compete with existing fossil fuels. Moreover, a transition to a more sustainable energy system differs essentially from past energy transitions.[5] First of all, the duration is estimated to be around fifty years instead of ten to twenty years. Secondly, whereas the current transition takes place in a liberalized market setting involving many actors, in the past, transition took place in a regulated setting with few actors, among which the government assumed the dominant position. Finally, the current transition has a very diverse set of technologies and complex solutions in mind, whereas past transitions had comparatively simple technical

[5] Kok 2004: 4.

goals. Consequently, the general public was more supportive about past transitions than the current one, which is often considered to be rather vague in terms of goals, policies, and results. Combined, these reasons urge for the development of new ways to govern such a transition.

It is expected that even by 2030 the role of renewables in the global energy mix will be marginal at best. The World Energy Outlook 2006 of the IEA states that while renewables and alternatives today cover 19% of global primary energy supply (if one includes nuclear and biomass; renewables alone account for 3%), by 2030 this will still be only 19% (and 4%). This is not due to a lack of development of renewables, but simply because global oil, gas, and coal consumption will also continue to rise. Nevertheless, as oil and gas become increasingly scarce, developing innovative technologies is the only long-term alternative.

International conflicts over the control of global oil and gas become more likely as consumption and imports rise, availability of oil and gas decreases, prices of these commodities increase, and environmental conditions deteriorate. These energy security challenges urge the EU and China to respond. The obvious questions remaining are how they (should) do so and what the impacts of their energy security policy strategies on each other are. To that end, this volume analyzes the effects of competition for access to oil and gas resources among the main global consumer countries and its implications for the security of energy supplies of the EU and China. The volume also analyzes domestic energy demand and supply patterns, policies to increase energy efficiency, and the prospects for the exploitation of renewable energy resources. The aim is to give insights into both the geopolitical and domestic aspects of EU-China energy relations in order to understand where the possibilities and impediments for cooperation and competition lie.

2. EU's Energy Security Issues

The EU is a net importer of energy due to the combination of high consumption, few resources, and little domestic production. According to the BP Statistical Review of World Energy 2008,[6] the EU's 27 member

[6] British Petroleum, *BP Statistical Review of World Energy* 2008, http://www.bp.com.

states combined consumed in 2007 14861 thousand barrels of oil daily (tbb/d), which is 17.8% of global consumption. This figure represents a 2.6% decrease compared to the previous year and interrupts therewith a longstanding rising trend. Oil production has declined over the last year and now stands at approximately 2394 tbb/d, or 2.9% of global production. Even worse for future prospects are the EU's oil reserves, which stand at 6.8 billion barrels, a staggeringly low 0.5% of global reserves and declining.

The statistics on natural gas are not much more optimistic. In 2007 gas consumption stood at 481.9 billion cubic meters (bcm), or 16.4% of total world consumption, which is a 1.6% decrease from the year before and again interrupts an overall increasing trend. At the same time, gas production fell 6.4% to 191.9 bcm over the previous year, continuing the decreasing trend and making EU production 6.5% of global production. Again, reserves further aggravate the problem, standing at 2.84 trillion cubic meters[7] or 1.6% of global reserves and decreasing. Thus the EU possesses insufficient reserves to meet future demand.

Expectations are that the EU's import-dependence will continue to grow. The European Commission has repeatedly warned since 2000 (Green Paper) that the EU's net energy import-dependency will rise from 50.5% in 2005 to 70% of the EU's total energy requirements by 2030. More precisely, where 45% of oil is currently imported from the Middle East, by 2030 this will be 90%. The situation for gas is similar. Currently 40% is imported from Russia, 30% from Algeria and 25% from Norway. By 2030 this will be 60% from Russia and an 80% overall import dependency.[8] It is especial cause for alarm that the EU import-dependency on gas from Russia is likely to grow and that gas is increasingly replacing oil as the main energy source.

2.1 *EU's Energy Security Strategies*

To answer these challenges to energy security, the EU opened a debate on future EU energy policy with its Green Paper of March 2006. Concerns about rising oil and gas prices, dependence on a limited number of non-European suppliers and climate change, especially CO2 emissions, led to the formulation of three energy security principles:

---

[7] 1 cubic meter = 35.3146667 cubic feet.

[8] Euractiv 03-03-2009.

security of supply, competitiveness (affordability), and environmental sustainability. Although reiterating the need for cooperation on such issues as competitiveness and the internal market, supply security and solidarity, creating an efficient, sustainable and diverse energy mix, tackling climate change, and innovation in energy technology, it especially highlighted the need for a coherent and common external energy policy.[9]

In January 2007, the Green Paper was followed up by an energy and climate change package with which the Commission aimed to reduce EU dependency on imported fuels and abate climate change. Its endorsement in the following EU summit (March 2007) led to a two-year action plan to establish a common EU energy policy targeting the diversification of sources and suppliers and the establishment of a common external energy policy on the one hand, and greenhouse-gas emission reduction, boosting of renewable energies, and increasing energy efficiency by 2020, on the other. Just recently the final version of this package was agreed upon by the member states at the EU 2008 December summit and by the European Parliament. In the meantime the commission has put up an EU energy security and solidarity action plan that focuses on supply security and fossil fuels, interconnectors, and external energy policy.[10] EU energy ministers also recently endorsed this plan (February 2009). However, the success of these external and internal policies faces serious challenges and is by no means assured.

2.2 *Geopolitical challenges to EU Energy Security*

"As one of the world's largest importers of oil, gas and coal, the EU is a major player on the international energy market. However, it remains a dwarf on the political stage as member states keep the upper hand on foreign policy."[11] The EU Green Paper of March 2006 gave a new impetus to establish a common external energy policy, the need for which

[9] Euractiv 03-03-2009.

[10] The Commission points to a 5-point action plan: a) infrastructure needs and the diversification of energy supplies (unbundling and EU-Med FTA); b) external energy relations (EFP); c) oil and gas stocks and crisis response mechanisms; d) energy efficiency; e) making the best use of the EU's indigenous energy resources (renewable and coal).

[11] Euractiv 03-03-2009.

already was signaled in the Commission's 2000 Green Paper[12] on energy supply security and the risks of import dependency. This 'energy foreign policy' should go beyond the concept of energy supply security as it envisages the active use of foreign (and security) policies as an extension of energy policy while aiming at furthering the diversification of origin. It was argued that the "energy challenges facing Europe need a coherent external policy to enable Europe to play a more effective international role in tackling common problems with energy partners worldwide. It would be a break from the past, and show member states' commitment to common solutions to shared problems."[13] As such it is the EU's response to the 'energy weapon' of producer countries and a possible remedy to strong international competition from the US and the rapidly developing Asian nations as it allows the EU to speak with one voice in global energy geopolitics.[14] Although promising on paper, the common external energy policy has a number of obstacles to overcome if it is to play the role envisioned for it.

Currently, the EU is still wrestling with the establishment of a common energy policy. Although the European Commission laid down priorities clearly (establishing an internal market for oil and gas out of the fragmented national and sub-regional markets, creating links with energy-rich countries in its periphery and neighboring regions, especially Russia and Central Eurasia etc.), many EU organizational issues remain. Overlapping competences of policy-making institutions in the EU and its member states, complex government-business ties, and competing energy priorities all hinder the effective establishment

[12] Green Paper—Towards a European strategy for the security of energy supply, COM (2000) 769, November 2000, http://ec.europa.eu/comm/energy_transport/doc/2005_green_paper_report_en.pdf.

[13] EU, Green Paper, op. cit., 2006.

[14] The key goals were set to be to: 1) form a clear policy on securing and diversifying energy supplies; 2) establish energy partnerships with producers, transit countries and other international actors in order to a) by creating a dialogue with major energy producers and suppliers and b) develop a pan-European Energy Community; 3) be able to react effectively to external crisis situations; 4) integrate energy into other policies with an external dimension, and 5)use energy to promote development. To this end the Commission proposed to: a) identify European priorities for the construction of new infrastructure necessary for the security of EU energy supplies; b) develop a pan-European Energy Community treaty; c) create a new energy partnership with Russia; d) develop a new Community mechanism to enable rapid and co-coordinated reaction to emergency external energy supply security situations impacting EU supplies; e) deepen energy relations with major producers and consumers, and f) make an international agreement on energy efficiency.

and execution of a common energy policy. Moreover, the constitutional crisis has also put further developments on hold for a while. Finally, distrust among member states about which interests will prevail has led to caution, hindering the formulation of clear strategies that the common external energy policy should focus on in pursuit of the Commission's goals.

Internationally, the effectiveness of a common energy foreign policy also remains to be seen. The Asian industrializing countries of China and India have emerged as powerful actors on the global energy market with which individual member states of the EU will have difficulty competing in their pursuit of energy contracts, while the US and Japan remain competitors for access to energy sources. From the EU's perspective, Russia appears to be its most immediate geopolitical concern, considering its import-dependency on Russia and two Russian-Ukrainian gas crises fresh in mind. Questions arise about its trustworthiness as a supplier. The energy dialogue with Russia is crucial in this respect. Worries abound about the new partnership agreement, mutual understanding of each other's perspective on energy trade and investment, a common approach among member states towards Russia, and cooperation on climate change (Kyoto) and nuclear safety and decommissioning.

Other important producer regions for the EU are of course the Persian Gulf and North Africa. Considering the current political instability in the Middle East and its effect on oil prices, the EU's dependence on the region does not bode well for energy security aims. Although the EU has the ambition to become a significant actor in the Middle East through the peace process, when it comes to energy it is the US that dominates the region both politically and militarily. Fortunately, North Africa provides more promising results. Not only are there serious talks about a Mediterranean free trade planned for 2010 that would further integrate Algeria and Libya into the EU on energy matters (December 2007 common Euro-Mediterranean energy market plan), but there also is the possibility for pipelines from Nigeria through Algeria to the EU.

Central Asia and the Caspian Sea region have been identified as possible sources for diversification. The former Soviet states however lack capital, technology, and necessary infrastructure for exploitation and export of energy resources. They continue to be mostly reliant on the Russian pipeline system. Conflict and cooperation around oil and gas resources not only involve questions of who dominates production of resources but also who has control of existing networks and who influences possible export outlets. Interesting in this respect is the Baku-Tbilisi-Ceyhan (BTC) pipeline that officially opened on May 25, 2005 and bypasses Russia. In addition, there is the planned Nabucco project (East Turkey—Bulgaria—Romania—Hungary—Austria) that is scheduled to open in 2014. It faces serious obstacles, however, in the form of the South-Stream pipeline that is supported by Gazprom and involves many of the same European energy companies. Some also point to private investment problems, but Chancellor Angela Merkel stated that the bigger issue might be where the gas for the pipeline should come from. Other possibilities include the expansion of the BTC to Kazakhstan and Iran, both potentially large suppliers. However, this is subject to major uncertainties due to the strategic situation in the Caucasus, and also in relation to the Iranian nuclear issue. The role Turkey could play as a regional power for future EU supply security is interesting in this respect.

Rivalry for oil and gas resources might lead to the development of an energy market and security complex on the Asian continent that is (partly) independent of US involvement. Bilateral agreements between China and Russia, China and Iran, and Russia and Iran but also regional cooperation such as the Shanghai Cooperation Organization (SCO),[15] are stepping-stones for such a development. This rift between the US on the one hand and China and Russia on the other seems to become the main geo-strategic reality in Central Asia, with the EU, Japan, and India having a secondary role. The SCO is well on its way in becoming an integrated Asian economic, political, military, and energy network able to counterbalance US regional influence.

It must be kept in mind, though, that energy is not the only determinant of geopolitical relations, and that the pursuance of other vital interests will also impact energy relations around the world. For instance, the Cold War has been a very dominant factor in international relations and in the way certain regions were approached in the past. The Soviet gas campaign, at the time, met substantial resistance from the US, which feared Europe's vulnerability as a result of the structural import dependence, while similar dependencies on other suppliers did not meet with this resistance. Indeed, geopolitics is a

[15] The member countries of the SCO are Russia, China, Kazakhstan, Kyrgyzstan, and Tajikistan. Iran, India, and Pakistan have observer status in the organization.

multifaceted arena in which energy is only one of the many political, military, economic, and ideological agendas pursued.

2.3 *Renewables, Alternatives, and Efficiency Measures in the EU*

Securing energy is not solely a matter of acquiring oil and gas. Renewable and alternative energy sources like solar, wind, hydro, geothermal, hydrogen, biomass, and nuclear energy do not only offer possibilities for the diversification of sources, but also of origin, as they are often not dependent on a specific geographical location and can be produced domestically. An added bonus of the use of renewable energy sources is their contribution in reducing greenhouse gas emissions and other pollutants. Additionally, a more efficient use of energy, meaning more energy output in terms of light, heat, and mobility per unit of primary energy input, can also lower dependence on imports and increase sustainability.[16] This can be achieved by reducing the quantities of energy consumed or by technological advances that make energy production processes less wasteful.

The EU Commission's energy and climate package and accompanying two-year action plan set legally binding targets for 2020: a 20% greenhouse gas emission reduction, a 20% share of renewable energy in the EU energy mix, boosting energy efficiency with a target of saving 20% of the EU's total primary energy consumption, and an increase in the share of bio-fuels in the transport mix to 10%. However, this roadmap only provides the goals; member states should fill in themselves how to achieve them, whether through developing renewable or alternative resources or by increasing the energy efficiency of the existing infrastructure. Member states may, for example, even invest in renewables in third countries (like solar farms in the Sahara) and import the energy in order to meet the goals. Although the energy mix remains in principle a matter of national sovereignty (state or company), the EU is increasingly defining energy choices by setting these and similar targets. The EU energy policy has continued along the lines of the action plan through liberalization of the internal energy market (3rd package: September 2007) and plans for CO2 burden sharing/carbon trading (March 2008).

---

[16] EU, Green Paper—A European Strategy for Sustainable, Competitive and Secure Energy COM (2006) 105, March 2006, http://ec.europa.eu/energy/green-paper-energy/doc/2006_03_08_gp_document_en.pdf.

Renewables harbor great promise for dealing with current environmental and geopolitical concerns surrounding the affordability, availability, and supply of energy. Not only that, renewable technologies are cutting-edge technologies that could create new export products. Unfortunately, the current picture looks quite different. According to the IEA, in 2006 the global primary energy supply consisted of 34.3% oil, 20.5% natural gas, and 26.0% coal. The combination of combustible renewable and waste energy (10.1%), nuclear energy (6.2%), hydropower (2.2%), and geothermal/solar/wind power (0.6%) together made up almost 20% of global energy supply.[17] By 2030, the overall share of biomass and nuclear is expected to fall marginally, while the share of hydropower is expected to rise slightly. Relatively speaking, wind, solar, and geothermal energy is expected to rise the most, but as it has such a small base this rise seems very insignificant in the bigger picture. In the EU's 27 member states things do not look much better. Currently, renewables on average make up only around 6% of national energy mixes and rarely go above 10%. This is deplorable, since energy accounts for 80% of EU emissions.[18]

Compared globally, the prospects for the EU to start exploiting renewables look good. The basic necessities like product knowledge, information on the costs and availability of new technology and on energy consumption, proper training of technicians, and investments in networks and maintenance, are all present in the EU member states. In addition, per capita energy use, environmental consciousness, and energy intensity are relatively lower than in other developed nations. Germany's experiences with renewable energy resources suggest reasons to be hopeful: the deployment of renewable resources could spur the creation of new high-lvel professional jobs and translate into net savings.[19] Hope also stems from a spillover effect where the growth in demand for renewable energy in the US and EU will enable increased

[17] See International Energy Agency (IEA) http://www.iea.org/textbase/stats/pdf_graphs/29TPESPI.pdf.

[18] Euractiv 03-03-2009.

[19] Klusmann, B., Third International Conference of the Energy Program Asia, Beijing 24–25 June 2006. The Third International Conference of the Energy Program Asia (EPA) was held in cooperation with the Institute of Industrial Economics (IIE), the Energy Economic Research Centre (EERC), and the Chinese Academy of Social Sciences (CASS). The topic discussed was "*Global Cooperation Towards Energy Efficiency: Barriers and Opportunities.*" For more details please see: http://www.iias.nl/epa/node/5.

access to them for developing countries like China and India. Crucial is if both regions can get the corporate, industrial, and financial backing in order to tip the balance away from the 'dirty' resources of oil and gas and still-dominant energy interests.

Next to the well-known solar, wind, hydro, geothermal, and future hydrogen alternatives, the use of coal gasification with carbon capture and storage facilities might be an interesting option, as it has great potential in the short as well as long-term for providing clean, and compared to other alternatives, relatively efficient energy. Nevertheless, some environmental concerns remain about CO2 storage. In addition, removing CO2 adds another step in the production process and is going to raise prices for consumers. Also interesting is the renewed interest in nuclear energy in the EU and its member states. Although after Chernobyl many planned to phase out nuclear energy, it is now back due to high oil and gas prices, and promises to deal with CO2 emissions, supply security, and competitiveness all at once. For example, the UK is starting to build new reactors, Italy and France are making new agreements, and even Germany is reconsidering its phase out. Nevertheless, member states remain divided. Whereas France and Finland are pursuing nuclear, Austria, Ireland, and Sweden are not. In addition, nuclear energy faces popular opposition. It also remains to be seen how renewable nuclear energy is, as uranium is also finite. At the same time, the use of bio-fuels in the transport sector, once hailed as the way to go, is increasingly criticized because of its effect on global food prices. New technologies are required to make the use of agricultural land more efficient so that more remains for food production. The use of hydrogen as a motor fuel might become a more socially acceptable alternative.

Finally, the end-goal of an internal energy market plays a role in achieving efficiency gains. By organizing production, trade, transportation, processing, and distribution in such a way that the boundaries of the relevant market are decided on technical criteria and economies of scale and scope rather than on a national basis, efficiency gains are likely to follow. Unfortunately, the process is currently driven by political compromises. As long as national governments are competent in a policy area, they are responsible for defending their national interests as best as they can. This in turn creates fear of an oligopolistic EU energy market in which some former national champions are expected to play a dominant role. Yet further unbundling faces hostility from some governments and electorates, while others point to the need of a regulatory framework to ensure positive economic and financial effects. Nevertheless, the process of liberalization has already resulted in shifting patterns of supply and demand, despite the fact that infrastructural capacities between member states are far from complete.

Unfortunately, the most serious challenges remain unsolved. The costs of renewable energy technologies remain high, although they continue to fall, making competition with oil and gas difficult until at least 2020/2030. A challenge is to find the right balance between investing in renewables now or later, i.e. between needs and costs. Of course, including environmental costs in the oil and gas prices might make renewables competitive sooner, but this is politically difficult. In terms of energy efficiency, new technologies also often cannot compete with established oil and gas technologies. Moreover, if the 2020 targets are to be met, making the existing infrastructure more efficient is likely to produce better results than focusing on renewables. In addition, renewable industries need to be built from scratch. The large investments and sunk costs involved pose considerable obstacles, necessitating continued government support. Moreover, there exist many 'chicken and egg' problems when building an infrastructure and ensuring the complementary deployment of user applications. Consequently, policy makers need to develop technologies and deploy applications by creating collaboration among the relevant actors in an effort to coordinate the introduction of all parts of a renewable infrastructure.

A promising approach to address this complexity is transition management. Based on transition theory, it addresses technological innovation processes within their wider societal context, claiming that changes in one without the other will remain fruitless over the long run. It is a step forward from the energy policies of the 1990s based on bottom-up, market-oriented policies in that it not only looks at techno-economic aspects, but also tries to create a framework in which government policy makers, industry stakeholders, non-governmental organizations (NGOs), and scientific institutes actively pursue accompanying institutional changes to ensure the emergence of a new energy system. Realizing its potential, the Dutch government has embraced transition management as the official governance framework for a transition towards a more sustainable energy system in 2000 (Fourth National Environmental Policy Plan). Considering transition management's novelty as a policy perspective, however, it has yet to prove itself in practice.

Finally, although renewables, alternatives, and more efficient means to use energy are developing, they are not likely to overcome the dependency of the world economy on fossil fuels until 2030. Basically, while the consumption of renewables is growing, so is the use of oil and gas. Thus even if the EU were to reach its 2020 targets, due to expected increases in global fossil fuel consumption the worldwide energy mix would not change. Although this improves EU energy security from the point of view of import dependency, on the count of global CO2 emissions and sustainability not much is expected to improve: the US is still not a signatory of the Kyoto protocol, global energy consumption and related pollution continue to grow, and much of the infrastructure in producer countries is aging. Moreover, while discussing issues of energy and the environment one must not overlook that both China and India are embarking on a major expansion of their coal burning power stations on a scale that could negate any achievements under the Kyoto protocol.

In the end, old dilemmas remain. Where renewables are sustainable and secure, they are not economical and require large investments to set up. Fossil fuels, by contrast, are competitive and enjoy an existing infrastructure, though they are not secure and sustainable. Hence, on the one hand, it is increasingly clear that geopolitical solutions aimed at diversification and acquiring access to fossil sources, although valuable for ensuring energy security in the short and medium term, will not be able to address the core problem of scarcity. On the other hand, the sense of urgency in social and environmental spheres for the development and deployment of alternative and renewable sources and energy efficiency measures is not matched by economic and technical opportunities to support such energy transitions.

3. China's Energy Security Issues

China's energy issues have three remarkable characteristics. First, its per capita domestic resources are insufficient. China has diversified energy resources with an abundant aggregate volume. However, compared with its huge population, energy resources are inadequate. China's per capita coal and water resources equal 50% of the world's average, while per capita oil and natural gas resources take up only about one fifteenth of the average level across the globe.[20] Second,

---

[20] State Council Information Office of the People's Republic of China, white paper of *China's Energy Conditions and Policies*, 2007, http://www.scio.gov.cn.

energy demand is highly intensive. Due to factors concerning the industrial structure and production technologies, China has a relatively low energy utilization rate and highly intensive energy demand for economic growth. Since the beginning of the new century, the country's coefficient of energy consumption elasticity has been as high as around one for many years. Third, fossil energies constitute a heavy weight. In the structure of primary energy consumption in 2007, coal and oil accounted for 62.8% and 19.0% respectively, with other energy only making up 18.2% of the total.[21]

As a result of the above-mentioned characteristics, China's rapid economic growth over the past three decades has produced two big challenges. On the one hand, rapid economic growth results in steep rise in energy consumption while domestic energy production cannot meet the ascending energy demand. Therefore the status of energy self-sufficiency is disturbed and energy consumption increasingly depends on external supply. On the other hand, rapidly climbing energy consumption leads to soaring fossil energy consumption and has ever-more negative impact on the environment.

Dependence on external energy supply is reflected in the increase of oil imports. As early as in 1993, China saw an end to the energy self-sufficiency era. In 2006, it became the world's third-largest oil importer. In 2008, China's oil import volume exceeded its domestic output for the first time and the weight reached 51%, which meant import has become the main source of oil supply. High reliance on external oil supply makes the country's supply security vulnerable to external risks, such as insufficient spare production capacity in the international oil market, political turbulence in the oil producing areas of the Middle East, strong fluctuations of the international oil market, and consequent supply disruptions and price hikes.

Impact on the environment is mainly related to fossil energy consumption. With its current energy structure, China mainly relies on fossil energy supply to meet the increased demand for energy, thus leading to more greenhouse gas emissions and heavier environmental pollution. In 2005, although the nation's per capita CO2 emission was only 3.9 tons and far lower than that of developed countries, the gross volume reached 1.5 billion tons, ranking second in the world.[22] China's

[21] International Energy Agency, *World Energy Outlook 2007*, Chinese version, Paris 2007, p. 144.

[22] International Energy Agency, *World Energy Outlook 2007*, Paris 2007, p. 200.

SO_2 emission amounted to 25.888 million tons in 2006, 90% of which was caused by coal combustion. In addition, coal production has also affected the land resource that already suffers from shortages, and the groundwater resources.[23] The impact of increased fossil energy consumption on the environment constitutes negative factors impeding China's sustainable development.

3.1 *China's Energy Security Strategies*

China's energy security strategies are stated in the White Paper of China's Energy Situations and Policies released by the Chinese government in December 2007. This document emphasizes that the country should respond to the challenges of energy security from five different angles, namely energy conservation, domestic resource exploitation, environmental protection, science and technology development, and mutual-benefit cooperation. From the viewpoint of practices, specific measures taken by China to cope with the above-mentioned two major energy security problems are composed of two aspects. One targets external factors of energy security, especially for the safeguard of oil import security. The other aspect focuses on internal factors, especially policies adopted to achieve sustainable development.

Measures to safeguard oil import security mainly include establishing strategic petroleum reserves, diversifying supply sources, and promoting peace and stability in oil production areas. In terms of the first issue of establishing strategic petroleum reserves, as a nation largely relying on oil import, China did not construct strategic oil reserves until the beginning of the 21st century. Therefore, it was unable to meet any emergency due to international oil supply disruption. In recent years, China has accelerated the building of strategic petroleum reserve bases. Four petroleum reserves bases of the phase-one project have completed construction, located in Dalian in Liaoning Province, Huangdao in Shandong Province, and Zhenhai and Zhoushan in Zhejiang Province. The four bases have a total storage capacity of 17.6 million cubic meters and have already started injection. The phase-two project of petroleum reserves bases has begun construction.

[23] Research Institute of Economics and Technology of China National Petroleum Corporation, *2008 Report on the Development of Domestic and International Oil and Gas Industry*, January 2009, p. 32.

The second measure is the diversification of supply sources. Excessive reliance on a single oil supply source or transportation road is unfavorable for preventing the risk of oil supply failure. After years of effort, China has formed a diversified oil import structure. In the first ten months of 2008, China obtained 48.9% of its total imported oil from the Middle East, 31.2% from Africa, and 9.7% from Russia and Kazakhstan. As the China-Kazakhstan Crude Oil Pipeline went into operation in July 2006 and the agreement to build a pipeline from the Russian Far East oilfields to the Chinese border were signed in February 2009, China will undoubtedly enhance its capability to obtain oil from central Asia and Russia. Since the late 1990s, Chinese petroleum enterprises have made contributions to the diversification of oil supply sources by means of offshore investment. By the end of 2007, Chinese petroleum enterprises participated in oil and gas exploration and development in 31 countries, with offshore oil and gas output topping 87 million tons, of which 43% came from Africa and 34% from Russia, Kazakhstan, and Turkmenistan.[24]

The third measure is the promotion of peace in oil production areas. The peace and stability of oil production areas are critical to the security of international oil supply. Some Chinese diplomatic activities, though not fully motivated by concern for oil supply security, may produce this kind of effect. In 1998 and 2004, China resolved border disputes with Kazakhstan and Russia in a peaceful way, creating favorable circumstances for economic cooperation. In 2006, the Shanghai Cooperation Organization took up energy cooperation as one of its priorities and established a special energy work group. In Asia, China proposes to adopt the principle of laying aside disputes to facilitate joint development of resources in the South China Sea. China, Japan, South Korea, and ASEAN established the "10+3 Energy Ministers Mechanism" in 2004. In the Middle East and Africa, China insists on the political settlement of regional conflict and opposes the use of force. In 2002 and 2007, the Chinese government appointed a special envoy to the Middle East and another to Africa to mediate major regional conflicts, such as the Arab-Israeli conflict and the Darfur issue. By March 2009, 2,146 Chinese peacekeeping soldiers

[24] Research Institute of Economics and Technology of China National Petroleum Corporation, *2008 Report on the Development of Domestic and International Oil and Gas Industry*, January 2009, p. 283.

were deployed abroad, the majority in Africa and the Middle East, in order to implement peacekeeping tasks in the framework of United Nations peacekeeping programs.

Measures to realize sustainable development stress the improvement of energy efficiency, the development of alternative energy, and attach importance to energy conservation, emission reduction, and the development of clean energy. Regarding the stress placed on the improvement of energy efficiency, the *11th Five-Year Plan for Energy Development* released by the Chinese government in 2007 sets the goal of reducing energy consumption per unit of GDP by 20% by 2010 from 2005. In the same year, the Standing Committee of the National People's Congress approved the revised *Energy Conservation Law of the People's Republic of China*, which defines administrative measures on energy conservation in industry, construction, traffic and transportation sectors, public institutions, and key energy consuming enterprises, while formulating the relevant incentive measures and legal liabilities. In order to put this law into practice, the State Council published implementation regulations such as *Regulations for Energy Conservation in Public Institutions* and *Regulations for Energy Conservation in Civil Buildings* in 2008, and publicized the energy saving status of the one thousand enterprises that signed the *Accountability Agreement on Energy Conservation Goal*. It also released the results of an examination of energy conservation goal fulfillment among various provinces, municipalities, and autonomous regions, and established the auditing system.

The second dimension is the emphasis placed on the development of alternative energy. In 2008, the Chinese government published the *Medium and Long-term Development Plan for Renewable Energy*, setting goals for hydropower, biomass energy, wind power, solar energy, rural renewable energy, and nuclear energy development between 2010 and 2020. According to this plan, renewable energy will comprise 16% of total energy consumption by 2020. *The National Special Program for Development of Nuclear Power (2005–2020)* was approved in November 2007 and revised in 2008, targeting an increase in running installed capacity of nuclear power from the current 9.068 million kilowatts to 60 million kilowatts. The Ministry of Finance also offers a series of tax policies to encourage development of energy that uses substitutes for oil and coal. China is also working on some coal-to-oil and coal-to-gas projects.

Finally, the third aspect is the importance attached to energy conservation, emission reduction, and the development of clean energy. As early as in 1998, China signed the *United Nations Framework Convention on Climate Change and its Kyoto Protocol*. In 2005, it published the *Measures on Management of Clean Development Mechanism Projects*. In 2007, it released *China's National Climate Change Program*, which promotes reduction of fossil fuel consumption and drives renewable energy production. China will take these measures to raise the proportion of renewable energy to 10% of the total by 2010 and 16% by 2020, and contain emission of carbon dioxide.

3.2 *Geopolitical Challenges to China's Energy Security*

China's energy security strategies have achieved significant effects in practice. For example, China has made remarkable progress in the diversification of oil supply sources. Energy efficiency has consistently improved. Based on constant prices in 2005, energy consumption per 10,000 Yuan of GDP decreased from 3.39 tons of standard coal in 1980 to 1.21 tons in 2006.[25] However, China still faces severe challenges and has a long way to go before it can significantly reduce the risk of reliance on external supply and realize a harmonious development between energy and the environment.

On the one hand, a wide gap exists between current achievements and the energy security goals. Firstly, the strategic oil reserve capacity is insufficient. Compared with the average daily oil consumption of over one million tons and the average daily oil import of over 500 thousand tons in 2008, the total existing capacity of strategic petroleum reserve bases is like a drop in the bucket, lagging far behind the 90-day import volume reserves standard defined by the International Energy Agency, and even further behind the real oil reserve capacity level of developed countries today. It will take many years for China to build up adequate capability to cope with an emergency. Secondly, opportunities to invest overseas are few and far between. Compared with the well-established transnational oil companies, Chinese enterprises have limited investment and oversea output. Transnational firms have already accessed regions with favorable conditions

[25] State Council Information Office of the People's Republic of China, White Paper of *China's Energy Conditions and Policies*, 2007, http://www.scio.gov.cn.

for development and profit-making, while Chinese companies have to enter high-risk regions and sometimes may encounter geopolitical and strategic obstacles created by other countries. Thirdly, China's diplomacy is not powerful enough to remove complex geopolitical risk in the major oil producing areas. In fact, world oil supply capacities tend to concentrate towards regions with higher geopolitical risks, especially the Middle East. For example, it is forecasted that OPEC will increase its share in world oil output from 36.7% to 50.9% between 2005 and 2025.[26] However, it seems unlikely that the potential for conflict in this region can be removed in short term.

On the other hand, in the quest to achieve energy security, China faces many constraining factors that cannot be overcome in a short period of time. The first factor is connected with the development stage. China's economic growth patterns indicate some characteristics of traditional industrialization and its economic development is currently at a stage in which energy-intensive industries, such as electricity, iron and steel, nonferrous metal, construction materials, oil refining, and chemical comprise the main driving force for economic growth. China will require a long time to go beyond this stage via industrial structure upgrading.

The second restrictive factor relates to resource endowment. As energy resources are limited, China's crude oil production is predicted to reach its peak around 2015.[27] Although the country enjoys rich coal resources with a reserve-production ratio of 80 years, the exploitation conditions are increasingly complex and the problem of shortage of domestic production is emerging, making China a net coal importer for the first time in 2007. Therefore, despite efforts, it seems to be very difficult to reverse the trend of dependence on external energy supply.

The third restrictive factor is institutional obstacles. China has made some achievements in energy price reform, but water, electricity, coal gas, thermal power, natural gas, and petroleum products are still priced by the government or have to adopt the government guiding prices. As a result, the energy prices cannot fully reflect the scarcity of resources, supply-demand relationship, and the environmental

[26] OPEC, Oil Outlook to 2025, *OPEC Review Paper*, Blackwell Publishing, UK, 2004, p. 12.

[27] Huang Minxuan, *China Energy Development Report 2008*, Social Sciences Academic Press, 2008, p. 33.

cost. As institutional reform may affect the general balance of reform, development, and stability, it cannot be achieved overnight, and must adopt a longer-term and gradualist method.

The fourth factor is the cost of substitution. Like other countries, China has yet to end the era in which fossil energies are the leading ones. The development of alternative energies is challenged by its high cost. It seems that legislation and government incentives in favor of alternative energies may produce some effects, but they are not powerful enough to counterbalance market forces completely; therefore there is no prospect for radical change of the energy mix at the expense of fossil energies in the short-term. The progress of alternative energies will depend, to a large extent, on the real price level of fossil energies, especially that of oil.

The fifth constraining factor is related to technology. China faces technological challenges in realizing energy security. According to the *11th Five-Year Plan for Energy Development*, the nation currently needs to develop some key applicable technologies, including coal and oil exploitation technologies for use under complex geological conditions, technologies of clean coal utilization, advanced nuclear power technologies for million-kilowatt pressurized-water reactors, super power transmission, distribution and a secondary system of power grid technologies, low-cost and large-scale development, and utilization technologies for renewable energy.

In conclusion, China still has a long way to go to realize its energy security, especially in terms of solving the two key issues of energy import security and environmental protection. Over the next 20 years, China will remain a country whose energy production can meet the lion part of its demand, but the trend of reliance on external supply may continue. By 2030, the nation's ratio of oil import reliance will rise to 80% with an import volume equal to that of the European Union.[28] Dependence on coal imports will decrease at a slow pace with imported coal volume expected to account for 3% of energy consumption by 2030.[29] Chinese experts estimate that China's overall external energy dependency ratio may increase from 9.7% in 2007 to 25% by 2030. Higher reliance on external energy supply will further enhance the impact of geopolitical factors on China's energy security.

---

[28] IEA, *World Energy Outlook 2007*, Chinese version, Paris 2007, pp. 145 and 184.
[29] IEA, *World Energy Outlook 2007*, Chinese version, Paris 2007, p. 175.

The government's stimulus will be conducive to alternative energy development. However, it is unlikely to change the basic structure of primary energy, in which the share of fossil energy would be maintained or even slightly grow. It is forecast that by 2030, coal will retain its proportion of 62.8% and the share of oil will grow to 21.1%, while the weight of natural gas, nuclear power, hydropower, biomass energy, and other renewable energies will drop to 16.1%, though the alternative energy aggregate will rise.[30] Increased consumption of fossil energy will continue to challenge efforts at environmental protection.

In brief, for China the pursuit of energy security is a long-term process. During this process, China will be confronted with challenges arising from both geopolitical and domestic factors.

4. Organization of the Book

This volume consists of two parts. Part One, which includes chapters 2 to 6, is concerned with issues related to energy and geopolitics. In Chapter Two Frank Umbach analyzes the challenges to energy relations between the EU and China. The EU and China face the same twin challenges of energy supply security and climate change. Both have become increasingly aware of the interregional interdependencies in coping with these difficulties and have therefore intensified dialogues and agreements on both subjects. However, tensions may arise in the years and decades ahead, with the EU, China, and other great powers competing for the same energy resources in the Middle East, Russia, and Central Eurasia. For example, worries exist regarding China's rising energy consumption and its impact on climate change. Umbach analyzes the internal and external challenges to China and the EU's energy policies, identifies the common interests and opportunities for cooperation, and examines problems and constraints for dialogue on energy issues.

In Chapter 3, Cutler explores the current Russian energy policy in Eurasia in the context of geopolitical developments. The EU has begun to manifest a more active interest in Central Asia and the South Caucasus. The EU is developing energy pipelines circumventing Russia,

[30] IEA, *World Energy Outlook 2007*, Chinese version, Paris 2007, p. 144.

whereas Russia used to be the transit country for gas and oil to Europe. From Russia's perspective, the EU is encroaching upon its former sphere of influence. The author postulates that the choice of which pipelines to construct and that of the partners with whom to build them, are an indicator of political entente, if not alliance. The construction of new pipelines in Eurasia can therefore alter the balance of power between Russia, the EU, and Eurasian countries. Against this background, the objective of the article is to gain a deeper understanding of Russia's place in Eurasia's energy complex.

In Chapter 4, Eva Patricia Rakel examines Iran's foreign energy policies and relations with China and the EU. Iran is a country with huge oil and gas resources and simultaneously a transit country for energy resources from the Persian Gulf countries. As a result, Iran is positioned as a strategic and influential player in the global energy market. The EU and China are potentially important economic partners for Iran. The EU and China are both in search of diversification of energy supply, and Iran needs the EU's foreign direct investment and knowledge for developing its industry, and the EU and China for exporting its oil and gas. However, Iran's nuclear program has created difficulties, especially for economic cooperation with the EU, making Iran turn towards China. Other obstacles are Iran's unsafe investment climate and its hostile relationship with the US. Rakel addresses these problems and their consequences for Iran's relations with the EU and China.

Philip Sen discusses challenges of China's and India's energy strategies in Chapter 5. Much attention has been paid to China, with its demand for energy imports increasing rapidly. However, India's energy needs are also soaring and deserve more attention. India's energy needs are almost identical to China's, with both countries experiencing an industrial boom and thus increasing demand for energy resources. With stocks of energy resources dwindling, logic dictates that sooner or later there won't be enough for both of them. India's need for energy imports poses a threat to China. Sen's objective is to examine whether China-India relations tend to cooperation or competition against this background.

In Chapter 6, Chen Mo explores Chinese policies and measures for achieving greater oil supply security, through an 'easier access to oil' and a 'keep away from oil' perspective. 'Keep away from oil' refers to domestic measures to lower consumption of oil. 'Easier access to oil,' on the other hand, refers to foreign tactics to improve possibilities for acquiring oil by improving relations with supplier countries and powerful nations. With the latter perspective in mind, Chen Mo conducts a case study of Saudi Arabia and Sudan, two significant oil-exporting countries to China.

Part Two brings together research papers on various aspects of renewable energy and sustainable development in China, Japan, and the Netherlands.

Chapter 7, written by Shi Dan, is an analysis of China's legal provisions and policy measures regarding renewable energy. The goal of these measures is to eliminate rural poverty, optimize China's energy structure, and to increase the energy supply by establishing a renewable energy industry. Shi Dan's objective is to analyze the practicability of these targets, the problems that emerge in the implementation of these policies, and to put forward countermeasures and suggestions for the development of renewable energy sources.

In recent decades China experienced an impressive economic growth. The downside of this growth is China's increasing need to import energy in order to fuel its economy. China has become more and more dependent on foreign oil imports as the result of a steadily growing economy, and with increased consumption of energy, environmental issues also emerge. In Chapter 8, Edward Vermeer analyzes the results of a public opinion survey conducted among advanced students and researchers in Beijing. The objective of the survey was to gain insight into the concerns and opinions which students and researchers have about China's oil dependency and environmental degradation, and the solutions they bear in mind.

Nuclear energy is experiencing a renaissance as a result of high oil prices, oil supply insecurity, and climate change. Raquel Shaoul writes in Chapter 9 on Japan's nuclear energy policy and possibilities and opportunities for cooperation with China on nuclear energy development. Japan has since the 1973 oil crisis made a commitment to enhancing nuclear energy for electricity generation. Energy hungry China perceives nuclear energy as a viable alternative to fossil fuels to cope with environmental problems and energy security. Sino-Japanese relations have been characterized by diplomatic tensions in the past. However, cooperation would significantly improve the two countries' energy security equation. Japan's expertise on nuclear energy can be important to Chinese efforts in nuclear energy development. Japan's ability to maximize its nuclear capabilities depends largely on expanding cooperation with China, if it is to achieve the ambitious targets set in its 'New national energy strategy.'

In Chapter 10, Daniel Scholten investigates the Dutch government's experience with stimulating a transition to a hydrogen economy. Fossil fuel scarcity and deteriorating environmental conditions urge for a more sustainable energy system. Such a transformation not only requires technical and economic change, but also institutional reform. Scholten conducted research on Dutch efforts to create or alter institutions in a way supportive of hydrogen in the Netherlands. Transition management and transition theory are promising tools for creating the transition to a hydrogen economy. The Dutch government embraced 'transition management theory' as its official framework. Because of the novelty of this theory, the author investigated the concrete achievements the Dutch government has made in facilitating a transition to a hydrogen economy.

CHAPTER TWO

EU-CHINA ENERGY RELATIONS AND GEOPOLITICS: THE CHALLENGES FOR COOPERATION

Frank Umbach

Abstract: *Beijing's rapidly rising import dependencies on energy and raw materials have numerous consequences for foreign, security and defence policies, as its policies to the EU or to the Iranian nuclear question have demonstrated during the last years. The EU, China, India and others may compete for the same energy resources in the Middle East, Russia and Central Asia. Whether they follow a 'market strategy' or a 'strategic approach' may ultimately decide whether they are able to cooperate for regional and global energy security or whether they will increasingly compete.*

Keywords: European Union, China, Geopolitics, Cooperation

1. Introduction

During the last years, China has replaced the United States as the center of the world's raw materials market and as a price setter for these industrial raw materials (Hale 2004). In 2009, it is expected to even surpass Germany as the largest exporter of goods in the world. Since 2000, China has accounted for 40 percent of the world's crude oil demand. In 2003, it already displaced Japan as the world's second largest energy consumer, and surpassed even the United States and Japan as the second and third largest exporter (after Germany). While having the third largest coal reserves worldwide, China only became a net importer at the beginning of 2007. Domestically, China's heavy reliance on coal in its primary energy consumption has raised enormous environmental problems and costs that increasingly threaten its future economic growth. According to an analysis of the Environment Assessment Agency of the Netherlands, China has already replaced the U.S. as the world's largest emitter of greenhouse gases (GHG) in 2006 (Netherlands Environmental Assessment Agency 2007).

As a consequence of its hunger for energy and industrial raw materials, China has become ever more dependent on imports from distant, often politically unstable parts of the world. It was forced to conduct much more pro-active foreign and security policies on the regional as well as global levels—reflecting China's self-perception of its energy insecurity. In the last 15 years, the economic rise of Asia, and above all of China, has created an enormous regional and global energy demand that raises not only important economic issues, but also countless foreign and security policy issues for both regional and global stability (Umbach 2003: 122–150; Umbach 2004).

Like many other Asian countries (with the exceptions of South Korea, Japan, Singapore, and Hong Kong), China has long subsidized energy consumption. The result has been an increasing inefficiency: China consumes up to five times as much energy to produce each dollar of economic output—which is an often underestimated factor in the government's energy forecasts. China's energy (foreign) policy seems to be based on a strategic approach (but with an increasing market orientation), thereby focusing on guaranteeing the rising energy imports for its socioeconomic stability and, therewith on its supply security. Yet, until very recently, it rather neglected energy conservation, economic efficiency factors, and environmental costs. At the same time, China has experienced an acute shortage of energy since 2003, which severely disrupted its industrial output and electricity supply.

The emergence of PR China as the world's leading consumer (overtaking the U.S. in 2004) and as one of the largest importers of oil, gas, and many industrial raw materials, had been overlooked in Germany and many other EU member states until 2004. But in the years and decades ahead, the EU, China, Japan, India, and other great powers may compete for the same energy resources in the Middle East, Russia, and Central Asia. Although China's deepening access to Africa's oil and gas resources, for instance, is often not the result of highly-coordinated government strategies to ensure China's energy security (Downs 2007), Beijing's energy foreign policies have been perceived as undermining U.S. and European foreign and development policies to promote good governance, human rights, and democratic political systems as well as Western hopes of China becoming a "responsible stakeholder" in global order and governance. As a Chinese expert admitted in 2006: "China must now view energy security in terms of economic threats and market solutions rather than military threats and diplomatic responses" (Daojiong 2006: 181).

Even after 9/11, the EU member states—in contrast to the European Commission, which has addressed energy security since its first Green Book of November 2000—did not focus on global and regional energy security issues. It was only the Russian-Ukrainian gas conflict at the beginning of 2006 that raised increasing doubts in the European public, governments, and parliaments about whether Moscow would remain a reliable energy partner for the EU. As a consequence of the Russian-Ukrainian conflict, in March 2007 the European Council agreed on an integrated climate and energy policy with an 'Energy Action Plan' (EAP) for the next years (2007–2009)—currently the world's most ambitious energy and climate policy. The EU is now in the process of defining an implementation strategy for its 27 member states and their 27 very different national energy policies, energy mixes, and economic conditions. However, there is a growing awareness within the EU as well as China of the interregional interdependencies of energy security and climate change.

The EU and China established bilateral relations in 1975, and these have broadened and deepened through more than 24 sectoral dialogues and agreements. The bilateral energy discussions are one of the oldest sectoral dialogues, and first began in 1994. These energy dialogues take place in the form of annual working group meetings and the bi-annual Conference on EU-China Energy Cooperation. It includes discussions of energy policy and development strategies, the evolution of energy markets, and security of supply and sustainable development. The "Memorandum of Understanding on Transport and Energy Strategies" of September 2005 envisages cooperation in areas such as energy regulation, renewable energy (including alternative transport fuels), energy efficiency, natural gas, clean coal technology (near zero emissions), and other new technologies in the energy sector. Furthermore, a new EURATOM agreement with China focuses on research into the peaceful use of nuclear energy and grants researchers from both sides access to each other's nuclear facilities. Both sides are also participating in the international ITER programme for the construction of an experimental controlled fusion reactor. In 2005, the European Commission and the Chinese Ministry for Science and Technology (Most) signed an Action Plan on Clean Coal and the terms of reference for an Action Plan on Industrial Cooperation on Energy Efficiency and Renewable Energies (European Commission 2008a).

Most recently, both parties have broadened their energy dialogue by including climate protection issues (in addition to the newly established EU-China Partnership on Climate Change) in their governmental and track-two energy meetings, such as working groups and conferences. At the same time, both sides are now more willing to take over more global responsibilities for coping with these global challenges, as China's efforts to create low carbon zones and eco-cities, funded jointly by Chinese and foreign sources, or common research projects on carbon capture and storage, (CCS) highlight (Lee/Froggatt et al. 2007; Guijang 2008).

The objective of this chapter is (a) to analyse the internal and external challenges of China's and the EU's present and future energy policies, (b) to identify common interests and projects for cooperation, and (c) to examine the problems of and constraints on their mutual energy dialogues. For a comparative analysis, each sub-chapter on the respective Chinese and EU energy policies begins with a special focus on their energy resource mix and their oil, gas, and coal supply situation and development. They also offer perspectives on nuclear power and renewable energy sources. In addition, the comparative analysis will also address energy-related environmental and climate protection efforts as well as their respective energy foreign policies and diplomacies. The chapter thereby seeks to highlight the differing and common problems and challenges, which may offer more concrete possibilities for future interregional energy cooperation and mutual efforts to stabilize global energy security.

2. China's Energy Challenges in Perspective

2.1 *Domestic Challenges*

With its population of 1.3 billion and status as the second-largest energy producer in the world, China is also the world's second-largest consumer of primary energy, the third-largest energy producer, and after the US the largest contributor to global carbon dioxide (CO_2) emissions. Together with India, China may account for 45 percent of the total increase in world energy demand between 2005 and 2030, and even 82 percent of the increase in coal consumption over the same period, according to the IEA's Reference Scenario (IEA 2007: 42).

China's rapidly increasing demand for energy is not due solely to population growth, but above all to the accelerated pace of agricultural electrification, urbanization, rapidly increasing consumption (such as refrigerators, washing machines, televisions, air-conditioning), and the development of the transport and industrial sectors. In fact, total urban consumption has grown larger than the secondary or industrial sector (Allaire 2007). This imbalance between demand caused by economic and population growth and domestic energy production is increasing slightly. In general, the roles of increasing energy efficiency and energy conservation for the enhancement of national energy security have been underestimated. One major obstacle is the tremendous regional differences in regards to energy efficiency, which make a unified energy saving concept for different provinces very unrealistic. The least energy efficient provinces are in the coal resource-rich hinterland that depends heavily on coal consumption stemming from its secondary industry (Dan 2007).

Against this general background, I will firstly analyze China's domestic energy challenges before addressing the risks for its energy foreign policies.

Rising Oil Demand

Despite being the second-largest energy and sixth-largest oil producer in the world, China currently produces neither enough crude oil nor natural gas to meets its demand. Most of its fields of production have reached or even passed their peak. Newly discovered fields awaiting development do not offer large enough reserves to change the overall balance between its demand side and its domestically assumed production. Its crude oil production will decline from 3.67 mb/d in 2006 to just 2.7 mb/d by 2030. In 2008, China overtook even Japan as the second-largest importer of crude oil (Tu 2008a: 1). By 2030, China's net oil imports alone will rise from 3.5 m/d to 13.1 mb/d in the IEA's Reference scenario, to even 9.7 mb/d in the Alternative Scenario, and 17.2 mb/d in the High Growth Scenario (IEA 2007: 166).

China has only 2.43 percent of global crude oil reserves and 1.2 percent of the world's reserves of natural gas. As a result of its limited oil reserves and continued increase in oil demand, its net imports account at present for about half of China's total oil consumption, and may rise to 84 percent by 2030. At the same time, Chinas has become the biggest market for new cars worldwide. The number of light-duty vehicles is projected to increase from 22 million in 2005 to more than 200 million in 2030 (IEA2007: 122–123).

Table 2.1 China's Primary Energy Demand (Reference and Alternative Scenarios in million tons/mt).

	1990	2005	2015	2030	2005–2015*	2005–2030*
Coal	534	1.094	1.869 (1.743)	2.399 (1.842)	5,5%	3,2% (2,1%)
Oil	116	327	543 (518)	808 (653)	5,2%	3,7% (2,8%)
Gas	13	42	109 (26)	199 (225)	10,0%	6,4% (6,9%)
Nuclear Power	0	14	32 (44)	67 (120)	8,8%	6,5% (9,0%)
Hydro	11	34	62 (75)	86 (109)	6,1%	3,8% (4,8%)
Biomass and Waste	200	227	225 (223)	227 (255)	–0,1%	0,0% (0,5)
Other renewables	0	3	12 (14)	33 (52)	14,4%	9,9% (11,9%)
Total	874	1.742	2.851 (2.743)	3.819 (3.256)	5,1%	3,2% (2,5%)

* Average growth per year.

The statistical figures of the Alternative Scenario for the years 2015 and 2030, and the average growth figures per year are in brackets. Source: IEA 2007: 287 and 384.

Furthermore, China still has to cope with insufficient oil product volumes. In Guangdong, for instance, roughly one-fourth of the service stations were closed due to lack of gasoline and diesel for the expanding fleets of private cars and taxis. Others have resorted to rationing fuel to cope with the underestimated energy demand. Pressure from local governments to increase supplies in order to minimize social and economic fallout has increased losses for state-owned refineries as the result of subsidizing domestic gasoline.

Continued state control over energy prices has discouraged industry from improving energy efficiency and prevented reversal of the trend of energy-wasting by sending the wrong price signals to the economy as a whole. In the future, China may suffer more severe supply constraints and shocks as long as the government is reluctant to raise the gasoline and diesel prices to real market prices (Tu 2008b). Nonetheless, price pressures are already having an impact on the broader economy and beyond, such as on angry consumers of an increasingly assertive middle class or hard-pressed farmers.

At the beginning of 2007, the IEA criticized China for a lack of transparency and inconsistency in its oil data forecasts, something which could have worldwide consequences. Nobody seems to know, for instance, how much illegal diesel and gasoline is smuggled in and out of China (Su 2007: 32). The uncertain legal status of local and private refineries (LPR) has seriously distorted China's official oil industry statistics. Even official Chinese statistical agencies have been unable to provide accurate data regarding these facilities, often leading to an underreporting of the actual capacity of China's petroleum refining industry. These LPR operations have also fuelled oil-related criminal activities. According to reports, China had already closed more than 6,000 of those LPRs at the end of 2000. But thousands of those "teapot refineries" still appear to be operating illegally (Tu 2008b: 2–3).

Another major challenge is the annual 916.5 percent diesel import growth as the result of an escalating diesel demand for electricity generation, recently also linked with refinery underutilization due to Beijing's price regulations on oil products and the stockpiling of oil products before the Olympics. Ultimately, many of these problems are the result of China's economic growth and development strategy that is based on the exploitation of poor rural communities and resource hinterland in favor of developed urban centers and coastal regions in which the former provide heavy energy subsidies to the latter (Tu 2008a: 2–3).

Natural Gas

Slightly larger deposits of natural gas have been found both in China and the entire Asia-Pacific region. As of the end of 2006, China has proven natural gas reserves amounting to 3,720 billion cubic metres (bcm) or two percent of global gas reserves (IEA 2007: 328). However, the costs of constructing pipelines and liquifying plants are huge, due to the long transport routes. However, for environmental reasons, China has made increasing the use of natural gas a high priority, despite the massive investment costs and the fact that at present gas accounts for less than three percent of its total energy consumption (EIA 2006: 8). In 2004, China's gas consumption had almost doubled from five years before (EIA 2006: 8).It is believed that China's gas consumption will increase by almost four times by 2030 (from 32 bcm in 2000 to 60 bcm in 2006 to 103 bcm in 2015 and 118 bcm in 2020, but then falling back to 111 bcm in 2030) (IEA 2007: 126 and 330–331). The Chinese government hopes that gas will cover eight to ten percent of the country's entire energy consumption by 2010. Although demand for natural gas should rise by eight percent a year, even in the next decade up to 2020 it may only account for a maximum of 11 percent of China's total energy consumption.

China's natural gas imports will increase from zero in 2005 to 12 bcm in 2015, 28 bcm in 2015 and 128 bcm in 2030 (IEA 2007: 332–33). Chinese companies are planning to build up to sixteen LNG terminals. 10 of them might be operational by 2010. Although current imports are rather small in volume, China's share in the global gas trade will continually increase until 2030.

Coal

China currently meets around 69 percent of its total primary energy demand and 78 percent of its electricity supply with coal (Chinese State Council 2007; IEA 2007: 262; Weitz 2008). Coal contributes heavily to the final energy used by industry, commercial businesses, and households. China and India combined account for 45 percent of world coal demand.

In contrast to the period of 1997–2000, China's coal consumption has been growing during the last six years. In 2004, China consumed 2.1 billion short tons of coal, which represented more than one third of the global total and an astonishing 46 percent increase since 2002 (EIA 2006: 10).

Although China possesses the third largest coal reserves in the world at 126.2 billion short tons (behind the U.S.A. and Russia) and is currently the world's largest consumer and producer of coal, it will have to import greater quantities of foreign cleaner coal in the future. This is because of its inadequate road network, particularly in the interior of the country, which poses insurmountable transport problems. As a result, it is sometimes more expensive to transport domestic coal than it is to import from abroad. China's traditional reliance on coal, and particularly low quality coal for power generation and domestic use has lowered the energy efficiency of the country due to inefficient management, insufficient investment, outdated equipment, and poor safety records (Tu 2007). At the end of 2005, China had 28,000 coalmines. 2,000 of them were state-owned. By then, China had already closed down between 20,000 and 50,000 small coalmines as part of China's restructuring plan for its coal sector (EIA 2006: 11). Hence reduction of the use of low quality coal and improvement in the energy efficiency of coal-fired plants are indeed of vital importance to China's energy efficiency. Moreover, in Beijing's view, further increases in coal production must be limited as the environment in China is already suffering from excessive pollution levels that increasingly threaten economic growth.

China's coal reserves can therefore only play a greater role if clean and cost-efficient incineration technologies find widespread use (carbon capture and storage/CSS). However, as financial resources are inadequate for this, the Chinese leadership is increasingly relying on other sources of energy for the time being.

In 2020, the share of coal as a percentage of total energy consumption will decline, but not below 60 percent. Its production volume, however, will further increase because coal projects will probably remain much cheaper than natural gas or other sources in the years ahead. Global coal consumption is expected to increase by around two billion tons from 1996 to 2020, 85 percent of which will be attributable to China and India. If the present coal trends continue (IEA-Reference Scenario), China's production is expected to increase from 1,636 million tons of coal equivalent (Mtce) to 2,248 Mtce in 2010, 2,604 Mtce in 2015 and 3,334 Mtce in 2030 (IEA 2007: 336). Despite a new record in domestic coal production of 2,523 billion tons—39 percent of global coal consumption in 2006—at the beginning of 2007, China had already become a net importer of coal, which has had a large impact on international coal trade. Those imports will mainly meet demand in southern coastal regions remote from the major-coal-producing areas. Nonetheless, China will also become a net importer of steam coal and even need to stop any of its net exports of coking coal. Overall net coal imports may reach 129 million tons (mt) by 2030 (IEA 2007: 342).

China also plans to build several large coal liquefaction plants to convert Chinese coal into oil products in order to reduce crude oil imports. In 2006, China's National Development and Reform Commission (NDRC) announced that it would invest more than US$128 billion in alternative coal-based synthetic fuel production and chemical feedstock in order to reduce its national oil import dependence. Currently, about 20 coal-to-liquids (CTL) are concretely planned or seriously considered with a total investment of US$15 billion and an estimated capacity of 16 mt of oil (IEA 2007: 273).

In the long term, the State Development Planning Commission plans to reduce the percentage of coal production for the national energy requirement to 35 percent, while oil and gas should account for 50 percent, with hydroelectric power, nuclear energy, and other alternative sources of energy making up the remaining 20 percent by 2050 (Umbach 2008e: 50).

Renewable Energy Sources

In 2005, renewable energy sources represented about 15 percent of China's total primary energy consumption, but dropped to just 8.1 percent in 2006 due to the much faster growth of fossil fuel consumption (Dan 2008a). Primary renewable energy sources are biomass, mainly used in rural households for heating as well as cooking, and hydropower in the electricity sector, accounting for 16 percent of total generation (IEA 2007: 354). But given increasing resistance to gigantic construction projects (such as the Three Gorges Project), environmental constraints have limited the expansion of alternative sources of energy such as hydroelectric power.

In 2005, China adopted a National Renewable Energy Law, which envisages drawing 16 percent (presently 7%) of its primary energy production from renewable sources by 2020. For the electricity sector, the target is 20 percent of the capacity from renewables by 2020. It will include 30 GW of windpower, 20 GW of biomass power and 300 GW of hydropower capacity (National Development and Reform Commission 2007; Lewis 2007: 2).

China also wants to increase the use of windpower, in which case the regions Sinkiang and Inner Mongolia offer the best possibilities. Between 2005 and 2006, its power capacity doubled from 1.3 gigawatts (GW) in 2005. Despite having a large land mass and long coastlines providing abundant wind sources, future expansion needs long overdue development as windpower has so far only met one percent of the national energy requirement. It also needs to be accompanied by large investments in grid expansion and transmission upgrades. But China lags far behind other newly industrialised countries—such as India—in this area. The expansion target envisages 5 GW in 2010, 30 GW in 2020, and 49 GW in 2030, but would account for just 1.6 percent of the nation's electricity supply (IEA 2007: 355).

While the installed capacity of photovoltaic systems is expected to reach just 9 GW in 2030, China has already become the world's leader in solar thermal systems for heating and hot water supply with about 75 million m^2 of solar collectors—about half the world total. The national target foresees the expansion to 150 million m^2 for 2010 and 300 million m^2 for 2020 (IEA 2007: 356). However, at present, China's solar photovoltaic development does not match well with its domestic demand market. This has resulted in an overgrowth of output value of the renewable energy industry and limited utilization as long as the feed-in price of solar PV power is more than ten-fold of that generated by coal (Dan 2008: 17).

Moreover, wind and solar power as an integral component of a decentralized energy supply offer the most economic solution to energy problems in isolated regions of China, where large power plants and power grids are not economically feasible. By 2010, 20 million inhabitants of isolated regions could be supplied with decentralized wind and solar power, which would significantly improve their educational opportunities and chances of economic development.

In general, as China's "Law on Renewable Energy" of 2006 and administrative decrees demonstrated, the development of renewable energy sources helps to ensure its national energy supply security, widens and decentralizes its national energy mix, reduces the negative effects on the environment, and decreases rural energy poverty in China—a fact, while increasingly recognized by energy experts, still presents huge problems to overcome.

2.2 *Nuclear Power and Electricity Generation*

In 2004, 24 of China's 31 provinces and municipalities suffered power shortages. For years, China's government has underestimated the growth of electricity demand as well as the structural problems linked with this growth. Thus it was surprised at the beginning of 2007, when in its official statistics the annual increase of 18 percent (102 GW) of electricity generating power surpassed the entire British national grid (McGregor/Dickie 2007). In the same year, China produced 104.8 GW of electricity—as much as the entire African continent (Tu 2008c). However, in the next years, China may face an oversupply problem because of an overinvestment in the power sector in recent years. But in the long-term perspective, China needs to add 1312 GW to its generating electricity capacity—more than the total current installed capacity in the U.S.A. (IEA 2007: 317).

The planned expansion of China's civilian nuclear power is outstanding in international dimensions. By 2050 China is projected to become the world's largest producer of nuclear energy (Zweig/Jianhai 2005: 36). Despite the long-term increase in the number of nuclear reactors from 11 at present to at least another 27 by 2030, the share of civilian nuclear power as a percentage of China's total energy consumption, however, will only rise from 1.5 percent in the mid-1990s to no more than 4-6 percent in 2020. A further expansion of nuclear power is hampered by the high capital costs for construction in competition with coal-fired power plants, whereas gas—as the environmentally cleanest fossil energy resource—is the most expensive fuel for power plant operation.

2.3 *Rising Environmental Costs*

Energy-related CO_2-emissions presently account for 61 percent of global GHG-emissions. Three-quarters of the projected increase in energy-related CO_2-emissions until 2030 will come from China, India, and the Middle East, and 97 percent in non-OECD countries as a whole. The OECD countries alone—even if they were to reduce their emissions to zero—are unable to reduce the world's GHG-emissions to a level that increases the global temperature just to 2°C as demanded in the Stern Report in order to prevent more disastrous consequences of the world's climate change and for its worldwide security (IEA 2008: 11–14).

China is already the largest producer of GHG and carbon dioxide (CO_2)-emissions that are held responsible for global warming. Even in 1997 China was the largest producer of sulphur dioxide (SO_2) emissions, surpassing both Europe and the U.S.A., with an output of 23.46 million tons. The main reason for this is the usually high sulphur content of the coal that is responsible for 75 percent of SO_2 emissions, 70 percent of smoke and smog, and 85 percent of all CO_2 emissions in China. Not only that, 33 percent of all SO_2 rainfall in South Korea and even 50 percent of acid rain in Japan are thought to have originated in China (Umbach 2008e: 51–52).

If current trends continue, China and India are together expected to account for 75 percent of all global CO_2 emissions in 2020. The largest absolute growth in carbon dioxide emissions is projected for the next two decades. By 2025, China's CO_2 emissions may reach 18 percent of the world's total emissions. The World Bank has already blacklisted 20 cities in China among the world's 30 most polluted cities (World Bank 2007). Air pollution alone, primarily from dirty coal burning, might be responsible for 300,000–400,000 premature deaths per year as estimated by the World Bank and other institutions (Economy 2005: 282; Economy 2004; Economy 2007). Furthermore, the economic-financial costs alone of health damage from air pollution are predicted to rise to the equivalent of 13 percent of China's GDP by 2020, if the present energy trends continue (OECD 2007). According to environmental

Table 2.2 Energy Related CO_2 Emissions of the United States, EU, China and by Scenario 2005–2030 (billion tons).

		Reference Scenario		Alternative Policy Scenario		High Growth Scenario	
	2005	2015	2030	2015	2030	2015	2030
World	26.6	34.1	41.9	31.9	33.9	34.9	44.8
OECD	12.8	14.1	15.1	13.2	12.5	13.9	14.6
Transition Economies	2.5	3.0	3.2	2.9	2.8	3.0	3.2
Developing Countries	10.7	16.4	22.9	15.2	17.9	17.4	26.3
U.S.A.	5.8	6.4	6.9	6.2	6.0	6.3	6.7
EU-27	3.9	4.0	4.2	3.6	3.2	4.1	4.2
PR China	5.1	8.6	11.4	8.1	8.9	9.5	14.1

Source: IEA 2007a: 199

regulators in 2006, SO_2 emission caused over 500 billion Yuan ($60 billion) in direct economic damage in 2005 (South China Morning Post 2006).

A new Chinese White Paper, released by the Information Office of the State Council (China's Cabinet) on 5 June 2006, confirmed Western analyses by painting a very grim picture of its ecological problems in spite of unprecedented efforts to protect the environment (Chinese State Council 2007). It has instituted a revised Energy Conservation Law, adopted in October 2007, that places more emphasis on the behavior and performance of the government itself. The same law previously set targets for individual energy intensive industries in order to reduce energy intensity by 20 percent between 2005 and 2010. But that goal already seems impossible to achieve. In the next 15 years, reportedly, China's economic imperatives are threatened by a reduction of around 10 percent of the country's annual GDP through clean-ups of ecological disasters and other socioeconomic impacts (Umbach 2008e: 51). Meanwhile, high-profile ecological disasters have been crippling the Chinese people, economy, and prompting increasing sociopolitical unrest. In 2005, environmental pollution provoked over 50,000 environmentally-related protests.

Meanwhile, China has sharply increased taxes on large cars (such as SUVs) to limit the environmental damage from the rapidly increasing use of vehicles and even placed a five percent tax on disposable wooden chopsticks and wooden floor panels. Vehicles already account for a third of China's petroleum demand, though only about 1.5 percent of the population has purchased cars. But the national number of cars doubled in just four years, rising from 8.5 million in 2000 to 17.4 million by 2004. China has already become the word's second-largest vehicle market. The total number of 40 million vehicles, which explains two-thirds of China's increase in oil consumption, is expected to rise by more than six-fold to 270 million by 2030 (IEA 2007: 283). The newly introduced fuel economy standards for cars are already tougher than those in most American states.

Despite the progress, China faces mounting problems on the implementation side. It often lacks the enforcement power to shut down polluting factories, whereas systemic corruption still seems to be on the increase in the countryside. While China uses one-eighth of the amount of energy that the U.S. uses and sees therefore no reason to compromise its further economic development, it had to realize at the beginning of this year that it failed to reach its own targets for reducing major pollution in 2006. High growth again outpaced national control efforts and exceeded the limits mandated in the national environmental protection plan for the Tenth Five-Year Plan period (Dan 2008b: 139). SO_2 emissions still increased by 18 percent in 2006 (13% in 2005). But Beijing's goal was actually to reduce such emissions by 10 percent, or 8.4 mt, by 2010 under the present 11th Five-Year Plan for Energy. Due to its often underestimated energy demand during the last years, China's government needs to make much greater emissions reductions in order to achieve its targets by 2010. But this seems more and more unrealistic as long as more fundamental changes in its energy and economic policies do not take place.

Internationally, China can no longer blame just the West as a fast-growing polluter causing climate change, though its per-capita emissions, at 3.9 tons of CO_2 in 2005, are only 35 percent of those in OECD countries. The Chinese government is now beginning to realize that climate change can heavily affect China's future economic development and sociopolitical stability by causing crop shortages, increasing floods in the rich coastal river deltas, and higher energy use as the result of mounting heat. Climate change is no longer a rich man's problem. Furthermore, China's present environmental policies contradict many of its declared foreign policy objectives. In the IEA's Reference Scenario, China's GHG emissions will rise by 5.4 percent annually to 2015 and 3.3 percent over the period 2005–2030. The emissions will expand to 11.4 billion tons, reaching a level 35 percent higher than that of the U.S.A. by 2015 and 66 percent higher in 2030. Although China's per capita emissions won't reach even the current OECD levels in 2020, its share of global emissions will grow from 19 percent in 2005 to 27 percent in 2030 (IEA 2007: 313). However, in its international environment policies, China's participation may remain conditional and it may continue to be reluctant in accepting any emissions quotas and targets due to its concerns for sociopolitical stability, at least in the short-term future (Meidan 2007).

2.4 *China's Energy Foreign Policy and its High-Risks Diplomacy*

In order to solve many of China's rising socioeconomic, energy, and environmental problems, the rising great power is dependent on a peaceful environment both in the Asia-Pacific region and on a global scale. Internationally, Beijing seeks increased legitimacy for its rising great power status, which in turn bolsters its political stability at home. In cherishing its concept of non-interference in the domestic affairs of other countries, China has been surprised by the increasing criticism of its energy diplomacy not just in the United States, but also in the EU and even increasingly in developing countries themselves in Africa and Latin America (International Crisis Group 2008).

Given the new energy policy dependencies in the 1990s, China's foreign and security policy had to deal with regions and countries that until then had played either no or only secondary roles in its traditional foreign policy. In the future, the possibility of greater economic and political rivalry, in particular with Japan (Umbach 2006b and Umbach 2006c), India, the U.S.A., and, in the medium- and long-term with Russia in Central Asia for shrinking global oil reserves, cannot be entirely excluded. Furthermore, Chinese energy experts are often more skeptical about global energy reserves and do not even rule out a serious shortage of oil reserves in the next 20 years. For this reason, they frequently arrive at much more alarming analyses than Western experts.

Against the above background of a rapidly growing demand for energy and deteriorating prospects for major new energy discoveries in their own country, the Chinese political leadership and managerial elite have been keeping a sharper eye out for new energy resources abroad since 1996–1997. As early as 1990, China purchased 81.5 percent of its crude oil from only three foreign states, although only Indonesia exported more than a million tons of crude oil to China. In 1997 the number of countries exporting more than two million tons of crude oil to China had doubled from 1990: Indonesia, Oman, Yemen, Angola, Iran, and Vietnam. Even in 1997, China already imported oil from all Gulf States except Bahrain. Until October 1997, China had 126 contracts and agreements with a value of US$5.38 billion, signed with 67 companies from 18 countries (Shengliang/Xiaojie 2000: 83). By 2005, China's oil importing sources had expanded to more than 40 countries.

Although China's government plans to launch a new round of exploration projects inside China to reduce the country's growing dependence on foreign energy resources, its has been forced to seek more overseas drilling rights for Chinese companies. They have stepped up their investment abroad to acquire direct control or partial rights in some of the world's potential oil fields. Beijing has forged closer ties with almost all continents. It has become much more proactive in Africa (Sudan, Chad, Angola), Middle East (Saudi Arabia, Iran, Algeria), and Latin America (Bolivia, Venezuela, Ecuador, Columbia, Peru and Brazil) (Thompson 2005 and Storey 2005). Despite the fact that China has also secured new supplies of oil and gas resources in the Asia-Pacific region itself (i.e., with Australia and Indonesia), the Persian Gulf and Africa have steadily become more important not only for the energy policies of China and the other Asian states, but also for their national foreign and security policies. In 2006, 44 percent of its oil imports came from the Middle East (with Saudi Arabia as the biggest supplier) and 32 percent rom Africa. Saudi Arabia was the biggest supplier that year (IEA 2007: 325–326). These, steps, however, present new risks to China's future oil security. If its net imports will rise from 3.7 mb/d in 2006 to 5.1 mb/d in 2010, 7.1 mb/d in 2015, and 13.1 mb/d in 2030, they will be more than those of the United States and a third more than the present total crude oil production of Saudi Arabia as the largest oil producer, as well as more than the projected net imports of Japan, South Korea, Australia, and New Zealand combined. As its oil import dependence grows from 50 percent today to 80 percent in 2030, China will eventually have to import as much as the EU (IEA 2007: 326).

As Beijing's new 2008 White Paper on Diplomacy highlights in its first chapter, "The Issue of Energy Security During the Period of High Oil Prices," the global competition for energy resources has increased in 2007 and regional competition has fuelled and complicated the global energy security problem. But the White Paper also views energy security as part of globalization trends and thus as a global problem, in which China as well as other countries are forced to international cooperation to guarantee the future stability of global energy security (Hsiao 2008). But China's energy diplomacy has also frustrated many efforts of NGOs and donor nations as well as organizations in their efforts to instill good governance, accountability, transparency, and improved human rights in these countries. All these Chinese diplomatic activities in the energy field have produced an economic-security nexus that is determined by the Chinese political leadership's most fundamental interest: economic growth and domestic stability in order to ensure regime survival. However, these unilateral energy-security strategies have also undermined multilateral and regional cooperations and have fuelled already existing strategic rivalries such as with Japan, India, and the United States.

But China's energy strategies still tend to rely on strenuous efforts to achieve as much as energy autarky as possible, and, as the result of the role of the PLA in domestic and economic security policies, do not exclude military concepts of safeguarding energy supplies, instead of market economic concepts, globalization requirements, and an international division of labor and transnational energy co-operations organized by the private sector. China thereby often seeks energy relationships with countries well-known for their anti-Americanism such as Iran, Venezuela, Sudan, Libya, and Angola. China's Iran policies have not only complicated American policy towards Teheran, but also that of the EU, which has an equal strategic interest in a nuclear weapons-free Iran (Chang 2008 and Umbach 2007b). Whereas American and European energy companies have withdrawn their business activities and investments for the time being, Chinese state-owned energy companies have moved in and taken over energy businesses from the West. By delivering modern arms to Iran and blocking harsher sanctions against Iran in the UN Security Council, China is supporting de facto Iran's non-compromising policies in regards to its nuclear weapons ambitions, though on a diplomatic level China is not interested in nuclear weapons proliferation in the Persian Gulf. Indeed, China has also reduced its oil dependence on Iran and diversified its oil imports from other producers during the last years. Still, around one-third of its total oil imports are shipped through the chokepoint of the Hormuz Strait (around 18% from Saudi Arabia and 12% from Iran).

Given Iranian threats to close and take control of the Hormuz bottleneck strait by deploying the Iranian Revolutionary Guard and its armed forces along the strait with Chinese weaponry, China's energy diplomacy is creating increasing contradictions and new security dilemmas for China itself. This is highlighted by the recent agreement signed between China Petroleum Engineering and Construction Corporation (CPECC) and Abu Dhabi's International Petroleum Investment Company to build a 360-km oil pipeline with a capacity of 1.5 mb/d that would bypass the Strait of Hormuz—albeit only a fraction of the total daily shipping of 17 mb/d of crude oil would bypass the Hormuz Strait (Shichor 2008).

China has undertaken efforts to satisfy its energy demand by increasing exploitation of other energy sources, modernizing its own production and extraction plants with the goal of increasing its own capacity, floating shares in Chinese oil companies on international stock exchanges, and making global investments to secure foreign energy resources while at the same time diversifying imports of crude oil and natural gas. But these positive energy strategies have only been partially successful. For instance, China's acquisition strategy of equity stakes in exploration and production assets overseas, which began in the late 1980s, has waned, though this strategy can enhance the country's physical oil supply or shield it from the effects of higher prices in the event of a supply crisis. But today China's national companies control just 600.000 b/d. Only 40 to 50 percent of this shipped oil is shipped to China. Although total Chinese equity oil output from overseas production could grow up to one mb/d by the beginning of the next decade (and not all of this oil will be shipped to China for technical and cost reasons), it will be just about 10 percent of the country's total oil demand. In this light, Chinese experts are divided in their opinion of whether this "going out" strategy and the equity oil really contributes much to China's future energy security (IEA 2007: 178–179 and 327–328; Downs 2007).

More important are China's plans to build up a strategic oil reserve for emergency situations. Although the construction and filling of storage sites have been postponed several times, the completion of the first phase (to be equivalent to 24 days of net imports) was planned for the end of 2008. A second phase would increase the capacity up to 61 days by 2010, and a third phase a capacity of 75 days by 2015 (IEA 2007: 327). In this regard, the earthquake in Sichuan province last May could accelerate the second and third phases of implementing strategic oil storage as well as the development of a long-debated strategic gas storage (Tu 2008a: 3).

China's energy policies and "oil diplomacy" have often given bilateral relations clear priority over multi-lateral strategies and solutions designed to safeguard its energy supply. But China could find itself exposed to growing political pressure from the oil and gas exporting states in the Middle East. This pressure could result in either even greater Chinese arms exports, including sensitive dual use goods and technologies, or concessions by Beijing on other political issues that run counter to Western and EU policies and long-term strategic interests such as in the case of Iran.

On the other hand, increasing political and economic interdependencies could have a number of positive effects on the basic structures of the international system and regional political stability in the Middle East. The long-term interest of the Chinese government in political stability in the region could therefore increase, in turn opening up greater possibilities of cooperation not only for bilateral Chinese relations with the U.S.A. and the EU. Expansion of its political and economic, military, and military-technology relations with the Middle Eastern states will also give China an increasing degree of influence over them and strengthen Beijing's position at a global level (in the UN, for example). At the same time, however, these energy and foreign policy dependencies are also a major problem for Beijing as it will increasingly run the risk of being unwillingly drawn into local or regional political conflicts, but without having a political influence comparable to that of the U.S.A. on the potential conflicting parties.

3. The EU's Energy Policies at the Crossroads: Is a Common Energy (Foreign) Policy Emerging?

3.1 *The EU's Evolving Energy Strategies since 2006*

Although historically energy questions dominated the negotiations leading up to the treaties of Paris (1951) and Rome (1957), the specific institutional provisions were made just for coal and the nuclear industries (leading to the *EURATOM* treaty in 1957). In regard to oil, gas, and renewable energy sources, each EU member state of the European Union (EU) is free to decide their own national energy policies. These historical circumstances explain the lack of a common and coherent energy policy of the EU. Furthermore, three other factors have complicated the design of a common energy policy:

- Firstly, during the last 15 to 20 years, the EU's and its member states' energy policies have been increasingly determined by market forces and a separation of energy questions from political factors and strategic developments. Ultimately, energy policies have often been left to the industry. Their business interests, however, are primarily guided by short-term economic benefits in an increasingly competitive environment.
- Secondly, for historical, geographic, and economic reasons, the 27 member states of the EU often have a very different national mix of energy resources (oil, gas, coal, and renewables). As a consequence, they often perceive different challenges in terms of energy imports and supply security, and accordingly, define different strategies to cope with them.
- Thirdly, both energy companies as well as national governments such as Germany have neglected both mid- and long-term national interests of energy supply security. In addition, with the privatization and liberalization of the European gas sector, in which new companies emerge, there will be no single party to assume overall responsibility for the security of gas supply.

Even after 9/11, the EU member states—in contrast to the European Commission, which has addressed energy security since its first Green Book of November 2000 (European Commission 2001)—did not address and discuss global and regional energy security issues. It was only the Russian-Ukrainian gas conflict at the beginning of 2006 that raised increasing doubts in the European public, governments, and parliaments about whether Moscow will remain a reliable energy partner for the EU (Umbach 2006a). So far, however, the EU has been unsuccessful in persuading Russia to pursue a market-oriented energy policy or in deepening a real "Strategic Energy Partnership" between both sides, as the latest Russian-Ukrainian gas conflict and the supply cuts to the EU have highlighted again. For its part, Russia's reputation as a reliable energy partner is undermined by the imminent threat of a potential gas shortage—meanwhile confirmed by President Putin himself as well as other prominent Russian government members. But Putin's strategy for dealing with the gas shortage by rapidly expanding nuclear energy (building more than 40 new reactors in the next 20 years) and coal production for domestic gas consumption won't solve the gas crisis today and tomorrow, nor does it provide a sufficient and

Table 2.3 EU—Primary Energy Demand 1971–2030 (Mtoe).

	1990	2005	2015	2030	2005–2015*	2005–2030*
Coal	451	317	291	275	–0.8%	–0.6%
Oil	626	671	678	670	0.1%	–0.0%
Gas	295	444	509	610	1.4%	1.3%
Nuclear	207	260	239	159	–0.8%	–2.0%
Hydro	25	26	34	37	2.8%	1.4%
Biomass and waste	46	83	127	182	4.3%	3.2%
Other renewables	3	13	33	72	10.0%	7.2%
Total	1,653	1,814	1,910	2,006	0.5%	0.4%

* Average annual rate of growth.
Source: IEA 2007: 616

realistic mid- and long-term strategy (Riley/Umbach 2007; Umbach 2007a: 17–18 and Umbach 2008d).

In March 2007, the European Council agreed on an integrated climate and energy policy with an 'Energy Action Plan' (EAP) for 2007 to 2009. The EU's long-term strategy for energy supply security needs to cope with uninterrupted physical availability of energy products on the market at a price which is affordable for all private and industrial consumers. At the same, the EU needs to balance its future energy supply policies with growing environmental concerns, which have become an even more important objective in light of the Kyoto Protocol. To this end, the EAP favors a liberalized internal market for gas and electricity, enhanced measures for security of supply, and a common approach to an external energy policy with a global dimension (European Council 2007). The summit's conclusions are mostly in line with the previous recommendations of the EU Commission in January 2007 (European Commission 2007c), its Green Paper of 2006 (European Commission 2006) and its "Green Paper on Energy Efficiency" of September 2005 (European Commission 2005). In general the energy policy of the EU aims at a careful balance of all three parameters: security of supply, competitiveness, and environmental sustainability.

The German EU presidency negotiated the world's most comprehensive action plan (containing 17 individual measures) on climate protection and energy supply. The EU-27 were able to agree on a set of tasks and targets at the March summit of 2007. In this "20–20–20 initiative," energy conservation and enhancing energy efficiency play the key role:

- Energy efficiency should be increased by 20 percent across the EU;
- The goals of the Kyoto Protocol should be exceeded and carbon emission should be reduced by 20 percent by 2020 compared to the levels in 1990 (if other industrialized countries such as the USA, India, and China commit themselves to similar policies, the EU would be willing to reduce emissions by 30%);
- Additionally, a 20 percent share of the energy mix should be generated from renewable energy sources. Latvia, Sweden, Finland, and Austria have already attained this target, although the Swedish and Finnish successes are due to the use of nuclear energy.

Disagreements arose concerning the ambitious climate policy targets, such as the increase in the share of renewable energies in overall EU energy consumption by 2020 and whether nuclear energy can be considered a carbon-free energy source. Controversies especially erupted around the question of to what extent nuclear energy could be used to reach this target.

The new European Commission's "2nd Strategic Energy Review" and its new "EU Energy Security and Solidarity Action Plan" of November 2008 (European Commission 2008b) have identified major weaknesses and problems that need to be overcome on the way to a real common energy (foreign) policy and enhancement of the energy supply security of its 27 member states. It has proposed five key areas for joint cooperation and projects in the forthcoming years:

- Infrastructure needs and the diversification of energy supplies;
- External energy relations;
- Oil and gas stocks and crisis response mechanisms;
- Energy efficiency;
- Making the best use of the EU's indigenous energy resources.

The EU has identified the structural and political weaknesses and shortcomings of the EU's declared common energy policy, and the insufficient physical infrastructure preventing a more effective crisis supply management as during the last Russian-Ukrainian gas conflict in January 2009. It remains to be seen whether the EU and its member

Table 2.4 Total Primary Energy Demand of the EU 2005–2030 (Reference and Alternative Policy Scenarios).

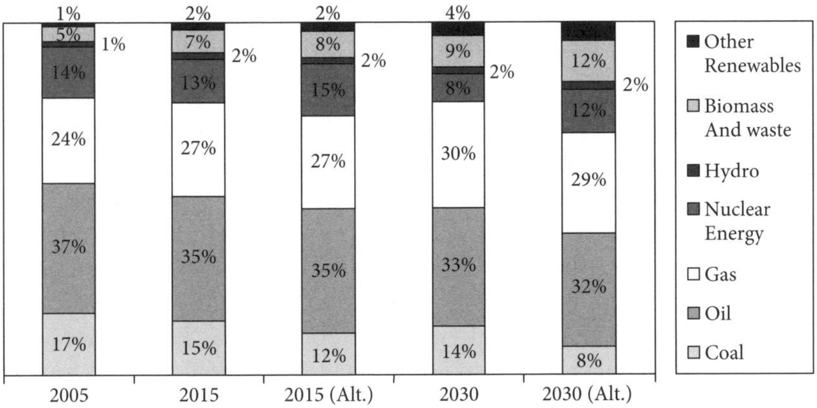

Source: IEA, WEO 2007: China and India Insights, Paris: IEA/OECD, 2007, 616

states have the sufficient political will and necessary financial resources, together with the European energy industry, to implement those projects and policies against all vested interests and political resistance. In general, however, and as many critics have recognized, the EU has accomplished a great deal in such a short period since 2006.

3.2 *Europe's Oil Supply Security in Perspective*

Since November 2000, the European Commision has warned in its first "Green Paper" that in the next 20 to 30 years up to 70 percent of the Union's energy demand (presently 54%) will have to be imported. In regard to oil, the EU's dependence could reach up to 93 percent, for gas 77 percent, and for coal up to 59 percent by 2020.

Indeed, EU members possess only about 0.6 percent of the world's proven oil reserves, 2.0 percent of the global gas reserves and, at least, 7.3 percent of proven coal reserves. Through its eastward extension, the EU has been able to increase its coal reserves substantially (by 41%), but not its oil and gas reserves. In 2006, the EU-27's total primary energy supply was generated by oil (37%), gas (24%), solid fuels (18%), nuclear energy (14%), and renewables (7%). 54 percent of Europe's energy is imported. These imports had cost an estimated €350 billion and €700 for every EU citizen until the summer of 2008 (European Commission 2008d).

Future capacity will still be predominantly generated by fossil resources, albeit with a rising percentage of gas, whilst the number of oil and solid-fuel power stations will continue to decline (European Commission 2008c). Accordingly, the European Commission's energy demand management strategy has always emphasized the diversification of energy supply, promotion of renewable energies, and a neutral look at the nuclear option.

But oil will remain the most intensively used product in the EU's fuel mix until 2030 in spite of having a decreasing share (see tables 1.3 and 1.4). Less than one-fifth of the EU's total oil consumption can be guaranteed by its own production. Currently oil also comprises most of the EU's total energy imports (60%) in contrast with gas (26%) and coal (13%). Its oil imports in 2006 of 608 million tons of oil equivalent (Mtoe) were delivered by OPEC countries (38%), Russia (33%), Norway (16%), and Kazakhstan (5%). In this regard, the EU's oil import diversity is much better compared to its gas imports, which are constrained to just four big suppliers. However, domestic gas production is currently better than oil production because domestic production satisfies about two-fifths of consumption needs (European Commission 2008c: 9; 15–17; 26–27). But given the decline in production in the North Sea from 6.8 million barrels per day (mb/d) in 2000 to some 5 mb/d in 2007 and 6 percent annually until 2020, the EU will become more dependent on imported oil supplies from much more politically unstable countries and regions in the next years and decades.

In order to improve coherence of the emergency strategic oil storage of the EU with the IEA, to increase the reliability and transparency of available stocks, to simplify compliance and verification as well as to clarify emergency procedures, the European Commission proposed in November 2008 a revision of the EU's emergency strategic oil stocks legislation because its member states have developed very different mechanisms and security standards. The existing mechanisms have proven effective for dealing with limited disruptions, but might not be effective enough for all future scenarios (European Commission 2008b: 10).

Table 2.5 European Natural Gas Imports from Russia in 2005 (in %).

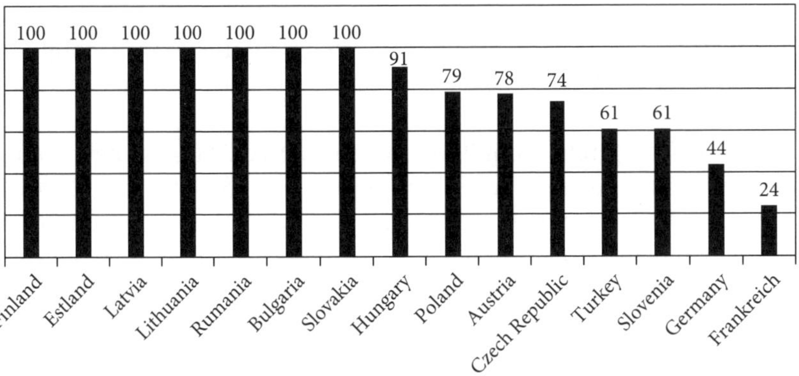

Source: EU, IEA, Eurostat, 2005

3.3 *Europe's Gas Challenge*

The expansion of natural gas as an environmental clean energy source will probably be the most problematic factor in the next two decades for EU member states. Europe is already today the largest natural gas import market and will continue to be the world's champion of gas importers until 2030. But today, almost half of the EU's gas consumption is imported from only three countries: Russia (23%), Norway (14%), and Algeria (10%). The new EU members and former allies of the Soviet Union in particular are still very much or even completely dependent on gas imports (see table 2.5) not just from one country (Russia), but even from a single company (Gazprom). Up to 2020, European gas production will decline between 3 to 4 percent annually, which is, however, taking place at a slower pace than oil production under various oil price scenarios and policy measures (European Commission 2008c: 17).

Given current trends, gas imports will increase to 80 percent over the next 25 years. In 2030, Europe will have to import 488 bcm (North America: 159 bcm and China/India 150 bcm). The projected additional EU imports of gas range from 50 bcm up to 300 bcm by 2030. The share of gas in total primary demand would simultaneously rise from 23 percent at present to 32 percent in 2030. Since the mid-1990s, low natural gas prices, relatively low capital cost, favorable technology innovations, particularly in regard to environmental emissions, and new investments in combined cycle gas turbines technology (CCGT) have been drivers of the increasing gas demand in Europe. In the future, a growing share of EU gas imports will be shipped as Liquefied Natural Gas (LNG), which would offer a better crisis stability for gas imports (IEA 2006: 117–118; Hirschl/Schlaak/Waterlander 2005; Götz 2007: 9–11).

Given the March summit results and agreed targets, new studies have predicted a much lower gas import for the EU until 2030. In 2005, the EU's combined gas production with Norway amounted to nearly 300 bcm, which will decrease to 200–250 bcm by 2030. Between 2005 and 2006, the IEA had already reduced its forecasts of the EU's rising total gas imports in 2030 from 530 to 488 bcm. In the future, they might be somewhere between 300 and 450 bcm (317 bcm in 2005 by EU-27) to compensate for the decline in domestic production and the increase in demand.

In order to strengthen its energy supply security, the EU has proceeded with a number of pipelines and LNG projects to import non-Russian natural gas sources (see table 2.6) in addition to the planned Nord-Stream Pipeline (2 × 27.5 bcm = 55 bcm). Furthermore, the Greensteam Pipeline from Libya to Italy became operational in 2008 and will transport 9 bcm. With increased production in the North Sea, Norway's exports will rise from 84 bcm to 125 to 140 bcm annually. In addition, the announced LNG projects would represent an additional import capacity of about 100 bcm per year after 2010 (European Commission 2007b: 8 and 24). Altogether, the EU will have an additional 150–200 bcm of non-Russian gas available in 2020. These alternatives will give the EU more leverage and bargaining power vis-à-vis Moscow.

Despite its engagement policy of "Verflechtung" (building interlinkages) with Russia, Germany has also strengthened its diversification of

Table 2.6 Main Greenfield Pipeline Projects.

Project	Supplier	From	To	Capacity (Bcm)	Investment (M €)	Foreseen Start-Up
Medgaz	Algeria	Hassi R'Mel	Spain	8–10	1300	End 2008
Galsi	Algeria	Hassi R'Mel	Italy	8–10	1200	2009–2010
ITG-IGI	Caspian	Greece	Italy	8–10	950 (IGI)	2011
Langeled	Norway	Ormen Lange	UK	22–24	1000	2006–2007
Nabucco	Caspian	Turkish Border	Austria	25–30	4600	2010
Total additional non-Russian gas supply capacity via pipelines to Europe: 71–84 bcm						

Source: European Commission 2007b: 24

gas imports. Even those companies like E.on Ruhrgas or VNG, which have close strategic ties to Gazprom, will import much more gas from Norway and are planning a LNG terminal in Wilhelmshaven. Furthermore, Germany's second largest energy company, RWE, has become an official member of the Nabucco-project (Umbach 2008a; Umbach 2008b; Umbach 2008c: 19–20).

Furthermore, the European Commission has also called for a new Security of Gas Supply Directive in 2010 for improving greater harmonization of security of supply standards and predefined emergency measures on the regional and EU levels. It also takes into account the limited strategic gas stocks within the EU, which cost at least five times more than oil stocks, but limit Europe's gas supply security and options for reactions particularly in a future gas supply crisis (European Commission 2008b: 11), as has been highlighted again during the last Russia-Ukrainian gas conflict in January 2009.

3.4 *The EU's Ambivalent Coal Policies*

After years of discrediting coal, the Commission has come to view coal as an important energy source in the future that can contribute to enhancing the security of supply in the EU. It decided to support technical progress in terms of the actual clean burning process of coal. Meanwhile, Europe, like the rest of the world, is experiencing a renaissance of coal due to its much lower economic costs (even with the cost of permits to pollute factored in) and as the result of the aim of reducing gas dependence (particularly in the new EU member states) and strengthening energy (supply) security. The Italian and German industries, for instance, see little choice but to build new, more effective, and environmentally clean coal plants to replace aging infrastructure, compensate for the phasing-out nuclear power, and to guarantee the base-load of electricity consumption. But here again, Germany seems rather to go for another "Sonderweg" in its energy policies in contrast with the other EU member states due to local resistance and a wider public protest in the form of a new anti-coal movement against the building of new coal power plants (Umbach 2008f: 4).

Currently, the sources of coal imports are also less concentrated and more diversified in contrast to those of gas imports. The largest suppliers are Russia (26%), South Africa (25%), Australia (13%), Colombia (12%), Indonesia (10%), and the United States (8%) (European Commission 2008d: 7). But imports will increase from under 40 percent today to 57 to 59 percent, depending on oil and gas prices as well as EU energy policies. While primary energy consumption of hard coal and lignite stagnated between 2000 and 2006 at around 18 percent of the EU's fuel mix, they may exceed their current levels by 6 to 7 percent in 2020 following projected soaring oil and gas prices. They will strengthen the relative competitiveness of coal, particularly in countries like Germany that maintain a policy of phasing out nuclear power (European Commission 2008c: 16–17).

Prior to the EU's '20–20–20 initiative' in March 2007, the EU's own energy production was projected to fall from 46 percent today to 36 percent by 2020 following a 'business-as-usual' or 'nothing to do'scenario. But with the EU's new Energy Policy, the share of its own energy production would be kept at around 44 percent of the EU's consumption. The Commission's call for "making better use of the EU's indigenous energy reserves" has also considered coal an essential component of Europe's domestic energy supply and an important alternative to oil and gas (European Commission 2008b: 14). Only higher CO_2 emissions are considered a main drawback, but this might be overcome by highly efficient plants and carbon capture and storage (CCS) technologies. But the EU's aim to have up to 12 commercial scale demonstration plants in operation by 2015 and the G8 commitment to launch 20 of those demonstration plants worldwide by 2020 will need greater incentives.

3.5 *The Renaissance of Nuclear Power in Europe*

The EU's first "Green Paper" in November 2000, being concerned over the projected increased external dependence over the next few decades, had already criticised the five (Germany, Sweden, Spain, Netherlands, and Belgium) out of eight EU member states (the other three are France, the United Kingdom, and Finland) with nuclear power who have adopted or announced a moratorium on nuclear power or decided to give up nuclear energy production. At present, nuclear energy does play a vital role—in 2006 it produced 29 percent of electricity in Europe compared with 15 percent from renewables—in the sustainable production of electricity. The share of nuclear power in primary energy consumption is considerable in many of those countries that have opted for it, such as France (42%), Sweden (35%), Lithuania (26%), Bulgaria (24%), Slovakia (24%), and Belgium (21%) (European Commission 2008c: 8). It is also seen as a major source of baseload electricity and the only industrially mature energy source with negligible greenhouse gas emissions which can be expanded. Moreover, EU uranium supplies are diversified within politically stable regions and countries such as Australia and Canada, which cover nearly 50 percent of EU imports (European Commission 2008b: 15). The cost of uranium has marginal impact on electricity prices or the building of new nuclear power plants. The Green Paper of 2000 warned that the EU would not meet its obligations under the Kyoto Protocol without nuclear energy. Annually, it avoids some 300m tons of carbon dioxide emissions—equivalent to half the amount produced by all cars in the EU (European Commission 2001). But in the next 10 to 20 years, the majority of nuclear power plants in the EU will reach the end of their originally designed lifetimes and the share of nuclear energy in power generation will decline accordingly if no decisions about lifetime extension, new investments, or replacement are made.

Against this background, some EU member states have begun to rethink the nuclear option as the EU Commission, the IEA, the World Energy Council (WEC), and numerous international energy experts have recommended for years. Even Germany's unilateral withdrawal from the use of nuclear power may not last in the coming years. Besides Russia and Ukraine as non-EU member states, Finland, France, Great Britain, and many new Central European members of the EU have already indicated that they do not want to renounce the nuclear power option. In fact, the construction of new nuclear power plants is being declared or at least seriously considered, as is the possibility that lifetimes of nuclear reactors be extended (like in Sweden, despite the recent nuclear accident, or in Great Britain and Italy, which have also announced that they will build new nuclear power stations). For economic, environmental, technological, and political reasons, the nuclear power option is also undergoing a renaissance in the United States, Russia, and especially Asia.

Since the G8 summit of 2007, Germany has isolated itself in regard to the use of civilian nuclear power and failed to assert itself at the European level. The Spring 2007 summit concluded that the lifetime extension of nuclear power plants can improve the CO_2 balance. Yet the Spring European Council's agreement was clearly a compromise and a common European response on the future of nuclear energy is still missing. France, a country that generates 40 percent of its energy supply from nuclear energy (and 77% of its electricity supply), thus has been able to comply with the required share of renewable energies without implementing any additional measures.

Given the EU's ambitious climate protection policies of reducing CO_2-emissions by 20 percent by 2020 and under projected future high oil and gas prices, falling power generation from gas will be compensated by higher electricity generation from nuclear power (+13%), renewables (+8%), and solid fuels (+4%). In this case, nuclear power would represent 13 to 14 percent of the primary energy demand in 2020 as it is today (European Commission 2008c: 17).

4. The EU's Climate Policies: Global Leadership Ambitions Versus Economic-Ecological Realities

The EU's March 2007 decisions on an integrated climate and energy policy recognize the fact that the current EU energy policy is "unsustainable." Given the projected 'business as usual' energy and transport trends, the CO_2 emissions of the EU would rise by 5 percent and by 55 percent worldwide by 2030. The Commission follows the recommendations of international climate change experts to stop the global average temperature from rising more than 2° C higher than pre-industrial levels.

On February 20, 2007, the EU concluded with a "firm independent commitment" to reduce its GHG emissions by at least 20 percent below levels in the 1990s. If other OECD nations follow, the output of GHG could be reduced by no less than 30 percent by 2020. In the view of the EU Council, developed countries should collectively reduce their emissions by even 60 to 80 percent by 2050 from levels in 1990. Even in the case that the Kyoto Process fails to materialize, the EU emission trading will continue. Until the March 2007 decisions, the EU agreed to reduce its emissions by 8 percent by 2012. Underpinned by a spirit of solidarity amongst member states with their "differentiated responsibilities and respective capabilities," and by emphasizing cost-efficient measures, the burden-sharing process for reducing the EU's carbon emissions by 20 percent by 2020 means that certain countries like Germany will have to carry a greater burden on account of their greater economic power.

The Federation of German Industries (BDI) and others, however, had warned prior to the summit that overly ambitious climate protection targets would jeopardize the competitiveness of German companies without meaningfully improving global climate conditions. Indeed, compared with the 15 percent of global CO_2 emissions today, the EU will only account for about 4 to 6 percent of global CO_2 emissions and less than 15 percent of new ones by 2030 (Umbach/Skiba 2007). However, not addressing the climate change problems may result in much higher costs later. In this regard, the EU sees itself as the leading political actor worldwide in international efforts to contain the effects of climate change and hopes to benefit from this role politically as well as economically. But it remains questionable whether the self-proclaimed leadership role of the EU on climate matters will be honored and followed by the United States, India, and other transition countries.

The present and future development of the EAP and the integrated climate policies are also hampered by the fact that until now only a few members have implemented attractive strategies for renewable energy sources. Hence even for the EU it is uncertain whether or not it will be able to live up to its obligations under the Kyoto Protocol, which envisions cutting greenhouse gas emissions by 8 percent between 2008 and 2012 from 1990 levels. At the end of 2005, only Sweden and Great Britain had fullfilled their obligations of the Kyoto Protocol.

With the increasing critical global debate on the first generation of biofuels as a replacement for petrol and diesel supply for the transport sector, the 10 percent binding minimum target for the share of biofuels in overall EU transport petrol and diesel consumption by 2020 is already been threatened and thus the overall objectives of its policies for mitigating climate change as well. But if the EU is able to implement and achieve its March 2007 aims by 2020 it would be using 13 percent less energy than today, which is equivalent to a savings of more than 100 billion Euros and a reduction in CO_2 emissions of about 780 million tons per year (European Commission 2007: 13). Yet as it seeks to achieve this, even Germany faces tremendous challenges if it won't undermine its national energy supply security and economic competitiveness in its energy triangle of objectives (Umbach 2008f: 2–4).

On March 14, 2008 the High Representative of the CFSP and the European Commission presented a paper on "Climate Change and International Security" (High Representative/European Commission 2008). It highlights the implications of climate change for international peace and security. The authors of the paper view climate change as an "irreversible and largely unpredictable" process, and at best as a "threat multiplier" exacerbating existing security trends, tensions, and instabilities. It envisions that already fragile and conflict-prone states and regions will be overburdened by threats brought on by climate change. The paper demands comprehensive policy responses by the international community and sees the EU itself as in a unique position to respond to the impacts of climate change on international security and to take up a leading role. The paper lists seven sources of direct threats of climate change can bring on worldwide peace and security:

- Conflict over depleting resources such as arable land, water, food, and fish stocks;
- Economic damage and risk to coastal cities and critical infrastructure, which could cost the world economy up to 20 percent of global GDP per year, whereas the costs of effective and timely concerted action can be limited to just 1 percent;
- Loss of territory and borders, such as small island states, and the potential conflict over resources in polar regions which will become exploitable as a consequence of global warming;
- Environmentally-induced migration, which the UN has predicted to be in the millions by 2020;
- Situations of fragility and radicalization, particularly in weak and failing states by overstretching the already limited capacity of governments. It can lead to tensions between ethnic and religious groups within countries and to political radicalization, even destabilizing entire regions;
- Tensions over energy supply arising from intensified competition over access to, and control over, energy resources;
- Pressure on international governance because the negative impacts of climate change may fuel resentment between those responsible for climate change and those most affected by it (having both a North-South as well as South-South dimension) at a time when the international security architecture is already facing increasing problems to cope with, as it faces very different circumstances compared with the time of its creation.

The EU has recognized that despite its claim to a leading role in the process of mitigating global climate change and its interrelated impacts on worldwide peace, security, and stability, it needs to first enhance its own common political will and capacities at the EU level. Those capacities range from observation, analyses, and monitoring to pre-warning and conflict prevention and finally to crisis management and disaster response instruments (both civil and military). Moreover, it needs to engage major powers and emitters such as the U.S.A. and China and commit them to a new ambitious climate agreement under the UN framework (Kyoto-2).

5. Speaking with One Voice?—The EU's Emerging Energy Foreign Policy and its Energy Dependence on Russia

As the EU has recognized in recent years, it cannot achieve its energy and climate change objectives on its own. By 2030, the EU may consume much less than 10 percent of the world's energy. But the enlarged European Union borders the main oil and gas producing areas such as Russia, the Caspian Sea, and North Africa, and is not far from the Middle East and the Persian Gulf. Given the close relationship between geopolitical stability and energy supply security, the rising dependence of the EU on energy imports and the growing interdependence between producer, transit, and consumer states, the Commission has recognized the need for closer cooperation with supplying partners as part of an proactive energy foreign policy. It seeks to encourage geopolitical and economic stability in supplier as well as transit countries and predictability in producer-consumer relationships.

Consequently, the EU has become more proactive in order to widen and deepen its energy dialogues with neighboring countries and other producers in Central Asia, North Africa, the Gulf-region (OPEC), Australia, Africa, Canada, and recently also with Brazil and other countries in Latin America and the Caribbean.[1] The EU has also strengthened its energy dialogues with other consumer states such as the United States, Japan, China, and India. While the EU in general still favors a strong market-approach in its international energy strategy, it has also begun to pay much more attention to energy supply security and its inherent challenges as well as geopolitical risks.

Meanwhile, the EU's original great hopes of a close strategic energy partnership with Russia have dissipated as the two Russian-Ukrainian gas conflicts of January 2006 and January 2009, as well as the Russian-Belarusian gas conflict in early 2008 have repeatedly highlighted. While the EU now views such a partnership more realistically as a long-term

[1] For more on the EU's and the European Commission's energy external relations see http://ec.europa.eu/external_relations/energy/index.htm.

vision, it has also become increasingly uncertain about whether Moscow will be able to increase its gas exports beyond 180 to 200 bcm after 2020 due to an emerging domestic gas crisis. In other words: the EU is forced to diversify its gas imports anyway (Riley/Umbach 2007; Umbach 2007a: 17–19).

Furthermore, the EU heads of state and governments have failed to agree upon a common strategy towards Russia, the bloc's most important energy supplier. Since the EU's March 2007 summit, the lack of coherence of the bloc's external energy policy has enabled Russia to continue successfully with the "bilateralization" of energy partnerships with Austria, Italy, Hungary, Greece, and Bulgaria at the expense of the other members of the EU and the EU itself (Umbach 1007a: 18–19). Thus Russia has remained successful by adopting its traditional politics of "Divide and Rule!" and playing individual European states and their "national energy champions" against each other. Russia's approach has been most visible in its efforts to undermine a common European policy toward Central Asia. Moscow tries to torpedo the Nabucco Pipeline project, which is of crucial importance for European energy autonomy and the diversification of energy supply with a rival pipeline ("South Stream"). It also tries to maintain and strengthen its gas pipeline monopoly not just from Central Asia, but also from all other real and potential gas suppliers to the EU member states (such as Iran, Qatar, and North African states) which would have far reaching impacts on the EU's liberalization of its energy (particular gas) markets and its CFSP (Loskot-Strachota/Pelczynska-Nalecz 2008; Umbach 2008a).

Norway has become a winner in the EU's worsening energy relationship with Russia. As a democratic country with a rule-based political system and a well-functioning market economy it is already integrated in the internal energy market as a member of the European Economic Area. It is expected to increase its gas and oil supplies to the EU and Germany. The EU-Norway energy dialogue will be further developed through new common projects such as offshore wind in the North Sea and new developments of Norway's oil and gas resources, such as through LNG projects.

The recent Russian-Georgian war in August 2008 and the European Commission's "2nd Strategic Energy Review" and its new "EU Energy Security and Solidarity Action Plan" (European Commission 2008a) have highlighted again the improved, but still insufficient, political solidarity between its 27 member states when speaking with one voice to external energy partners. This is also true of its international relations at a time when the overall power balance between consumers and producers have changed in favor of the latter, leading to the creation of a "sellers market". But by building new gas and electricity connections as part of enhancing energy supply security and the creation of common energy markets, these developments will ultimately lead to more common energy (foreign) policies. What the EU needs most of all in the coming years is the common political will to implement all the decisions they have agreed upon—which ultimately boils down to the issue of political credibility in its economic, energy, and Common Foreign and Security Policies (CFSP).

6. Summary and Perspectives

For the EU and China alike, integrated solutions to the energy-climate nexus are needed more than ever to balance energy priorities with economic and environmental objectives. Of course, national security objectives are always difficult to work out and to implement due to the fact they require difficult tradeoffs between domestic and foreign policy interests. But meanwhile climate scientists largely agree that the world's glaciers and northern ice caps are melting at accelerating rates and that sea level rise will threaten many coastal and low-lying areas.

As long as fossil fuels continue to dominate the global fuel mix, energy-related greenhouse gas emissions and increased reliance on oil, gas, and coal imports from politically unstable countries will increase concerns about climate change as well as energy security. Having no adequate and secure supplies of energy at affordable prices is being perceived as a major threat as soaring energy prices and consumption cause irreversible environmental damage for societies.

Hence policymakers need to address these twin challenges of energy security and climate change to ensure the security of our global energy system and to reduce greenhouse gas emissions as part of an overall strategy for a sustainable energy security concept. The energy-climate nexus has been highlighted by numerous events. Energy supply disruptions, for instance, are also the result of extreme weather conditions or accidents as hurricanes Katrina and Rita in August and September 2005 taught the international community. They had worldwide implications for global oil prices, energy policies, climate change, strategic oil stocks, and perceptions of supply security. The global energy security challenges, its interdependencies with climate change, and their implications for the future foreign and security policies of the West have already opened new policy options for closer EU-China cooperation on numerous energy and environmental challenges.

On the EU side, due to the fact that the energy sector gives rise to 80 percent of all its GHG-emissions, the EU is focusing on raising the level of renewable energy sources up to 20 percent by 2020 and increasing energy conservation and efficiency by another 20 percent. This would help to both improve the overall energy balance and to decrease import dependencies, as well as improve on the environmental objectives.

But given the difficult burden-sharing of the implementation process of the March 2007 decisions on the EU's integrated energy and climate policies, as well as the skyrocketing oil and gas prices, the French presidency had to focus on internal crisis management to maintain the political credibility of the EU's policies and to strengthen the political solidarity amongst its members, moving away from short-sighted free-riding attitudes and populist unilateral tendencies of individual member states.

Despite its huge progress on the way to adopting a common energy (foreign) policy as part of the Energy Action Plan, the EU is now confronted with two major challenges in the months and years ahead. Firstly, the public debate about the results of the spring summit primarily concentrated on the "historic agreement on climate change." This rather narrow focus, however, jeopardizes the balance within the energy triangle between security of supply, competitiveness, and sustainability. A one-sided orientation toward climate issues leads to a marginalization of the other two objectives, namely competitiveness and security of supply and, therewith, an inability and lack of attention to cope with many of the above described global and geopolitical challenges. On the other hand, with few exceptions like Great Britain, Sweden, and Germany, most of the EU-15 member states, which are obliged to reduce their emissions by eight percent between 2008 and 2012 from 1990 levels, will fail to fulfill their Kyoto targets (Die Welt 2008: 9).

In the mid- and long-term perspectives, the EU considers its energy (supply) security to have been primarily threatened by a rising gas dependence on politically unstable or unpredictable energy partners such as Russia, wherein many new EU member states are still dependent on just a single supplier and company (Gazprom). China, by contrast, views its energy security through its high dependence on increasing oil imports and coal as its primary energy source, which has raised environmental costs and threatens its economic growth—a precondition of its socioeconomic and political stability as well as regime survival. This is also true of LNG imports via Sea Lanes of Communication (SLOCs), which can be blocked by the US naval forces.

For China even more than the EU, the primary challenge in the years and decades ahead remains how to transition to a more secure and a low-carbon energy system by taking action without weakening economic and social development.

In the past, China largely responded to environmental crises on a piecemeal basis instead of using broader and comprehensive strategies to sustain a stable ecological system or in seeking sustainable development. Beijing's newly announced "scientific development concept," however, has announced its intention to implement an environmentally friendly approach to industrialization, which gives energy conservation a high priority. For the first time, Beijing has established compulsory targets on the efficient use of energy by 2010: (1) energy consumption per unit of GDP is to decrease by 20 percent, (2) water consumption per unit of industrial added value is to decline by 30 percent, and (3) industrial solid waste recycling and the conservation rate are to grow 60 percent (Jiang 2006). While this new approach will be welcomed in the U.S.A., EU, and Japan, doubts remain as to whether China is able to implement it effectively in its vast landscape due to systemic factors in its political system—a "development autocracy" (Minxin Pei)—and its present economic development philosophy.

China's energy foreign policy in the Middle East, Central Asia, Africa, and Latin America has produced an economic-security nexus that is determined by the most fundamental core interest of its political leadership: economic growth and domestic stability in order to ensure regime survival. However, the resulting unilateral energy-security strategies have often undermined multilateral and regional cooperations as well as fuelled already existing strategic rivalries such as with Japan, India, and the United States.

On one hand, China's long-term interest in political stability in the Middle East and other regions of the world could increase and open up greater possibilities for cooperation with the U.S. and the EU.

Expansion of its political, economic, military, and military-technology relations with these energy producers will also give China an increasing degree of influence over them and strengthen Beijing's position at a global level (like in the UN). On the other hand, however, these energy and foreign policy dependencies are also an increasing risk for Beijing, as it will be unwillingly drawn into local or regional political conflicts, but without having political influence comparable to that of the U.S.A. on the potential conflict parties.

In the future, China's energy policy must comprehensively address the issues of energy conservation and pollution control in order to avoid and control rising environmental, economic, and social costs. Although Beijing has set new priorities for increasing energy efficiency, such as adjusting electricity supply structure for higher efficiency or importing modern coalmining technologies with high efficiency and clean burning technology, they are insufficient and need to be an integral part of an overall comprehensive strategy which would include all sectors of the economy and private households.

Furthermore, though China does not have to fulfill any obligations of the Kyoto Protocol, international pressure on Beijing will increase to improve energy efficiency in the years ahead. Hence China needs to design more radical incentive strategies to promote energy efficiency comprehensively, thereby including renewable energy development as well as enhancing its cooperation with the international community. Due to the fact that the EU members and Japan are the leading and most experienced countries with high energy efficiency standards, the EU's sectoral dialogues with China on energy and climate change challenges need to be based on a more comprehensive concept and integrated strategy that takes into account more systematically both China's domestic challenges to its energy-climate nexus as well as its energy foreign policy directions, which are increasingly interrelated and cannot be de-linked as in past EU-China energy dialogues.

The EU's integrated energy and climate strategy of March 2007, its 2nd Energy Security and Strategy Energy Review of November 2008, China's new "White Paper on Energy" of December 2007 (Chinese State Council 2007), as well as its "National Climate Change Programme" of June 2007 (National Development and Reform Commission 2007) offer new, good starting points for a closer, deeper and better-structured EU-China energy cooperation and prospects for improving China's critical energy efficiency and climate protection policies in the future. Moreover, given China's heavy dependence on coal for its primary energy consumption and the global climate change challenge, the "Memorandum of Understanding in Near-Zero Emissions Power Generation Technology" of February 2006 as well as joint projects on renewables open a wide area for joint energy technology projects and an exchange of experiences, as the joint CCS projects demonstrate.[2]

In this light, the Western aim of encouraging China's integration into the international global cooperation structures, while insisting, in return, that Beijing abide by the same rules as everyone else, will remain the major strategic goal and challenge for the years to come. Along the way, whether the EU has the political will, attention, and resources, and whether China will offer more transparency, openness, and strengthen the market orientation of its energy policies and develop integrated concepts for its energy and environmental/climate protection policies will be decisive. In order to have a greater influence on the direction of China's energy and environmental/climate protection policies, it will be important for the EU to take China's views, perception, and political-economic priorities more seriously and to develop some understanding of them. Otherwise, the EU will undermine its own future leverage and influence on China's energy and environmental/climate protection policies.

[2] See, for instance, the "Support to Regulatory Activities for Carbon Capture and Storage (STRACO2)—Project", which is designed to support the development of a regulatory framework for CCS in the European Union. By supporting a CCS regulatory framework inside the EU, STRACO2 will be instrumental for establishing best practice standards globally. By incorporating the Administrative Centre for China's Agenda 21 (ACCA21), the project is part of the EU-China partnership on Climate Change and ensures that the developed solutions are applicable to rapidly developing economies outside Europe which will be crucial in fighting CO2 emissions and climate change. The programme started in January 2008 and will run for at least 18 months—http://www.euchina-ccs.org/index.php.

CHAPTER THREE

RUSSIA'S EMERGING PLACE IN THE EURASIAN HYDROCARBON ENERGY COMPLEX

Robert M. Cutler

Abstract: *The geo-economics of Central Asian energy and its transit are a natural focus for examining Russia's emerging place in the Eurasian hydrocarbon energy complex. In the years 1989–1994, Russia jockeyed for position with other regional actors and also the US in both the South Caucasus and Central Asia. From 1995 to 2000 this competition deepened in Central Asia and the EU began to manifest a more active interest. China, too, began to project its influence into Central Asia up to the Caspian littoral. From 2001 to 2006, as the EU's interest deepened partly due to the Ukraine energy imbroglio. Kazakhstan and Turkmenistan implemented increased energy cooperation with China and began working with the EU to develop oil and gas transit lines circumventing Russia. Complex-systems analysis, including focus on trilateral and quadrilateral relationships (which cannot be decomposed into aggregates of bilateralisms), provides categories for understanding these developments and projecting their continuation into the future.*

Keywords: Russia, energy, geopolitics

1. Introduction

This article addresses Russian energy policies in Eurasia, including the Caspian Sea region, in relation to China and the EU, in the context of current geopolitical developments. Necessarily, it also addresses where appropriate the role of the US in the structuration of geo-economic relations in Eurasia. However, it focuses mainly on the stakes concerned in the negotiation and construction of major Central Eurasian energy pipelines over the last two decades, with special attention to the twenty-first century and its most recent years. The reason for this is that it is in such a domain that the geo-economic position of Russia in Eurasia has most definitely asserted itself.[1]

[1] The analysis draws also on years of the author's discussions and interviews with diplomats, industry figures, and others, as well as on press analysis ('open-source

In the early part of this decade, the national energy trust Gazprom moved aggressively to invest in fields outside Russia rather than promote development of the country's own resources (Stern 2005). Likewise, the expropriation through juridical means of Yukos from its chief Mikhail Khodorkovsky and his imprisonment, together with the gradual forced transfer of management and assets of BP-TNK to formal and informal Russian state representatives, has slowed down and discouraged foreign direct investment in the Russian oil market. Meanwhile, Russian exports of energy to China will not begin until the middle of the next decade, if then, while its control over transit of Central Asia oil and gas to Europe will hardly be alleviated by then, either. These are the reasons why a focus on Russia as a transit country for Central Asian energy is the key aspect for understanding Russia's emerging place in the Eurasian hydrocarbon energy complex.

Treating the evolution of the Eurasian hydrocarbon complex as a 'complex system' draws special attention to pipeline projects, especially undersea pipelines, because they unify elements of adjacent subregions in new ways that can alter the inherited balance-of-power geopolitics. A complex system is a system of which the behavior cannot be predicted by studying the behavior of its constituent parts: for instance, the post-Cold War international is a complex system. It is a self-organizing network wherein bottom-up structuration contributes significantly to systemic configurations and intermediate levels of emergence and coherence influence subsequent evolution more than ever (Bar-Yam 1997).

The construction of international energy pipelines in the twenty-first century is comparable to that of national railroads in the nineteenth (see Gerschenkron 1962). Since then, the policies of national energy trusts in geo-economics have become a principal means of advancing national interests in geopolitics. They are not only axes for economic development but also instruments of national power projection. Consequently, the choice of pipelines to build and of the partners with whom to build them is an index of political entente, if not alliance.

---

intelligence') that he has produced for both public and private consumption. He will be happy to supply documentation for any statements of fact to the degree that promises of anonymity to interviewees can be respected.

2. Geo-Economics and Complexity Science

The 'geo-economics' of the Eurasian hydrocarbon energy complex refers to the international economic and political factors influencing hydrocarbon resource development, and their consequences. Central Asia's oil and gas deposits have attracted the attention of many international powers. Kazakhstan's large energy resources drew significant Western attention to Central Asia soon after the disintegration of the Soviet Union. Turkmenistan, the other energy export giant of Central Asia, has yet to live up to its real potential. Uzbekistan's large population accounts for the country's high domestic consumption and lower level of exports, despite its very high production levels. Because of this, the present chapter does not address Uzbekistan in detail.

International regions today enjoy an increased relative autonomy of the general international system in comparison with the bipolar Cold War system. Not only have new international regional subsystems emerged, but also new categories of such regions as well; littoral basins, for example, have become more important, and regional international systems are more and more densely linked to one another. The key aspect of such littoral basins is international energy pipeline construction, but their growing significance is also due to international public policy issues in the broad sense, including ecological security, applicable legal regimes, and the need to set cross-sea trade by the littoral states on a firm and regular footing (Cutler 1999).

'Complex-system' analysis draws special attention to pipeline projects, especially undersea pipelines, because they unify elements of adjacent subregions in new ways that can alter the inherited balance-of-power geopolitics. Complex-system analysis reveals three successive phases in the transition to the international system now unfolding. These correspond to the complex-system categories of emergence, autopoiesis, and coherence. (Autopoiesis is the epigenetic stage reached by an entity that is able to begin, relatively autonomously of its environment, to direct the future course of its own evolution (see Luisi 2003; Mingers 1994; Varela, Maturana, and Uribe 1974)). In the Central Eurasian theater of energy geo-economics, these three phases have manifested since 1989 respectively as the 'bubbling-up' of possibilities for new patterns of international relations, free from bipolar constraints, the 'settling-down' of unsustainable patterns of structuration of regional subsystems (including the beginning of their relatively autonomous self-direction of their own evolution as regional subsystems of international relations), and the 'running-deep' of reciprocal relations among those new subsystems(including their incipient coherence).

The present period represents the emergence of a new international system after the end of the Cold War. An evolutionary approach that historically situates this present period is extremely useful, if not necessary. For both narrative and analytical purposes in the all-Eurasian theater, and with particular reference to Central Eurasia, the 'bubbling-up' phase may be assigned the dates 1989 to 1994, the 'settling-down' phase 1995 to 2000, and the 'running-deep' phase 2001 to 2006. During the first of these, with respect to energy geo-economics, the Russia-Southwest Asia-Europe triangle is central, during the second the Russia-Central Asia-Europe triangle, and during the third the Russia-Central Asia-China triangle. This is not to say that each set of triangular relations exhausts itself within six years; rather, they overlay one another, progressively inflecting their reciprocal development. The six-year periods corresponding with each triangle represent the launch of the 'emergence' phase of the development of the given regional triangular energy geo-economic relations within the overall Eurasian space.

Since I have just mentioned 'Southwest Asia' instead of the South Caucasus, it is imperative to define this term. Southwest Asia is sometimes taken to include the Arabian Peninsula. However, because there is no Russia-China competition for energy resources there, and since the geo-economics of that region have not undergone substantial restructuring as a result of the disintegration of the Soviet Union and transformation of the Cold War system, the Arabian Peninsula is excluded from the scope of 'Southwest Asia' as treated here. Rather, Southwest Asia is here construed in the first instance to comprise Turkey and the three South Caucasus countries, and in the second instance as including also the North Caucasus and eastern Black Sea littoral as appropriate to the particular scale of the energy project being discussed (compare Lewis 1992).

Such an approach is justified because that is the region where geo-economic conflict over energy between Russia and non-Russian powers originated immediately after the end of the Cold War. This has continued into the present day. Even if the US was the initial driving diplomatic force behind the projects for pipeline construction westward from the western coast of the Caspian Sea, nevertheless executives of the UK-based international energy company British Petroleum (which now simply calls itself BP) were in fact the first to sketch these pipelines on a map.

3. Emergence and Autopoiesis, 1989–2000

This section addresses together the phases of emergence and autopoiesis in Central Eurasian energy geo-economics, i.e., 1989–1994 and 1995–2000. The phase of emergence (1989–1994) was characterized mainly by a focus on oil, principally in the South Caucasus but also with reference to the North Caucasus and offshore Caspian and Black Seas, plus Turkey as a transit country. Turkey's role here anticipates its still higher profile that became evident during the autopoietic phase, 1995–2000, which was marked by the deepening of plans for connections between Southwest Asia and Southeast Europe (compare Balkir and Williams 1993; Fuller and Lesser 1993). Those connections include projects for an oil pipeline between Bulgaria and Greece, now under construction, and a natural gas connector between Turkey and Greece, both with plans for onward transshipment through the Balkans and into Central Europe. The autopoietic phase also saw the first sketches of projects heading eastward from Central Asia into China, which became more definite during the phase of coherence (2001–2006) and are treated later.

3.1 *Emergence in the South Caucasus, 1989–1994*

The energy geo-economic dynamic in the South Caucasus radiated first of all from the existing route through southern Russia for export of Azerbaijani oil. This pipeline already ran from Baku to Russia's Black Sea port of Novorossiisk through Chechnya, which during the Soviet period had been a major producer of highly refined petroleum products such as various aviation oils. However, this route became problematic even before the Chechen Wars definitively closed it, for a 'cottage industry' arose whereby local residents would siphon from the pipeline and sell literally by the roadside either the product 'refined' in homemade stills or the crude itself. This practice became more widespread and 'industrial' under the fiefdoms of various local Chechen warlords before, during, and after the two Chechen Wars.

As the situation in Chechnya became intractable, Azerbaijani crude was loaded onto railcar tankers and taken through Dagestan, between Chechnya and the Caspian Sea in southern Russia, northward so as to enter the pipeline to Novorossiisk from a different feeder line. That feeder line also served other production stations, in Russia itself. Nevertheless, payments contracts from Moscow to Baku continued to be honored for political reasons, as Russia continually attempted to stave off the 'southern route' for Azerbaijani oil that turned into the Baku-Tbilisi-Ceyhan (BTC) main export pipeline, although in the end no one really knew exactly how much of the crude arriving at Novorossiisk really came form Baku (Cutler 1999; LeVine 2007). During these years of emergence, Russia was still competing for Azerbaijani crude with the Baku-Supsa 'western early oil pipeline' (in that it carried smaller quantities, and earlier, than planned for the BTC) through Georgia to the Black Sea.

The BTC itself would not have been built but for the Western sponsorship of the refurbishing of the Baku-Supsa pipeline terminating on Georgia's Black Sea coast. Azerbaijan's 'early oil' was essential because it was able to justify, economically by itself, the construction of the westbound leg of what later became integrated into the BTC route. The international energy industry's general preference at the time was for a route through Armenia, instead of Georgia, as this would have been a bit shorter and more economical in the long run. However, the lack of settlement of the conflict over Nagorno-Karabakh made such a route impossible.

Azerbaijan, too, actually preferred a pipeline through Armenia in the early 1990s, conditional upon settling the Karabakh conflict, in order to ensure sufficient energy supplies to its exclave, Nakhichevan. The failure to settle the Karabakh conflict led to the exploration of the Georgian route, despite Turkey's hesitations over any route to a Black Sea port, since the Turkish Straits, the only egress for oil to world markets from the Black Sea (so far), are highly trafficked and difficult to navigate, posing legitimate fears for ecological security. That stumbling block may be eroding for future projects. In particular, the Russian invasion of Georgia in 2008 has had the effect, under conditions of Turkish border controls with Armenia, of cutting Armenia off from its sole overland route to and from world markets, through the now damaged Georgian port of Batumi. As this changes the calculus of interests of the parties concerned, the possibility of a route through Armenia for trans-Caspian oil and gas has once again become the subject of discussions within the Turkish-Armenian-Azerbaijani triangle.

Any domestic turmoil in Azerbaijan in the early 1990s would have delayed the development of Baku's oil industry, so Moscow would have sought to maintain rather than resolve political instability in the region. Russia's military and security apparatus was probably involved in the governmental crisis in Azerbaijan, which lasted all through the summer of 1993 and into the autumn, and put Heydar Aliev back in power (Goltz 1998). In November 1993, Russia signed agreements with Azerbaijan to ship its oil through Russia but was unable to help resolve the Karabakh conflict. So in 1994, the Azerbaijani and Turkish presidents began to examine whether Turkey could help solve the conflict. At the same time, they began to explore the feasibility of an oil pipeline for the export of Azerbaijan's oil through eastern Turkey to the Mediterranean.

Under Yeltsin's presidency, Russia sought to prevent other powers from asserting themselves in the South Caucasus by playing a balancing game. Russia initiated a rapprochement with Turkey directed at settling the Karabakh problem. At the same time, Moscow encouraged Teheran's pro-Armenian policy as a counterbalance to Ankara, until Teheran's meddling extended to Central Asia and the civil war in Tajikistan (for background, see Mesbahi 1997). In the early 1990s it was widely thought that a new diplomatic assertiveness, together with a certain cultural history, would make Turkey the dominant regional power in Central Asia. In the end, Turkey's inability to provide large amounts of investment capital disappointed the Central Asian states, which turned to other sources for international support. However, through much of the first half of the 1990s, Western assessments of Central Asia tended to frame the situation as one of competition between Iran and Turkey, oddly excluding Russia.

3.2 *Autopoiesis in Central Asia, 1995–2000*

The project of the Caspian Pipeline Consortium (CPC) was the only oil pipeline with implications for Central Asia that was largely completed during the autopoietic phase of the mid- and late 1990s. This phase primarily witnessed the inauguration of negotiations or work on pipelines that came to fruition during the first years of the twenty-first century. The Kazakhstan-China oil pipeline, notably, was agreed upon between the two countries in 1997, but arriving at final terms took years and construction of the first section from Atyrau was not finished until 2003.

In the mid-1990s Russia officially supported cooperation with Turkey on Caucasus and Caspian oil but still assumed that the Novorossiisk route through southern Russia would be built first. In mid-January 1995, Russian, Kazakhstani, and Omani companies agreed on a CPC pipeline plan for both Kazakhstani and Azerbaijani oil to go across southern Russia to Novorossiisk. By then, the military conflict in Chechnya had made 'northern route' proposals through the Caucasus much less plausible because a main segment of the pipeline went through Grozny. However, the CPC was unable to obtain funding before (with the continuing and low-profile assistance of the US embassy in Kazakhstan) it was restructured so as to decentralize decision-making and streamline oversight structures.

All throughout 1998 and 1999, despite ongoing intensive negotiations, many people considered the BTC project to be dead and were only waiting to bury it. Then, in mid-October 1999, BP-Amoco (formerly British Petroleum) shifted its stance slightly by qualifying the BTC pipeline as strategic and declaring in a public statement that it should be built. This shift was accentuated by the company's newly declared willingness to take the initiative with governments and international institutions so as to find adequate and appropriate financing. Next, in mid-November 1999 at an OSCE meeting in Istanbul, following years of strong US support plus stamps of approval from major international financial institutions, and with BP-Amoco from the UK taking the lead, agreements were signed among Turkey, Georgia, and Azerbaijan, providing the framework for the construction of the BTC pipeline. They also provided the political security necessary for the cementing of four further agreements in which the industrial investors participated: a cost guarantee agreement, an agreement between investors and the transit states, the main export pipeline agreement itself, and the construction contract. As a further guarantee, the three state signatories to the Istanbul intergovernmental agreements subsequently incorporated the pertinent terms into national legislation for approval by their respective parliaments, explicitly transforming the language of those agreements into governing domestic law (Babali 2006).

When it was thought that Azerbaijani offshore oil might not be plentiful enough to fill the BTC to maximum capacity, and as Kazakhstan's offshore Kashagan deposit still seemed on-track for early development, oil from Kashagan was considered a prime candidate for topping off the BTC. That is because the Kazakhstani government had been counting on Russia to make good on its promises to double the volume of the pipeline of the CPC from 615,000 to nearly 1.3 million barrels per day (bpd), so as to accommodate increased production at the onshore Tengiz field. Yet despite such promises, as well as repeated public statements by Russian leaders at the highest level, Russia's commitments to expand CPC pipeline volume have never been realized. Regardless of whether this failure is due to internal Russian bureaucratic and interregional squabbling or to the unwillingness of the Russian leadership to act on its words, the result for Kazakhstan has been the same.

Two gas pipelines in Southwest Asia followed the oil pipelines there during this phase. The South Caucasus Pipeline (SCP, also called Baku-Tbilisi-Erzerum) literally followed the path of the BTC oil pipeline. When the Western oil majors began to explore the Azerbaijani offshore in the early 1990s, they did not have natural gas on their minds, but only oil. Questions about natural gas entered the picture only later in the decade, when PSG International, a joint venture between Bechtel and the GE Capital unit of General Electric, contracted with Turkmenistan to develop the Trans-Caspian Gas Pipeline (TCGP). This project would have taken that country's gas to Turkey via Azerbaijan and Georgia, alongside the route of the BTC for export of Azerbaijan's offshore oil. While the TCGP was still being negotiated and planned, unexpectedly large volumes of natural gas were separately discovered in the offshore Shah-Deniz deposit, where Azerbaijan had given BP-Amoco the rights to explore for oil. Azerbaijan's president Heydar Aliev, unable to find agreement with Turkmenistan's president Saparmurat Niyazov concerning expansion of the TCGP or allocation of its volume between the two countries, then decided to go ahead with an export pipeline exclusively for his country's own natural gas from Shah-Deniz (compare Cutler 2001; Winrow 2004).

In mid-December 1997 the Turkish and Russian prime ministers signed an agreement for the Blue Stream gas pipeline under the Black Sea. At the time it was projected that the Blue Stream pipeline would increase Turkey's dependence on Russian sources of natural gas from the current two-thirds level to about four-fifths, so there were suspicions that economic rationality was not the main driving force behind the deal. In spring 2002, the Turkish energy minister was indicted for corruption but won a vote blocking the revocation of his parliamentary immunity. Europe opposed the TCGP project supposedly on environmental grounds but really because it did not want the US in the region. However, environmental concerns did not prevent Europe from arranging the bank credits to make possible the construction of the Blue Stream pipeline, even though it was technically a more difficult project, deeper than any previous undersea natural gas pipeline and fraught with ecological hazards. The president of the European Commission at the time was Romano Prodi and the lead European company in the project partnering with Gazprom was the Italian firm Eni. The Blue Stream pipeline entered into service in November 2005.

Finally, it must be mentioned that the late 1990s saw important progress on agreement over the demarcation of national subsurface zones in the Caspian Sea basin. In February 1998, Azerbaijan and Turkmenistan agreed on the 'median-line principle' of division in a joint statement, although this did not decide their still unresolved dispute over the disputed Kyapaz/Serdar oilfield. That same year, Russia and Kazakhstan reached an agreement to divide the north Caspian seabed by the 'modified median-line principle' while the waters remained subject to joint use and such issues as the environment, shipping, and fishing were to be cooperatively resolved. The modified median line approach usually does not provide for joint development of disputed fields. Rather, the median line is modified precisely in order to take into account such considerations, allocating fields to national sectors on the basis not only of geography but also based on previous industrial development, historical precedent and so forth. Nevertheless, this agreement between Russia and Kazakhstan has become the basis for subsequent successful negotiations over cooperative joint development of a number of offshore north Caspian fields (for background, see Raczka 2000).

4. Coherence Around the Caspian, 2001–2006

The CPC pipeline began operations in late 2001. At present, after arriving at the Black Sea, the oil reaches world markets through the Turkish Straits. Other transshipment routes under development include those across the Black Sea that then pass via either Ukraine or a Bulgaria-Greece pipeline now under construction to end on the Mediterranean. Other than the CPC route, the export options that Kazakhstan developed during this period of 'coherence' in the Central Eurasian hydrocarbon energy complex included the existing Atyrau-Samara pipeline (also through Russia) and a combined rail-and-barge route across the Caspian Sea into the South Caucasus.

About 80 percent of Kazakhstan's oil has nowhere to go today, other than through Russia's pipeline system. Until the Russian invasion of Georgia in 2008, half the remainder was exported through the Georgian Black Sea port of Batumi, the seaside capital of the formerly rebellious Georgian province of Ajaria; the rest goes to China. Consequently, Kazakhstan has now decided to construct a 590-mile pipeline, for Kashagan oil in particular, running from Eskene, where Kashagan's onshore processing facility will be located once full-field development gets under way, to the port of Kuryk, near Aqtau, on the east coast of the Caspian Sea. Starting at 500,000 bpd, its volume will later be increased to 750,000 bpd; to this, another 400,000 bpd may be added by doubling the capacity of the Aqtau port itself.

TengizChevrOil already exports about 120,000 bpd of Kazakhstani oil across the Caspian Sea by barge to the Sangachal terminal at Baku, whence the oil is transported overland by railcar to Batumi. A new export terminal at Kulevi, on the Georgian coast near Poti on the Black Sea, was planned before the Russian invasion to handle about 100,000 bpd of 'late oil' from Azerbaijan to start with. This capacity could be doubled within two years, and then doubled again if necessary to handle additional oil from Kazakhstan. It is not out of the question that Kazakhstan will later adds further elements to the Eskene-Kuryk pipeline so as to decrease its dependence on the CPC pipeline from the Tengiz into Russia and then across to Novorossiisk on the Black Sea.

Tankers from Batumi and Kulevi/Poti can take Kazakhstan's as well as Azerbaijan's oil to Ukraine over the Black Sea for insertion into the Odessa-Brody pipeline (OBP). This pipeline in Ukraine, running from southeast to northwest, was finished in 2001 but lay empty for three years because Russia refused to allow transit of oil from Kazakhstan to fill it, oil that would have been destined for Europe. So instead of flowing southeast-to-northwest, the OBP has since 2004 flowed northwest-to-southeast, carrying Russian oil domestically inside Ukraine. When Kazakhstani oil realizes its alternative route to the OBP, its flow will be reversed back to the originally intended direction, although intra-elite Ukrainian political conflicts may affect this (Kudelia 2007). From Brody, the oil would continue to Plock, Poland, if higher world prices make this continuation of the OBP economically justifiable to construct; from Plock, an existing pipeline going to the port of Gdansk and thence world markets could be refurbished (for background, see Klid 1997). Since Kashagan is not planned to enter production until 2013, this project can, if necessary, use oil from Tengiz or even the Azerbaijani offshore in the meantime.

Prices paid to Turkmenistan for delivery of gas at the Russian border have gradually increased over the years, even before Russia began increasing its own prices for export (or re-export of Turkmenistan's gas) to third countries such as Ukraine and Belarus. Turkmenistan's new president Gurbanguly Berdimuhammedov has assured Russia publicly that Turkmenistan will honor its contractual obligations to

Moscow until 2028, the terminus date of a 25-year contract signed by his predecessor Saparmurat Niyazov in April 2003. This contract provided for sales to Russia at a price of $44 per thousand cubic meters (tcm). By late 2005, Gazprom had ceded to Niyazov's insistence for a rate increase, agreeing to a price of $65/tcm for deliveries during the first half of 2006. Almost immediately after the start of the new year, in the wake of the Ukraine gas imbroglio, Niyazov began agitating for yet another price hike. Moscow resisted until the end of the summer but finally caved in and agreed to a price of $100/tcm with Turkmenistan.

At the time of his death in December 2006, Turkmenistan's Niyazov had given no impression of relenting in his efforts to drive up the price. The simple fact is that Gazprom has managed its Russian resources so poorly and failed so greatly to make any significant improvements or important capital investments, that Russia is dependent upon gas from Turkmenistan not only for exports to Ukraine and, through Ukraine, to Europe, but also simply needs Ashgabat's gas for domestic Russian consumption during the cold winter months. There are other non-Russian outlets theoretically available to Turkmenistan: through Kazakhstan to China, through Afghanistan to Pakistan, and through Iran to Turkey, to name but three. Each of them has, like a renovated TCGP project, its own particular problems (although construction on the first began in mid-2008). The Turkmenistan-Iran-Turkey route may have the greatest practical obstacles, despite an oral agreement in principle between Turkish and Iranian representatives to pursue the idea.

The TCGP project, which failed in the 1990s due to Niyazov's insistence on personally leading Turkmenistan's negotiations and his inability to grasp the technical details of project planning and financing, is now integrated as an aspect of the Nabucco pipeline led by Austrian concerns. The route of the Nabucco pipeline would run from Turkey's eastern border (with Georgia and/or Nakhichevan, in two non-mutually exclusive variants) through Ankara and Istanbul into eastern Greece, then northwards through central Bulgaria and western Romania, finally snaking across Hungary from the southeast to the northwest, and terminating at Austria's Baumgarten gas hub.

5. Post-emergence: 2007 and Beyond

At the beginning of the present decade, Russia contracted to upgrade Kazakhstan's gas pipeline infrastructure in the west and north of the country, linking this to the prospect of that gas flowing through those pipelines to be exported to Russia. An oil pipeline from Karachaganak was eventually constructed to Atyrau so that its liquids could be conducted into the pipeline of the CPC to Novorossiisk on Russia's Black Sea coast. In mid-2007 the two countries reached agreement for further investment in Karachaganak with the intention of more than doubling current production levels from 7.5 to 16 billion cubic meters per year (bcm/y), all still going to Orenburg with the exception of a separate and much smaller Uralsk Gas Pipeline for local customers (Yenikeyeff 2008: 25).

The development of the natural gas production industry in northwest Kazakhstan has historically depended upon the capacity of the Russian gas processing industry, much as the development of the oil production industry there has depended upon the capacity of Russian oil pipelines. The Karachaganak gas deposit, for example, where today fully half of Kazakhstan's gas is produced, has ever since Soviet times depended upon the capacity of the trans-border Orenburg processing complex in Russia, which has been its only outlet. Throughout the 1990s, Russia was able to use its monopsonistic position to limit possibilities for developing Karachaganak's gas production (Ögütçü 2006). Various projects were elaborated over the past decade and a half for the westward transit of at least part of Karachaganak's production towards the Caspian Sea basin for trans-Caspian export. However, whenever such projects became sufficiently well-defined to appear technically and economically feasible, Russia would revive negotiations with Kazakhstan for raising the prices and/or the quantities it would accept from Karachaganak, so as to render other routes uneconomical by comparison, only to alter the terms on offer yet again once the momentum of the alternative project had dissipated.

Meanwhile, the (re)construction of the Caspian coastal (Prikaspii) pipeline, announced for Turkmenistan's cooperation with Russia and Kazakhstan with much fanfare, was on hold for a long time, at least from Turkmenistan's side. Despite great publicity at the time about the project for a 'gas OPEC' headed by Russia, it was rarely if ever noted, in all the commentary at the time, that the Prikaspii agreement represented nothing other than yet another intergovernmental MoU. It did not establish an international consortium to undertake the work; rather, the three governments became responsible for funding and assuring the execution of the work on its national segment. This allowed Berdimuhammedov to placate Russia, which is promised as much as 50 bcm/y by agreements inherited from the Niyazov era, while gaining breathing space during which to pursue further possibilities. In Turkmenistan and Kazakhstan, there have been delays. Not only is the TCGP of greater benefit to both of them since it would provide an additional outlet for the gas produced, but it is also more economical for them because it would be funded and constructed by the European companies concerned. Under the terms of a trilateral December 2007 agreement, the Caspian coastal pipeline is planned to carry 20bcm/y, beginning in 2012. There has been some delay, but some of the more specific terms for this work were presumably agreed upon in December 2008 and embodied in the three countries' respective national legislation (Cutler 2009).

Uzbekistan does not usually come to mind when one thinks of Central Asian energy producers, because it does not export as much as either Kazakhstan or Turkmenistan. However, it is one of the top ten gas-producing countries worldwide, and it is third in gas production in the Commonwealth of Independent States. It has not gained corresponding attention on the world energy map because its population of approximately 28 million consumes over 80% of the gas produced, leaving comparatively little for export.

When Bulgarian president Georgi Parvanov visited Tashkent in late 2008 to try to induce his Uzbekistani counterpart Islom Karimov to dedicate volumes of gas to the Nabucco pipeline project planned to run from Central Asia to Central Europe, the latter declined, saying that his country exports gas through only one pipeline (it dates from the Soviet era) and therefore exports only to Russia. Parvanov was, in fact, a bit late. In September 2008, Karimov had already agreed to build a new pipeline to Russia parallel to the existing Central Asia-Center and Bukhara-Urals pipelines. With a capacity of 54bcm/y, it will originate in Turkmenistan and also cross Kazakhstan before terminating in Russia. About half the volume will come from Uzbekistan, increasing its exports to Russia by about 50 percent from 2006 levels, with Gazprom investing about $1.5 billion in development of the gas condensate fields in the country's Ustyurt region.

In addition, Uzbekistan's three-year agreement to supply 3.5bcm/y to southern Kazakhstan will terminate at the end of 2010. Kazakhstan has been 'repaying' Uzbekistan via a swap arrangement by sending 3.5bcm/y from its own Karachaganak complex in the northwest of the country to the transborder Orenburg gas processing plant in Russia. However, Karachaganak's production capacity is being further ramped up and the Orenburg plant is being expanded, so those volumes from Kazakhstan will not be lost to Russia in the future. All this suggests that despite Uzbekistan's on-again, off-again relations with the Eurasian Economic Community, the country's relations with Russia are not suffering. Indeed, Putin's efforts to form a Central Asian 'gas club' within the Shanghai Cooperation Organization (SCO) may bear fruit: just not inside the SCO itself. The signing of a trilateral agreement for yet another pipeline in December 2007, beginning in Turkmenistan and running to Russia through Kazakhstan alone, is further evidence of this energy diplomacy.

As Russia and China seek to augment their influence over the development of Kazakhstan's energy production, Astana looks for other routes to overcome the constraints. The reinvigoration since 2007 of prospects for a new TCGP with Turkmenistan's participation creates the possibility for Kazakhstan, which already cooperates with Azerbaijan on trans-Caspian oil shipments, to also participate with gas exports. Delays in the development of the offshore Kashagan field make associated gas from the onshore Tengiz oilfield the first candidate for such exports. At the time of Kashagan's proving, it was considered that its associated gas might be piped under the Caspian Sea to Azerbaijan so as enter the SCP and eventually, through Turkey, reach Europe.

Due to delays with Kashagan, and with Karachaganak still dedicated to Orenburg for the foreseeable future, associated gas from the onshore Tengiz oil deposit is now the best candidate to supply Kazakhstani gas in a revamped TCGP project. Industry practice has been to flare Tengiz gas into the atmosphere, but now this must cease by 2011. The government in Astana considers the practice to be environmentally unsound (there is legislation against it) and, moreover, wishes to recover the gas for domestic use and revenue enhancement through export. Since the agreement of the multilateral 'Road Map' in Astana in November 2006 by the Second Energy Interministerial Conference of the Littoral States of the Black and Caspian Seas (a process set into motion under the EU-sponsored 2004 'Baku Initiative' but which acquired true momentum following the January 2006 suspension by Russia of natural gas exports to Ukraine), Kazakhstan has been working together with Azerbaijan and the EU to address concretely the realistic prospects for Kazakhstan's gas to reach Europe and the available techniques for this. The undersea portion of the TCGP as now conceived would run from Kazakhstan's Caspian Sea coast at Aqtau (whither gas from Tengiz would be brought overland) to Baku, connecting there to the SCP and eventually on to Europe. At the same time, a spur from this main line to the port at Turkmenbashi would connect Turkmenistan's gas fields to the TCGP. At present, the pipeline is projected to have an initial capacity of 20 bcm/y, possibly increasing to 30 bcm/y. Its total length would be almost 1600 kilometers, of which only 300 would actually be underwater.

Since the death of Turkmenistan's former President Saparmurat Niyazov at the end of 2006, Ashgabat is in principle no longer an obstacle to such plans. Following a visit by Turkmenistan's new president Gurbanguly Berdimuhammedov to Berlin and Vienna in late 2008, the major German energy firm RWE, together with Austria's OMV, formed a joint venture so as to move the TCGP project ahead. Berdimuhammedov had already visited Brussels for high-level EU discussions in late 2007, and by mid-2008 an agreement had been reached that 10 bcm of gas from Turkmenistan would reach Europe in 2009. This is being accomplished through interconnecting Turkmenistan's rigs with Azerbaijani gas rigs in the Caspian offshore, which are in turn connected to the SCP.

Thus in November 2008, Azerbaijan's state oil company SOCAR and Kazakhstan's state monopoly KazMunaiGaz signed an agreement setting out the main terms for a transport system to convey Kazakhstani oil across the Caspian Sea for entry into the BTC pipeline, as another step forward in the realization of the Kazakhstan-Caspian Transportation System (KCTS). The system, while long discussed, has now become Astana's response to Russia's unwillingness and/or inability to implement the long-promised doubling of the capacity of the CPC line. It is also entirely possible that the oil could come from the offshore Kashagan deposit now under development, or even from both. The new document is said to specify quantities of 500,000 bpd by 2012, rising to 750,000 bpd later (Guliyev and Akhrarkhodjaeva 2008: 19). This bilateral agreement was signed as an aside to a larger meeting in Baku earlier this month that saw another significant agreement in which Azerbaijan agreed to supply Georgia's natural gas consumption requirements for five years. This agreement represents Azerbaijan's rebuff of Russia's recent commercial offer to purchase of all of Azerbaijan's gas production. While the commercial basis of the offer was excellent, current president Ilham Aliev averred non-commercial interests that must be considered. Since then, however, Azerbaijan has agreed to sell Russia 500 million cubic meters per year from the existing Shah Deniz (Phase One) development as from the beginning of

2010 and to give it important consideration as a buyer of some of Phase Two's production.

That most recent Baku meeting marked the second anniversary and fourth ministerial level follow-up to the November 2006 Baku Initiative, itself in turn a follow-up to a 2004 conference that set the stage for a new cooperation among the European Commission (EC), Armenia, Azerbaijan, Belarus, Georgia, Kazakhstan, Kyrgyzstan, Moldova, Tajikistan, Turkey, Ukraine, and Uzbekistan. It is noteworthy that the original initiative in 2004 preceded Russia's ostentatious and highly publicized cutting-off of gas supplies to Ukraine at the start of 2006. This move affected the member—states of the EU to varying degrees and drew mass and elite attention in Europe to the precariousness arising from dependence upon Russian energy supplies. In 2006, Russia supplied 40% of the EU's natural gas consumption and 30% of its oil consumption. These proportions are at present projected only to rise in the future, particularly as the EU implements its policy decision of a few years ago to increase the proportion of natural gas in its energy consumption mix for environmental and ecological reasons. Russia was present at the Baku Initiative meetings in 2004 and 2006 as an observer but did not respond to an invitation to send a representative to the most recent one.

It is worth noting that the original initiative preceded Russia's ostentatious and highly publicized cutting-off of gas supplies to Ukraine at the start of 2006, a move that also affected member—states of the EU to varying degrees and drew public and political attention in Europe to the precariousness arising from their dependence upon Russian supplies. In 2006, Russia supplied 40% of the EU's natural gas and 30% of its oil. These proportions are at present projected only to rise into the future, particularly as the EU implements its policy decision of a few years ago, to increase the proportion of natural gas in its energy consumption mix for environmental and ecological reasons.

In November 2008, the EU also adopted a project for a 'supergrid' that would permit member-states to share electric power generated from different sources, including wind farms in the North Sea, solar energy in Spain, and the projected Nabucco pipeline from Central Asia to Central Europe. Over the past two years, Russia has moved swiftly to pose obstacles to the realization of the Nabucco project. It has proposed its own 'South Stream' pipeline, to branch westwards under the Black Sea from the Russia-Turkey Blue Stream Pipeline. It has played the national interests of EU members off against one another and sought to entice European energy companies into special relationships that would prevent the all-European cooperation necessary to realize Nabucco.

At present, Turkey appears to be a stumbling block, independent of Russian moves. According to press reports, Ankara, as a transit point for Nabucco gas, is insisting on purchasing at below-market prices, a portion of all gas transiting via Nabucco for domestic consumption. Former Azerbaijani president Heydar Aliev, the current president's father, was able magnanimously to forego transit fees in order to seal the deal with former Georgian president Eduard Shevardnadze for construction of the BTC back in the late 1990s. However, it is doubtful that his son is either willing or able to accord such a privilege to Turkey. At an EU-Turkey summit in Prague in early May 2009, a breakthrough agreement was reached over questions of pricing and legal regulation for Nabucco gas (which would come from Shah Deniz Two in the first instance), although the June 25 signing summit in Ankara was postponed until July 13 due to last-minute hitches.

The Turkmenistan-China gas pipeline, on which construction has already started in transit countries Uzbekistan and Kazakhstan is formally an extension or add-on to earlier Kazakhstan-China negotiations (Saurbek 2008). The route of the pipeline has not been made public, but the most reasonable scheme involves expanding the volume of the Bukhara-Tashkent pipeline within Uzbekistan and then taking it through Almaty to Alashankou on the border, where the existing Kazakhstan-China oil pipeline from Atasu also crosses into China.

Towards the end of the first half of the present decade, the Chinese National Petroleum Company (CNPC) started negotiations for gas imports from western Kazakhstan with the country's national energy trust KazMunaiGaz. The first phase of that project was assigned the figure of 10 billion cubic meters per year (bcm/y), and the second stage of the Kazakhstan-China gas pipeline was to have increased Kazakhstan's own exports to China to 30 bcm/y. This gas could come either from the Karachaganak deposit, where production has been subject to Russian limitations on volumes receivable by the trans-border Orenburg processing plant in southern Siberia. It could also come from the associated gas in the offshore Kashagan deposit, where China was rebuffed a few years ago when it tried to purchase a share of the consortium directing Kashagan's development. In the beginning, however, it will probably come from Chelkar, in the Aqtobe region of western Kazakhstan, where CNPC has already been active for nearly a decade.

From there, a feeder pipeline would logically descend southwards to Kzyl-Orda and then to the major city of Shymkent (South Kazakhstan province), thereafter rejoining the extension of the Bukhara-Tashkent pipeline to Almaty and beyond.

The idea was first sketched on maps as early as 1993 when Western companies began to survey possibilities for energy development in Central Asia following the disintegration of the Soviet Union. However, the sheer scale of the project, together with the self-imposed isolation of the country under its former president Saparmurat Niyazov, who died in December 2006, made it a non-starter for a long time. Nevertheless, it was Niyazov who, in April 2006, signed a framework cooperation agreement in Beijing with Chinese president Hu Jintao. By July 2007, there was an agreement signed by the CNPC and witnessed by Niyazov's successor Gurbanguly Berdimuhammedov.

Chinese experts conducting the geological exploration and development of the Bagtiyarlyk fields have already reported that they hold 1.6 trillion cubic meters of gas. A first phase will be opened for up to 10 billion cubic meters per year (bcm/y) from the already operating Samantepe and Altyn Asyr fields, which are together expected to supply 13 bcm/y to the completed project. After this quantity reaches 10 bcm/y, the second phase will be inaugurated, adding 17 bcm/y from deposits that the two countries will develop under the terms of the July 2007 production sharing agreement.

It is worth noting that China has explicitly linked this geo-economic investment to its geopolitical aims, by eliciting from Turkmenistan the statement, at the time of the 2007 signing, that Chinese interests would not be threatened from its territory by third parties. This implies a promise that Ashgabat will think hard and long before allowing a US military presence in the country under the new Berdimuhammedov regime.

Both Azerbaijan and Turkmenistan have new and younger presidents than they have had for most of the post-Soviet period. In May 2008 they held their countries' first bilateral summit in over a decade. Symbolically, the day the summit began, a ship from the Azerbaijani state oil company SOCAR delivered equipment to a Turkmenistani oilrig located in Turkmenistan's Caspian offshore. In a concluding joint press conference, the two presidents declared that all issues were resolved between their countries. Turkmenistan reopened its embassy in Baku, and Azerbaijan paid off its $44.8 billion gas debt to Ashgabat. It looks very possible that Azerbaijan and Turkmenistan will reaffirm their 1998 adoption the 'median line' rule to divide their Caspian Sea subsurface sectors. The next step would be adoption of the 'modified median line rule' already in force under bilateral agreements between Azerbaijan and Russia, between Russia and Kazakhstan, and between Azerbaijan and Kazakhstan. Even if formalized only on an ad hoc basis in the beginning, such a procedure would allow Azerbaijan and Turkmenistan to settle their dispute over ownership of the Kyapaz/Serdar field in the Caspian Sea (probably providing for its development under a joint venture), not to mention resurrection of the TCGP project. The major international energy companies have known for ten years that there is no technical obstacle to construction of an ecologically sound gas pipeline tracing a relatively shallow east-west undersea ridge between Turkmenistan and Azerbaijan.

On a visit to Baku in mid-2008, Gazprom's chief Alexei Miller unexpectedly offered to buy natural gas from Azerbaijan at European market prices, minus transport costs. It is likely, although unconfirmed, that this was intended to fill the Russian-Italian sponsored 'South Stream' pipeline project, on which Gazprom is partnering with the Italian firm Eni, the same team that built the 'Blue Stream' pipeline under the Black Sea from Russia to Turkey. However, the South Stream's route is still quite vague: first it would cross, under the Black Sea, the continental shelves of Ukraine and Romania (whose agreement would also be necessary and is not assured), reaching Bulgaria, whence it would either continue through Greece and under the Ionian Sea to Italy or instead join another Turkey-Greece-Italy ('trans-Adriatic') pipeline already planned by the Swiss energy trading company Elektrizitäts-Gesellschaft Laufenburg with Norway's StatOilHydro. Alternatively, it could take a northern route through Serbia, Hungary, and Slovenia to Austria's Baumgarten hub, unless from Slovenia it passed instead into northern Italy, or else via Bosnia-Herzegovina and/or Croatia to Trieste.

A potential problem from Baku's standpoint was that Gazprom insisted on long-term contracts at fixed prices, whereas the price of natural gas in Europe is projected to rise over time. Baku announced instead its readiness to participate in the rival Nabucco pipeline project, which takes a route through Turkey, Bulgaria, Romania, and Hungary, also terminating at Baumgarten in Austria. Moreover, in mid-2008, Azerbaijan agreed to supply the first real order for physical gas through Nabucco: Bulgaria will buy more than one bcm/y of Azerbaijan's natural gas beginning in 2013, when the Nabucco pipeline is projected to open. This amount represents over one-sixth of Bulgaria's annual consumption and about one-eighth of the pipeline's first-phase capacity.

6. Review and Analysis

6.1 *Review*

At the beginning of this article, I mentioned how the three chronological tranches that organized the narrative analysis here represent the complex-scientific categories of emergence, autopoiesis, and coherence. The significance of the period beginning in 2007 comes into focus in reference to the complex-scientific principle of scaling, which can also be understood as 'nesting.' The three phases 1989–1994, 1995–2000, and 2001–2006 may be properly understood as being nested within a longer cycle, of which they together represent only the phase of emergence. In other words, they respectively represent the emergence of emergence, the autopoiesis of emergence, and the coherence of emergence. From this perspective, we may say that a 'superphase' of autopoiesis has started that will likewise have its own subphases of emergence, autopoiesis, and coherence.

The analysis of Russia's emerging place in the Eurasian hydrocarbon energy complex therefore largely means exploring the possible triangularizations of bilateral Russia-Turkmenistan and bilateral Russia-Kazakhstan relations, where 'triangularization' refers to potentials for decreasing the dependence of the respective Central Asian states on Russia. In the years 1989–1994 it was actually the United States that stood in such a triangularizing role, even against EU protestations of American 'meddling' in the region. For the period 1989–1994, among the EU, the US, and China, only the US was present in Southwest Asia and Central Asia in a manner likely to impinge on Russia's energy interests. The only significant exceptions were the British firm BG Group and Italy's Agip (now called Eni) in the existing Karachaganak gas deposit in northwest Kazakhstan; the British firm BP (then known as British Petroleum, and subsequently, following merger, with an American company, as BP-Amoco) in the offshore Azerbaijani deposits, and Chevron in northwest Kazakhstan in the Tengiz deposit.

European opinion, both elite and popular, was under the illusion that the EU might reach a twenty-first century version of a great-power entente with Russia over the management of international relations in western Eurasia after the disintegration of the Soviet Union.

That was at a time when questions of energy exploration, development, and marketing were not so pressing as they are today. At that time, export routes to China were not yet a realistic possibility, so the search for new Russian routes targeted southern Russia and the Black Sea (exiting to world markets through the Turkish Straits), while the search for non-Russian routes targeted the South Caucasus, whence they moved to the Black Sea, or later, to Turkey's east Mediterranean coast. The US, at the same time, successfully promoted the restructuring of the CPC, resulting in the pipeline from Tengiz to Novorossiisk, and encouraged international financial institutions.

The autopoiesis of emergence, 1995–2000, saw the construction or the decisions to construct the CPC pipeline (Caspian Pipeline Consortium, oil from Kazakhstan to world markets across southern Russia and through the Turkish Straits), the BTC pipeline (Baku-Tbilisi-Ceyhan, Azerbaijani to world markets via Georgia and Turkey), the SCP (South Caucasus Pipeline, Azerbaijani gas to Turkey and perhaps eventually beyond), and Blue Stream pipeline (gas from Russia to Turkey under the Black Sea), and the failed exploration of the possibility of TCGP pipeline (Turkmenistani gas to world markets under the Caspian Sea to Azerbaijan and out through Turkey).

There are two oil pipelines to the world market, one from the eastern shore of the Caspian Sea and one from its western shore; similarly, there are two gas pipelines, one from Russia through Turkey and the other from Azerbaijan through Turkey. There is also one failed pipeline from the eastern shore of the Caspian Sea through Turkey. To these may be added other failed pipelines, most notably the TAP (Turkmenistan-Afghanistan-Pakistan) natural gas pipeline, which attracted new attention due to its possible extension to India. However, these other pipelines (such as the newly agreed but as yet unbuilt bilateral Iran-Pakistan gas pipeline, a rump project of the Iran-Pakistan-India which India in the end declined to pursue) take us further afield than the scope of the present article allows, as they would necessitate the introduction of Iran into the equation, which is not part of 'Central Eurasia' as defined here. Suffice it to say that nearly every successful westbound pipeline conceived during the autopoiesis of emergence transited Turkey.

In Central Asia during this period, Russia continued to reassert its influence through the national energy trusts, particularly Lukoil in Kazakhstan and Gazprom in Turkmenistan. But Russia's attempts, beginning in the late 1990s but still more evident in the first half of the present decade, to enforce its interests in Europe (including the new

Eastern Europe, viz., former Soviet republics west of the Urals) has led to cooperative initiatives by other countries and the EU to ensure their own respective interests. The energy producing Central Eurasian states have seen how Russia uses its geographic position to restrict how much of their own energy reaches the world market, limiting the quantities that they can produce and thus their revenue and possibilities for national economic development (compare Saivetz 2003; Stulberg 2005). Kazakhstan in particular was injured by the failure to double the volume of the CPC pipeline as agreed, as well as by Russia's veto over transit through its territory to the Odessa-Brody pipeline for export of Kazakhstani oil on towards Europe. Indeed, only after announcement of the agreement to implement the KCTS for Tengiz oil did Russia remove its blocking action from the prospect of doubling the CPC pipeline, though not by laying more pipe but rather by adding further compressor stations. Kazakhstan and more recently Turkmenistan have cooperated with China in the construction of eastward export pipelines; and Azerbaijan hesitates to accept Russia's offer to purchase the whole of the country's natural gas production because of avowed non-economic concerns.

According to the logic I have laid out, we have already entered the superphase of autopoiesis, specifically the emergence of autopoiesis. This would mean (and it is the case) that in the last two years there has occurred the autonomous consolidation of patterns of energy geoeconomic structuration that cohered in the first half of the present decade, after having emerged as mere possibilities at the time of the Soviet collapse and then survived as simple contingencies through the 1990s.

6.2 *Analysis*

As suggested above, a second complex-scientific insight is to rely on trilateral rather than bilateral international relations as the building block of the analysis. This approach is sustained by recent advances in sociological network analysis, which has conclusively demonstrated, for example, that the dynamics within a triad differ qualitatively from the aggregated dynamics of all dyads within the triad. The practical reason for this approach is that trilateral energy arrangements are becoming ever more typical, particularly in Central Eurasia, because the exploration and development of many new deposits requires transit through third countries to the world market. The BTC pipeline in the South Caucasus, for example, was the first project implemented in the history of the world hydrocarbon industry that included not just a producer country (Azerbaijan) and a consumer country (Turkey, actually only the country of final exportation), but also a transit country (Georgia).

The principal issues structuring the basis of bilateral and multilateral international energy relationships in Central Eurasia have been the directions of export of Turkmenistani natural gas and Kazakhstani crude oil. It is clear in retrospect that the key bilateral energy relationship without which the Eurasian hydrocarbon complex would not have acquired its present structure, is the Turkmenistan-Russia connection, particularly the latter's effective monopoly of significant export options for Turkmenistan's natural gas. Most of the evolving significant international energy networks in Central Eurasia branch out from that relationship, or else from Kazakhstan's oil export policy. The multidirectionality of Kazakhstan's policy has undergone changes in emphasis since 1991, but Russia remains such a key player that we may say the Kazakhstan-Russia-Turkmenistan triangle is the generative nucleus for the structuration of multilateral geo-economic relations in Central Eurasia and beyond.

Thus in the concentration of multilateral networks around the key bilateral Russia-Turkmenistan relationship, two triangles stand out. As we have seen, the first includes Kazakhstan as its third vertex; the other triangle has the European Union for its third vertex. Furthermore, on the basis of the latter Turkmenistan-Russia-Europe triangle, one may analytically construct a quadrilateral, adding the United States. This energy-based quadrilateral is analytically decomposable into four trilateral relationships, each of which omits one of the four members. As explained above, this is how they should be regarded, rather than as aggregations of bilateral relationships. Likewise, the Kazakhstan-Russia-Turkmenistan triangle is the basis for two strategic quadrilaterals, one of which adds China to the triangle and the other the US.

To illustrate the significance of this abstraction, let me now condense parts of the foregoing narrative s to illustrate the significant insights from the analytical perspective and to project possible developments for the future. As mentioned earlier, the relations among the members of the Russia-Turkmenistan-Kazakhstan triangle, which are irreducible to the aggregate of their pair-wise bilateral relations, are the basis for understanding Russia's role in Central Asian energy transit and therefore its emerging place in the Eurasian hydrocarbon energy complex.

What appears from the course of events is that that triangle is successively transformed into three quadrilaterals by becoming respectively linked up with each of three other vertices. Moreover, these quadrilaterals bridge the phases of emergence, autopoiesis, and coherence. Specifying this conceptually in relation to events on the ground clarifies the principle and its significance.

1. *The phases of emergence (1989–1994) and autopoiesis (1995–2000) are bridged by the Russia-Turkmenistan-Kazakhstan-US quadrilateral, which began to manifest during the years 1993 to 1997.* In addition to the basic Central Asia-centric triangle of Russia-Turkmenistan-Kazakhstan, the other triangles in evidence were therefore Russia-Kazakhstan-US (competition and cooperation manifested in Tengiz), Russia-Turkmenistan-US (competition manifested in the failed TCGP project), and Turkmenistan-Kazakhstan-US (also manifested in the failed TCGP project).

US-Russian relations remained the principal systemic factor still structuring the evolution of Central Asia. Competition between the two countries was focused mainly through the lens of hydrocarbon resources in the region. The Russia-Turkmenistan-Kazakhstan-US quadrilateral came into evidence beginning about 1993, as the ruble zone disappeared in Central Asia and American energy interests made themselves felt on the ground in Turkmenistan and Kazakhstan. Of the subsidiary triangles within this quadrilateral, the most significant at the time was US-Kazakhstan-Russia, as the question of export pipeline routes for the already obviously impressive Tengiz field in northwest Kazakhstan was under active study by different government and industry groups. The first half of the 1990s thus saw the Kazakhstan-Russia leg of the fundamental Kazakhstan-Turkmenistan-Russia triangle pointing towards the U.S., because American offshore terminals in the Gulf of Mexico were the first intended targets for Kazakhstani oil shipments.

Western interest in Turkmenistan during these years was exclusively American, and it concentrated mainly on ameliorating Ukraine's payments situation as a gas importer. This was a perennial problem that led to periodic closures of Turkmenistan's taps and to corresponding shortages in Ukraine. Thus, for example, high-ranking U.S. officials attended the November 1994 talks in Ashgabat that settled (at least for a time) the problem of Ukraine's debt to Turkmenistan for natural gas supplies since the end of 1991. Western concerns about Turkmenistan's energy exports had not yet clearly focused on trying to make a trans-Caspian gas pipeline happen, even though American diplomatic activity and financial interest were the moving force behind this export axis. The Turkmenistan-Russia leg of the triangle pointed towards Europe, through Ukraine, but only vaguely so, since Europe had not yet begun significant imports of natural gas through Ukraine. The Kazakhstan-Turkmenistan leg of the fundamental triangle remained undeveloped.

The phases of autopoiesis (1995–2000) and coherence (2001–2006) are bridged by the Russia-Turkmenistan-Kazakhstan-EU quadrilateral, which began to manifest during the years 1999 to 2003. In addition to the basic Central Asian triangle including Russia, the other triangles in evidence were Russia-Turkmenistan-EU (with the EU's belated interest in the failed TCGP project), Russia-Kazakhstan-EU (manifested in the Kashagan deposit and other offshore North Caspian developments), and Turkmenistan-Kazakhstan-EU (also manifested in the failed TCGP project).

The EU was highly allergic to the notion of US-Russian competition for Caspian oil at the time and opposed American power-projection into the South Caucasus and Central Asia. The final years of this phase, however, saw the international UK-based oil company BP declare themselves in favor of the Baku-Tbilisi-Ceyhan pipeline from the Azerbaijani offshore to the eastern Mediterranean. This inevitably brought in not only other European companies but also other international and transnational European actors. The Karachaganak operating consortium includes not only the American ChevronTexaco, but also the British BG Group, the Italian Eni, and the Russian Lukoil. In 1995, the Kazakhstan government signed a memorandum of understanding on production sharing with Eni and BG Group, and in 1997 Chevron and Lukoil joined the international consortium. Towards the end of this period, then, the EU injected itself into the South Caucasus and, by inevitable extension even against its declared intent, into at least the western and Caspian offshore regions of Central Asia. Also with the declared American interest in a Trans-Caspian Gas Pipeline (TCGP) from Turkmenistan that could also conceivably have picked up natural gas from the Karachaganak deposit in western Kazakhstan, the Russia-Turkmenistan-EU-US quadrilateral came into evidence and established its significance in these years.

It is worth noting that the US-Russia relationship was not so highly competitive at that time. For example, the CPC pipeline from northwest Kazakhstan through southern Russia to the Black Sea would not have been built without active American diplomatic facilitation. Indeed, it is possible to summarize the nature of Russo-American relations in Central Asia proper, during the second half of the 1990s, as defined by the most important countries there. It could be schematically argued, that Russo-American relations in Kazakhstan were characterized by cooperation and in Turkmenistan by conflict over routes for Turkmenistan's gas exports to world markets. Russia's desire to keep Turkmenistan's gas production exclusively for its own pipelines, whether for domestic consumption or re-export at world prices, was opposed by U.S. policy, which consistently sought to find other routes to market for energy supplies from the newly independent states. In the instance of Turkmenistan, this could have been via the undersea TCGP project to Azerbaijan that failed to materialize, or the Turkmenistan-Afghanistan-Pakistan pipeline project that was put on hold after the Taliban took power in Kabul.

3. *The phases of coherence (2001–2006) and post-coherence (since 2007) are bridged by the Russia-Turkmenistan-Kazakhstan-China quadrilateral, which began to manifest itself in 2005, and by the foregoing logic will have established itself by 2009.* In addition to the basic Central Asian triangle, the other triangles in evidence here are Russia-Turkmenistan-China (competition for Turkmenistan's gas resources as in the Caspian coastal pipeline vs. the Turkmenistan-China project), Russia-Kazakhstan-China (competition for Kazakhstan's resources as in the battle between Russian and Chinese companies for Petrokazakhstan, the pipeline that this company owned), and Turkmenistan-Kazakhstan-China (cooperation over the Turkmenistan-China gas pipeline).

Turkmenistan's agreement with Russia for gas sales extends out to 2028 but the prices are not set past the near future. In 2007, Russia, Kazakhstan, and Turkmenistan negotiated, and in December of that year signed, an agreement to refurbish and expand the capacity of the Caspian coastal pipeline, a Soviet era gas line from Central Asia to Russia. It was Russia's attempt to prevent Turkmenistani negotiations with the EU for the TCGP from restarting. Knowing how heavily Gazprom relies on Europe for revenue, Ashgabat bargained hard and succeeded in raising prices from $100/tcm to $130 during the first six months of 2008 and $150 during the second half of the year. Russia had thought that it was succeeding in purchasing the vast majority of Turkmenistan's potential production over the long-term but in October 2008 the results of a British expert firm's audit of the country's gas reserves (commissioned by a Niyazov's successor Berdimuhammedov) revealed that the new Yoloton-Osman field alone contains a minimum of 4 and a maximum of 14 trillion cubic meters, over and above current exports to Russia and Iran as well as planned exports to China:

and that was the result for this one new field only. Suddenly, Turkmenistan's available resources far outstripped Russia's attempts to corral them.

Kazakhstan's announced intent is now to target the world market as a whole with energy exports but focus attention on Russia, China, Central Asia, and the Caspian and Black Sea regions. Indeed, Kazakhstan is counting less on them than on state-driven integration of Central Asian energy grids and, probably, take-or-pay natural gas delivery contracts. The centrality of natural gas to ecologically friendly energy generation is surely why Kazakhstan has put a new accent on formulating a comprehensive national natural gas development and export policy (as well as exploring possibilities for developing nuclear energy). Nazarbaev is seeking to talk up the idea of a Central Asian system of state energy networks with Kazakhstan's neighbors, looking to create a Council of Energy Security that would create a market system providing regional and international energy security. Despite the contradiction, what emerges clearly is Kazakhstan's intent to assert itself as strongly as possible as a proactive player in Central Eurasian energy and not just be a field on which other states (and industrial trusts) play out their opposing interests. Whether history will follow the grand vision Nazarbaev has described, however, will depend on variables beyond Kazakhstan's control, including the attitude of its neighbors to the form of practical implementation of that design.

7. Conclusion

From Russia's perspective, the US as well as China and the EU encroach upon its former domain in Central Asia and the South Caucasus. China has established access to oil from eastern and probably in the future western Kazakhstan, as well as to gas from Turkmenistan. The EU has in the last few years become very serious about finding non-Russian energy resources and non-Russian routes for receiving them; and the US has from the beginning of the post-Soviet era consistently promoted the search for routes for energy resources to reach world markets by circumventing Russia, starting with oil in northwest Kazakhstan in the early 1990s. Putin's focus on Central Asia was qualitatively new in the post-Soviet period. Yeltsin's foreign minister Evgenii Primakov attended mainly to traditional Soviet allies such as Iran, India, and China. A 'Eurasianist' strand entered post-Soviet Russian foreign policy in the mid-1990s, and the growing focus on Central Asia was reflection of the shake-out from the post-Soviet disorder in the region, together with a reaction against US and European (and NATO) interest in the South Caucasus, although Putin's rapprochement with Central Asia in the first years of the twenty-first century would probably have occurred in some manner, irrespective of NATO's increased involvement in the South Caucasus.

It is useful to underline that if the US has encouraged these countries in such a direction, still they would not have so acted without the belief that doing so is in their own interest. Indeed, American policy does not even need to be Russo-phobic in order to engage in such behavior: this is, rather, a result of the European history that contributed to the formation of American national character. Briefly put, the dynamism of US economic growth was from the late nineteenth and throughout the twentieth centuries driven by the meritocratic promise to the 'wretched of the earth' who had been condemned to impotence, if not annihilation, by war and diplomatic entente and who, having nothing left to lose, decade after decade left the Old World for the New. Popular sympathy for small nations under threat by larger neighbors does not contradict elite *Realpolitik* calculations seeking to avert hegemonic domination of the Eurasian landmass, or any portion of it, by one power or by an alliance of them.

In recent years it is the EU acting in its own interest, and not the US, that has provided the greatest assistance to these countries seeking to orient their export policies in a non-Russian direction. This is due in significant part to Russia's miscalculations over how its use of the 'energy weapon' against Ukraine, Belarus, and other countries would be interpreted in Europe and how it would make an indelible impression on both popular and elite opinion.

The on-course westward extension of China's current oil pipeline from Kazakhstan, including the forced buyout of the Canadian firm Petrokazakhstan that controlled a key segment of existing pipeline, is an especially impressive piece of political-economic engineering, and almost paradigmatic of the aforesaid nature of the development of a complex system. Chinese energy geo-economic penetration into Central Eurasia is confirmed not only by the entry into service of the oil pipeline from Kazakhstan but also by the construction now under way of the gas pipeline from Turkmenistan, through Uzbekistan and Kazakhstan to Xinjiang in western China. These realizations are testimony to Chinese strategic planning which began fifteen years ago, when its national energy trusts first implanted themselves ever so delicately in the Caspian littoral.

Both Tengiz and eventually Kashagan oil could conceivably reach China. Already a pipeline runs to the Caspian port of Atyrau from Kenkiyak in the Aqtobe region of western Kazakhstan, where China has industrial interests in the country's hydrocarbon industry. Also, the Kazakhstan-China pipeline finished in 2006 runs from Atasu in the center of the country to the Dushanzi refinery in China. Between Aqtobe and Atasu, an existing pipeline already runs roughly halfway, from Kumkol to Atasu. For China to receive Tengiz oil, then, it would remain only to build the missing segment from Kenkiyak to Kumkol, and reverse the Aqtobe-Atyrau pipeline so that it flows from west to east. The result could eventually boost Chinese imports of Kazakhstani oil from 100,000 to 400,000 bpd, but whether it happens, or how fast, depends crucially on the accessibility of oil from Kashagan. However, Kazakhstan's decision in favor of the KCTS and its westward route for Kashagan suggests that the Kazakhstani leadership may not be too keen to repeat with China its mistake of depending too much on Russia.

Such are the most important of the now-established patterns that we should expect to become still more marked in years to come. In view of the post-Soviet turmoil throughout Central Asia, it would be possible to sustain the argument that this region's development is chronologically differentiated from that of the South Caucasus: for Central Asia, the second half of the 1990s would be the period of 'emergence' in the domain of energy geo-economics, 2001–2006 the period of autopoiesis and 2007–2013 that of coherence.

CHAPTER FOUR

THE ENERGY POLICY OF THE ISLAMIC REPUBLIC OF IRAN TOWARDS THE EUROPEAN UNION AND CHINA[1]

Eva Patricia Rakel

Abstract: *The Islamic Republic of Iran (IRI) is a strategically important country in the Middle East. Firstly, due to its position at the Strait of Hormuz, it is the main transport corridor for the export of oil from the Persian Gulf countries. Secondly, it has huge oil and gas resources. Both the European Union (EU) and China are potentially important economic partners for Iran. The country needs the EU because of its longings for foreign direct investment, technology, and knowledge transfer, which the EU member countries could provide. For the EU, Iran is a possible future supplier of oil and gas and a significant factor for stability in the Middle East and in its own backyard. Moreover, China's national oil and gas companies have signed several deals in recent years on the import of oil and the exploration of major Iranian oil fields to secure China's rising energy demand in the future. What does the strategic significance of Iran's energy resources tell us about the country's current and future role in international relations in general and its relations with the EU and China in particular?*

Keywords: Iran, foreign policy, energy policy, European Union, China

1. Introduction

Since the disintegration of the Soviet Union in 1991, Iran's position as a strategic player in the global oil and gas business has strengthened. Iran is one of the five Caspian littoral states and is thus a strategic link between the Persian Gulf and the Caspian region. Due to Iran's position adjoining the Persian Gulf and the Strait of Hormuz, it is the main transport corridor for the export of oil from the Persian Gulf countries. Iran has huge oil and gas resources. It ranks second in the world in proven oil reserves, behind Saudi Arabia, and ranks second,

[1] This article is partly based on my book 2009. *Power, Islam, and Political Elite in Iran-A Study on the Iranian Political Elite from Khomeini to Ahmadinejad.* Leiden, Boston, and London: Brill Academic Publishers.

behind Russia, in proven gas reserves. With oil demands rising across East Asia in general, and in China and India, in particular, Iran tries to strengthen its position not only among regional producer countries but also in world oil markets. At times, it might even set the main consumers—the United States (US), the European Union (EU),[2] China, and India—against each other. Both the EU and China are potentially important economic partners for Iran. Since the Islamic revolution it can be assumed that Iran needs the EU because of the country's longings for foreign direct investment, technology, and knowledge transfer, which the EU member countries could provide. For the EU, Iran is not only a possible future supplier of oil and gas but also a significant factor for stability in the Middle East and—considering eastward EU enlargement—in its own backyard. Closer economic cooperation between Iran and the EU, however, has been hampered in recent years due to Iran's nuclear program. The crisis around the nuclear issue has been to China's advantage to some extent. China's national oil and gas companies have signed several deals in recent years on the import of oil and the exploration of major Iranian oil fields to secure China's energy supply in the coming years.

This article first describes the theoretical framework of foreign policy practices and applies it to the IRI. It then discusses state institutions and the political factions in Iran, as well as the formal foreign policy decision-making process. It further gives an overview of continuity and change of foreign policy of Iran since the Islamic revolution until the presidency of Mahmoud Ahmadinejad. In the following, the article looks at the oil and gas complex in Iran. It then analyses foreign energy relations between Iran, the EU, and China.

2. Rethinking Foreign Policy Practices

This article argues that the notion of geopolitics should be integrated into the concept of foreign policy practices. The term 'geopolitics' finds its origins in the realist school of International Relations (IR), according to which states are the sole actors in international relations and struggle for dominance in an anarchic world. The Swedish political scientist,

[2] The term 'EU' is used throughout the whole article to refer to the European Union, thus it is also used for the time when the EU was referred to as the European Economic Community and European Community.

Rudolf Kjellén, used the term for the first time in the late 19th century to describe the interconnection between geography and politics. The British geographer, Halford Mackinder, then further developed it in the early 20th century. Confronted with the decline of British hegemony, he predicted that land powers would overtake sea powers and that the 'Eurasian landmass' would rise as the world's heartland, if not checked by Britain and its allies (Mackinder 1904: 421–42; Mackinder 1919). In the 1920s and 1930s, German geographers used the term to justify Nazi Germany's eastward expansion. After the Second World War, the geopolitical discourse was applied both by governments and citizens to define their state's position in the world in the Cold War context. The term geopolitics, however, was avoided, because of its association with Nazi Germany. Instead, terms like 'national security,' 'containment,' and 'deterrence' were used. During the Cold War, the international system was generally understood as a bipolar world with a conflict between the two superpowers, the US and the Soviet Union, carried out in a traditional balance of power politics (Amineh 2003: 18–19).

To study the foreign policy of a state does not imply only looking at foreign policy practices but also at how certain representations of space are incorporated into foreign policy practices. Agnew and Corbridge (Agnew and Corbridge 1995: 47–48) argue that the description of a foreign policy situation alone is in itself an act of geopolitics. The geographical identification of a place, and the labeling of it in a certain way, brings about specific visions and ideas about that place and the policies it pursues. To categorize a geographic area as 'Islamic' or 'Western,' for example, also implies certain presuppositions about its foreign policy practices.

The critical theory[3] and critical geopolitics[4] approach of IR and IPE might help us to solve this problem, as they reject the agency-structure dualism, and ahistoric structuralism and determinism in the analysis of world politics and the historical structure of IR. The representatives

[3] See for example Cox, R. W. 1987. *Production, Power, and World Order-Social Forces in the Making of History*. New York: Columbia University Press.

[4] See for example Agnew, J. 1998. *Geopolitics: Re-visioning World Politics*. London: Routledge; Agnew, J. & S. Corbridge 1995. *Mastering Space-Hegemony, Territory, and International Political Economy*. London & New York: Routledge; ÓTuathail, G. 1996. *Critical Geopolitics: The Politics of Writing Global Space*. Minneapolis: University of Minnesota Press; ÓTuathail, G. and S. Dalby 1998. *Rethinking Geopolitics*. London: Routledge.

of the critical theory/critical geopolitics approach believe in the transformative character of the world and assume that any social structure has its limits. This stands in contrast to (neo-)realism,[5] institutionalism/pluralism,[6] structuralism,[7] and traditional geopolitics, all of which have a static view on social relations, considering certain components of a social system as unchangeable, e.g. the state. Critical geopolitics and critical theory go beyond IR and IPE theories in that they analyze 'dynamic systemic change;' thus they take an historical perspective on international relations. The most important aspect of the critical theory approach is that it believes in the transformative abilities of human beings and that collective human action leads to historical transformation. Critical geopolitics adds to this the geographic dimension in analyzing complex systemic realties. It not only looks at the "material spatial practices" that constitute the global political economy but also the way they are represented and contested (McHaffie 1997). The representatives of critical geopolitics take state-society relations as the unit of analysis. According to this view, state-society complexes come into interaction through their (foreign) polices. By these interactions they create a "system level of social order" (Amineh and Houweling 2004/2005: 11).

Actors in state-society complexes legitimize foreign policy by presenting the public with certain assumptions of other states and regions beyond their borders (Taylor 1993; Dijking 1998). These assumptions of other states and regions can also be called 'geopolitical visions,' that is: "any idea[s] concerning the relation between one's own and other places, involving feelings of (in)security or (dis)advantage (and/or) invoking ideas about a collective mission or foreign policy strategy" (Dijking 1996: 10).

Assumptions about other states and regions emerge out of how the political elite of a society has defined its own state and its role in the world. This is what Campbell has called "geographical imagination": "A geographical imagination [...] can be defined as the way in which influential groups in the cultural life of a state define that state and

[5] See for example Waltz, K. 1979. *Theory of World Politics*. Reading, Mass.: Addison Wesley; Waltz, K. 1959. Man State and War. New York: Columbia University Press.

[6] See for example Keohane, R. O. 1984. *After Hegemony: Cooperation and Discord in the World Political Economy*. Princeton NJ: Princeton University Press.

[7] See for example Wallerstein, I. 1974. *The Modern World-System: Capitalist Agriculture and the Origins of the European World-Economy in the Sixteenth Century*. New York: Academic Press.

nation within the world. It addresses the primary acts of identification and boundary-formation that population groups within a state engages" (cited in Ó'Tuathail 2004: 84).

Geographical imagination is the basis of the "geopolitical culture" of a state. Geopolitical cultures are the product of the cultural and organizational processes that shape foreign policy in a state. But a geopolitical culture is not homogenous. Conflicts among the political elite, based on different political and/or economic interests: "produce a geopolitical culture that is powered by division and contradictory impulses and drives" (Ó'Tuathail 2004: 85, 87).

Geopolitical cultures are also characterized by "geopolitical traditions" that compete in the interpretation of a state's position in international relations: "A geopolitical tradition is a historical canon of thought on state identity, foreign policy, and the national interest, which is usually defined in opposition to alternative traditions" (Ó'Tuathail 2004).

Furthermore, foreign policy practices are also influenced by external elements, namely by the geopolitical visions of other countries as well as other countries' foreign policy practices.

The factors influencing foreign policy formulation discussed above are visualized in figure 4.1. In the following they are applied to the case of Iran since the Islamic Revolution and are visualized in figure 4.2.

In Iran geopolitical traditions have, over the last 200 years, been greatly influenced by the country's experiences of foreign intervention. The Tobacco Revolt (1881–1882),[8] the Constitutional Revolution (1905–1911),[9] the Oil Nationalization Movement of Prime Minister Mosaddeq (1951–1953)[10] and, finally, the Iranian Islamic Revolution

---

[8] In the 19th century the Qajars granted concessions to Britain for tobacco. Mirza Hassan Shirazi, the *marja-e taqlid* at the time, issued an edict that forbade Shi'ite Muslims in Iran from smoking tobacco. Because of great public pressure, the government finally withdrew the concessions. For more on the role of the clergy in the Tobacco Movement see further Keddie, N. R. 1966. *Religion and Rebellion in Iran: The Tobacco Protest of 1881–1882.* London: Frank Cass.

[9] On the Iranian Constitutional Revolution see among others: Afary, J. 1996. *The Iranian Constitutional Revolution 1906–1911: Grassroots Democracy, Social Democracy and the Origins of Feminism.* New York: Columbia University Press; Algar, H. 1969. *Religion and State in Iran.* Berkeley and Los Angeles: University of California Press; Martin, V. 1989. *Islam and Modernism: The Iranian Revolution of 1906.* Syracuse: Syracuse University Press.

[10] In the beginning of the 1950s, Prime Minister Mohammad Mosaddeq nationalized the British owned and operated Anglo-Iranian Oil Company. He was removed from power by a coup in 1953 by Mohammad Reza Shah, in cooperation with the

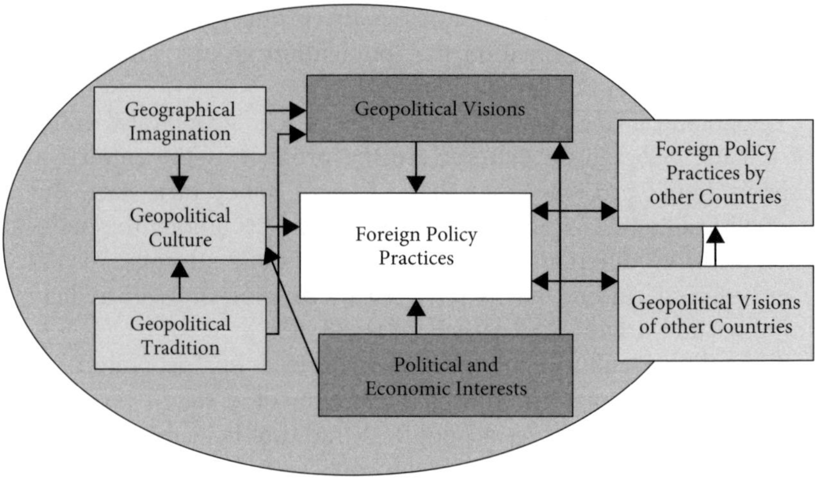

Figure 4.1 Factors of Foreign Policy Practices.

(1978–1979),[11] were all events that can be partly explained as reactions to the domination of Iran by foreign powers and exploitation of the country's wealth and resources by foreign powers and companies. These events were closely connected to Iran's historical experience of foreign influences and penetration: first, the rivalry with other empires (e.g. the Ottoman Empire), and second the interference into its internal affairs during the last 200 years by France, Russia, Britain, and the United States (US).[12] They were also affected by multiple

British and US intelligence agencies, see further Gasiorowski, M. and M. Byrne 2004. *Mohammad Mosaddeq and the 1953 Coup in Iran-Modern Intellectual and Political History of the Middle East.* Syracuse: Syracuse University Press; Katouzian, H. 1990. *Musaddeq and the Struggle for Power in Iran.* London and New York: I. B. Tauris.

[11] The origins, circumstances, and outcomes of the Iranian Islamic Revolution have been studied at length elsewhere and will not be repeated in this study. Important works on the revolution include: Arjomand, S. A. 1988. *The Turban of the Crown: Islamic Revolution in Iran.* Oxford and New York: Oxford University Press; Abrahamian, E. 1982. *Between two Revolutions.* Princeton: Princeton University Press; Keddie, N. R. 1981. *Roots of Revolution: An Interpretative History of Modern Iran.* London and New Haven: Yale University Press; Amineh, M. P. 1999. *Die globale kapitalistische Expansion und Iran- Eine Studie der Iranischen Politischen Ökonomie 1500–1980.* Münster, Hamburg, London: Lit Verlag, 1999; Amineh, M. P. and S. N. Eisenstadt 2007. "The Iranian Revolution: The Multiple Contexts of the Iranian Revolution." in M. P. Amineh (ed.) *The Greater Middle East in Global Politics: Social Science Perspectives on the Changing Geography of the World Politics.* Leiden, Boston and London: Brill Academic Publishers, 117–145.

[12] On the involvement of western powers in Iran since the 19th century see among others: Curzon, G. 1892. *Persia and the Persian Question.* London: Longman, Green;

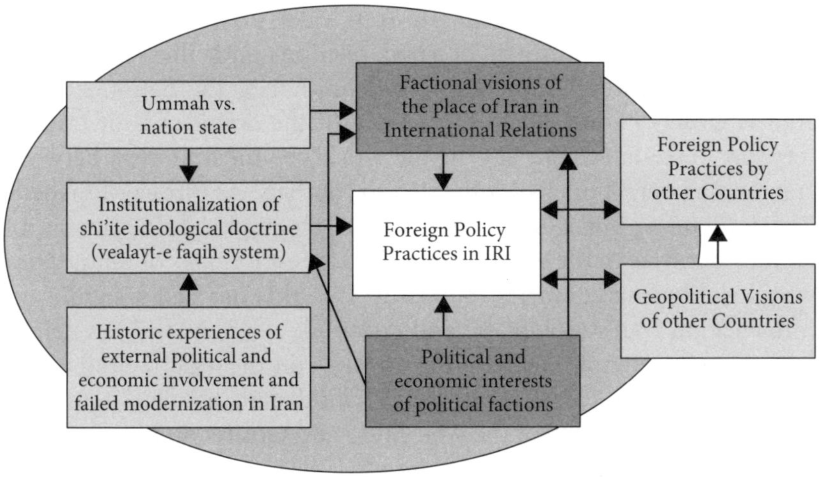

Figure 4.2 Factors of Foreign Policy Practices in the Islamic Republic of Iran.

failed attempts at modernization, starting in the 19th century, including those of the Qajar Shah,[13] and later, after the disintegration of the Persian Empire and the establishment of Iran as a nation state, by those of the two Pahlavi Shahs (Reza Shah 1921–1941[14] and Mohammad Reza Shah 1941–1979).

Furthermore, the geopolitical culture in Iran has been influenced by the question of what the Iranians should identify with: the *ummah* (Islamic community), as was proclaimed by Ayatollah Khomeini, or with Iran as a nation state, as former presidents Rafsanjani and Khatami, and probably also current President Ahmadinejad, see it. This has been termed geographical imagination. Khomeini's geopolitical visions of Iran manifest themselves in the two main ideological foreign policy principles of the Islamic revolution: "Neither East nor West" and the "Export of the Revolution." They can be summarized as: Iran should refrain from relations with the west and support those Muslims who are suppressed by the west or their un-Islamic rulers. Those who

Lenczowski, G. 1978. *Russia and the West in Iran (1918–1948): A Study in Big Power Rivalry.* Ithaca, N.Y.: Corneall University Press; Cottam, R. 1988. *Iran and the United States: A Cold War Case Study.* Pittsburgh: Pittsburgh University Press.

[13] On modernization during the Qajar Empire see among others: Bakhash, S. 1978. *Iran: Monarchy, Bureaucracy & Reform under the Qajar (1858–1896).* London: Ithaca Press.

[14] On modernization under Reza Shah Pahlavi see among others: Banani, A. 1961. *The Modernization of Iran: (1921–1941).* Stanford: Stanford University Press.

advocate Iran as a nation-state see Iran a key player in international relations. They are in favor of good relations with the west as well as with neighboring countries. The definition of the nation-state in Iran is closely connected to the setting of the boundaries of Iranian territory. Like anywhere else in the world, boundary issues between Iran and its neighbors have often led to conflicts. As Pirouz Mojtahed-Zadeh (2006: 9) notes, Iran is in a unique geographical situation as, of all the countries in the world, it has the highest number of boundaries with neighboring countries (currently 15). This has had a significant influence on Iran's diplomatic and economic relations with its neighbors. Recent examples include the disputes on the legal regime of the Caspian Sea with Russia, Azerbaijan, Kazakhstan, and Turkmenistan, and on the territoriality of the Abu Musa and Greater and Lesser Tunb Islands with the United Arab Emirates (UAE).

Thus, the geopolitical culture in the IRI is the product of past experiences, questions of identity, and territorial boundaries. After the Islamic revolution, the geopolitical culture in Iran manifested itself in the institutionalization of the *velayat-e faqih* system, as developed by Ayatollah Khomeini in the 1960s. Additionally, as will be shown below, the geopolitical visions and foreign policy practices of other countries and regions, such as the European Union and China, have had a great influence on foreign policy practices in Iran. Furthermore, the rivalry between different political factions for power has had a great influence on foreign policy practices in the IRI as will be discussed below. Each political faction has different views and interests on politics, the economy, sociocultural issues, and foreign policy. These views and interests have also changed over time. Consequently, each faction develops different visions of the place of Iran in international relations, based on these views and interests.

3. State Institutions and Political Factions in the Islamic Republic of Iran

With the Iranian Islamic revolution 1978/79, the secularly oriented authoritarian regime of Mohammad Reza Shah Pahlavi and its political elite were overthrown and replaced by the theocratically oriented political regime of Ayatollah Khomeini and its corresponding political elite, mainly clergy and religious laypersons institutionalized in the Islamic Republic of Iran (IRI). The nature of the political system of the IRI is unique as it is based on a combination of state institutions that derive their legitimacy from Islamic law—the religious supervisory bodies (the Council of the Guardian [*Majles-e Khobregan*],[15] the Expediency Council [*Majma'-e Tashkhis-e Maslahat-e Nezam*],[16] and the Assembly of Experts [*Shora-ye Maslahat-e Nezam*])[17]—and republican institutions legitimized by the people (the legislative [*majles*, parliament], the executive, and the judiciary bodies). The basic principle of the political system is the *velayat-e faqih* (the Government of the Jurist) system as developed by Ayatollah Khomeini, according to which the supreme leader (*vali-e faqih*) is the head of the political system. The supreme leader, who is not elected by the people, may overrule any bills passed by the legislature. His power, however, is not absolute, but checked by the religious supervisory bodies. Together, the supreme leader and the religious supervisory bodies oversee the republican institutions. This creates a continuous tension between the supreme leader and the religious supervisory bodies on the one hand, and the republican institutions on the other.

Like in all other domains, it is the supreme leader who has the final say in foreign policy decision-making. He approves or disapproves foreign policy initiatives. Though the president and his office constitute the main foreign policy-making organ in the IRI since 1989, foreign policy decisions have always to be made in accordance with the supreme leader. The foreign minister reports directly to the president. Foreign policy initiatives of the foreign ministry are always monitored through the president's office. The *majles* may not interfere in the executive foreign policy decision-making process. But the *majles* discusses foreign

[15] The Council of the Guardian consists of twelve jurists (six clerical and six non-clerical). The six clerical members are selected from among the ranks of the clerical elite and appointed by the supreme leader. The six non-clerical members are appointed by parliament at the recommendation of the head of the judiciary. The Council of the Guardian determines whether laws passed by parliament are compatible with the *shari'a* (Islamic law). It has supreme oversight of the elections for parliament, the Assembly of Experts, and the presidency (Schirazi 1997: 89).

[16] The Expediency Council was established in 1988 to act as a mediator between the *majles* and the Council of the Guardian, and to advise the supreme leader (Tellenbach 1990: 54). The Expediency Council has 31 members that are appointed by the supreme leader from among the ranks of the Iranian political elite (Buchta 2000: 61).

[17] The Assembly of Experts is a council of 86 clerics who are elected by the Iranian people for an 8-year term. The Assembly of Experts elects the supreme leader from its own ranks and may dismiss him if he does not fulfill his duties (Algar 1969), the latter of which is very unlikely to happen.

policy issues and individual members can make public statements on regional and international issues. The government needs the *majles*'s approval to enter into international agreements, treaties, memoranda of understanding, and so on. This division of competencies regarding foreign policy issues has provoked disagreement between the supreme leader, the president, and the foreign minister several times.

Legal political parties do not exist in the IRI. Political factions represent the different approaches to domestic (economic and sociocultural) and foreign policy issues. These different approaches are based on diverse ideological and material interests of the members of the factions and the groups that support them. The main political factions are the Conservative faction, the Pragmatist faction, and the Reformist faction.[18] The political factions are not homogenous but loose coalitions of groups and individuals with similar views. They have no coherent organizational structure and no official program. Sometimes, different opinions within factions can even cause disruptions and result in alliances with other factions or the decline and emergence of new factions. The extent to which a political faction participates in policy formulation or political discourse depends on what faction or alliance of factions control (semi-) governmental institutions in a certain period of time. However, though the domination of the republican institutions has shifted between the factions several times, the religious supervisory bodies, the military, and the religious foundations have, since the Islamic revolution, been under control of the Conservative faction. In other words, the Conservative faction has continuously ruled over key state institutions and, consequently, has had a decisive influence on domestic and foreign policy formulation.

The different approaches of the political factions to domestic and foreign policy have an ideological and a material component. Firstly,

[18] The division of the political elite into factions is not the author's own categorization, but is based on several works published on factional politics or factionalism in the IRI. See for example Moslem, M. 2002. *Factional Politics in post-Khomeini Iran.* Syracuse: Syracuse University Press; Seifzadeh, H. S. 2001. "The Landscape of Factional Politics and Its Future in Iran." *Middle East Journal* 57 (Winter): 57–75; Buchta, W. 2000. *Who Rules Iran-The Structure of Power in the Islamic Republic.* Washington: Washington Institute of Near East Policy; Bakhtiari, B. 1996. *Parliamentary Politics in Revolutionary Iran—The Institutionalization of Factional Politics* (Gainesville: University Press of Florida.This distinction is also used by the Iranian political elite itself and the Iranian public, as can be seen from academic publications, newspaper articles, and speeches.

there are diverse opinions between and within factions on whether Islamic jurisprudence should be the only or main basis of the legal system in the IRI. The Conservative faction considers Islamic jurisprudence—with varying interpretations—an important constituent of the juridical system in Iran. For the Pragmatist faction, Islamic jurisprudence is of relevance on the sociocultural level but less on the economic level. The Reformist faction considers Islamic jurisprudence insufficient to address all issues in Iranian society. Secondly, the political factions defend the material interests of their members and of economic groups that support them. The Conservative faction represents the interests of the traditional economic sector (the *bazaaris*), as well as ultra-orthodox clergy, and the highly religious public. It receives its major income from official economic sources based on fiscal instruments (taxes, fees, and borrowings) and oil/gas revenues (sources of foreign currencies), as well as from religious sources (the mosques, the Shiite holy shrines and sites) and the religious foundations outside the fiscal instruments. The other two main factions, the Pragmatist and the Reformist factions, rely only on official fiscal sources. The Pragmatist faction is supported by (religious) technocrats, parts of the middle class, and segments with liberal tendencies. The Reformist faction represents the interests of a wide range of (secular) social groups among which women, students, and intellectuals. The Conservative faction is in favor of trade liberalization but objects to large-scale privatization policies, which could counter the interests of the traditional economic sector and the religious foundations. The Pragmatist and Reformist factions represent liberal tendencies or support economic liberalization policies with a limited role for the state.

The varying approaches of the political factions to foreign policy are grounded in different geopolitical outlooks. The Conservative-dominated group is more ideologically driven in its foreign policy outlook, while the Pragmatist and Reformist factions have a pragmatic foreign policy approach. In general, two main positions of the political factions on foreign policy can be distinguished. The first group is represented mainly by the Conservative faction. It sees Iranian society as part of the *ummah* (Islamic community) and emphasizes the significance of the Islamic revolution and the return to Islamic values. In order to fulfill these prerogatives, the IRI has to have a good partnership with Islamic countries, but not necessarily with their governments, and also refrain from rapprochement with the United States (US). During the first decade after the Islamic revolution this position was dominated by the two main ideological foreign policy principles of the Islamic revolution: "Neither East nor West," and the "Export of the Revolution." Today, due to the huge economic problems in Iran, segments of the Conservative faction have softened their position on these two foreign policy principles. The second group is represented by the Pragmatist and Reformist factions. These factions are convinced that Iran has to play a key role in international relations, as international trade and international diplomatic relations are preconditions for economic development in Iran. Since the late 1980s, the Pragmatist and Reformist factions have been the driving forces behind the IRI's international economic policies and improvements in diplomatic relations with the Persian Gulf countries, European Union (EU), China, India, Central Eurasia, Russia, and, also, the US. Several events had a great impact on reversing the foreign policy approach: (a) the death of Ayatollah Khomeini, (b) the rise of the Pragmatist faction to power in 1989, (c) the end of the Iran-Iraq War in 1988, (d) the collapse of the Soviet Union in 1991, (e) the larger US military presence in the Persian Gulf since the Kuwaiti crisis (1990–1991), and (f) the economic problems in the IRI. Parts of the Conservative faction have complicated Iran's foreign relations by financially supporting subversive militant groups in Muslim countries, suppressing of Iranians in exile, accusing western citizens of blasphemy, or not recognizing of the state of Israel. Nevertheless, despite these different approaches to foreign policy the Conservative faction also agrees that Iran should be a key player in international relations. Since 1989, this foreign policy goal has been more or less independent of what composition of factions controls the republican institutions and religious supervisory bodies in a certain period of time. This means that in contrast to the economic and sociocultural levels, factional rivalries are of less importance on the foreign policy level. The interaction with other countries and regions plays a much greater role in foreign policy formulation than factional rivalries.

4. The Foreign Policy of Iran since the Islamic Revolution

Iran's foreign policy approach since the Islamic revolution can be summarized by and divided into four phases. During the first ten years after the revolution, when Ayatollah Khomeini was the Supreme Leader, it was dominated by two main ideological principles, manifested in two slogans: "Neither East nor West but the Islamic Republic," away especially from Western (US) influences in Iran, and "Export of the Revolution," to free Muslim countries and non-Muslim countries from their "oppressive and corrupt rulers." The latter served as a means of mobilization of the Iranian people to support the eight-year war with Iraq (1980–1988). Supreme Leader Khomeini (1979–1989) and his followers saw the Islamic revolution as a model that would trigger further revolutions in other Muslim countries. However, the guideline of the Export of the Revolution was strongest only in the first ten years after the revolution, and even then not as an ideological or revolutionary pursuit, but rather as a survival strategy in the war with Iraq (Bakhash 2001: 248). Most of the armed groups, which received financial support from Iran during the 1980s, were Shi'ite organizations opposed to Saddam Hussein in Iraq or to other rulers in the Persian Gulf, or groups active in Lebanon, Afghanistan, and Pakistan (Ehteshami 1995; Roy 1996/1999: 191). Above all, the almost unqualified support of Iraq by the Arab states and the west played a determining role in Iran's support of armed groups in the Middle East and beyond.In the 1990s, Iran supported Sunni groups such as the Islamic Salvation Front (FIS) in Algeria, the National Islamic Movement in Sudan, Hamas and Islamic Jihad in Palestine, the Muslim Brotherhood in Jordan, the al-Nahda Party in Tunisia, and the Jihad Group in Egypt. Furthermore, they also supported the Muslims in Bosnia in the 1990s and the Islamic MORO Movement in the Philippines in the 1980s (Ehteshami 1997: 30; Taheri 1987). Xinjiang province in western China was also subject to the IRI's aim to "Export the Revolution." During the first years after the revolution the Iranian government funded the construction of mosques and religious schools in Xinjiang province without the permission of the Chinese government and invited students from Xinjiang to study at one of the religious schools in Iran. Still, Iranian involvement in Xinjiang was rather limited due to the fact that most of Xinjiang's Muslims are Sunnis and of Turkish origin. To them, Persian Shi'a Islam was not attractive. Furthermore, other Middle Eastern countries carried out as many, and even more, subversive activities in Xinjiang province, Saudi Arabia among them (Garver 2006: 132–133). The two following presidents, Hashemi Rafsanjani and Mohammad Khatami, followed a pragmatist approach toward foreign policy. President Rafsanjani (1989–1997) adopted a more pragmatic foreign policy orientation focusing on post-Iran-Iraq war economic reconstruction and the country's reintegration into the international economy. A priority of President Rafsanjani's foreign policy was to improve relations with the Persian Gulf countries, especially Saudi Arabia, to strengthen Iran's role in the region. He even suggested the establishment of a regional security organization, which did not come about as the Persian Gulf countries preferred alignment with the US above Iran. During Rafsanjani's presidency the IRI also improved relations with Russia after the fall of the Soviet Union as well as the CEA countries and the EU. Iran's relations with the US remained frozen. During Khatami's presidency (1997–2005) important changes in Iranian foreign policy took place, especially in the form of improved relations with the EU. Khatami continued Rafsanjani's foreign policy with regard to the Persian Gulf as well as Russia and CEA. Due to Iran's huge oil and gas resources Iran also intensified cooperation with China and India in the energy sector. President Khatami tried to open a dialogue with the US. He aimed at combining democratic reforms at home with a pragmatic foreign policy abroad but failed, mainly because of resistance to these reforms by the Conservative faction. This means that even during the presidencies of Rafsanjani and Khatami, Islamic ideology, embedded in a nationalist desire to reject any "Westernization" of the country and the Iranian people, still prevailed among some elements of the Iranian political elite. With the election of Ahmadinejad as president in 2005, some shifts in foreign policy orientation can be noted—away from the pragmatic approach under Rafsanjani and Khatami—to a rhetorically more hostile attitude, especially towards the west and Israel. At the same time, Ahmadinejad has found companions among leaders of other developing countries in Latin America (Venezuela, Ecuador, and Nicaragua). In the short term, Ahmadinejad complicates Iran's foreign relations, especially towards the west. At the same time, the US, through its economic sanctions, the inclusion of Iran into the 'axis of evil,' and branding the Islamic Revolutionary Guards Corps (IRGC)[19] a terrorist organization, makes rapprochement between the two countries very difficult. The situation has further worsened due to the nuclear issue, affecting also the relations between Iran and the

[19] Together with the regular military the IRGC is formally subordinate to the Ministry of Defense and Armed Forces Logistics (MODAFL). They are responsible for defending Iran's borders as well as for the provision of internal security.

EU, and possibly preventing large-scale investments in the Iranian oil and gas sector.

5. Oil and Gas Complex in Iran

The oil industry has been Iran's main source of income for the last century. In 1901 the Englishman and entrepreneur William Knox D'Arcy received a 60–year concession from the Persian Shah Muzafar ed-Din to search for oil in Iran. In 1908, he found oil and set up the Anglo-Persian Oil Company (APOC) (Karbassian 2000: 629; Fateh 1358/1979). Due to the change from coal to oil on British ships, and because of British strategic interests in oil, the British government bought majority interests in the APOC. Britain thereby gained direct control of APOC (Fateh 1358/1979: 262–265). When Persia was renamed Iran in 1936, the APOC became the Anglo-Iranian Oil Company (AIOC), today known as British Petroleum (BP). The Iranian oil industry was nationalized in 1952 by Prime Minister Mosaddeq and received the name National Iranian Oil Company (NIOC).[20] In 1953, Mosaddeq was overthrown by a coup with the help of the US and British intelligence services. Before the nationalization of the oil industry, about half of the AIOC had been owned by Britain. After the coup, between 1954 and 1979,[21] the Iranian oil industry was operated in conjunction with a consortium of international oil companies (Karbassian 2000: 629). The profit share of this consortium was 75 percent to 25 percent to the advantage of Iran. Additionally, the international oil companies had to pay the exploration costs. Thus, the former profit share system of 50/50 percent was abolished (Alam 1993: 107). After negotiations with the oil consortium, the agreement of 1954 was changed on March 20, 1973, giving Iran control of its own oil industry. This meant that under the terms of the new agreement, the Iranian political elite controlled the production, marketing, and setting of prices on oil. The change from the 1954 agreement in favor of Iran was made possible through the collective agitation of OPEC, which had been established in 1960

[20] For more information on the NIOC, see its website http://www.nioc.org.

[21] On the oil industry in Iran during the Mohammad Reza Shah period see the following important books: Karshenas, M. *Oil, State, and Industrialization in Iran,* (Cambridge: Cambridge University Press, 1990); Fesharaki, F. *Development of the Iranian Oil Industry: International and Domestic Aspects,* (New York: Praeger, 1976).

by Iran, among others (Amineh 1999: 240). After the Islamic revolution, the post-revolutionary political elite embarked on reforming the oil sector through reducing oil production and export and terminating, or at least decisively reducing, the activities of foreign oil companies in the Iranian oil sector. The international oil consortium was abolished and all its tasks were transferred to NIOC. Moreover, all joint-venture oil companies were handed over to NIOC, which was put under the supervision of the Ministry of Oil, created in 1979. The war with Iraq, during which important oil installations were destroyed, resulted in a reduction of oil production and export and, therewith, a decline in the country's oil revenues (Amirahmadi 1990: 72–75), diminishing the financial resources available to the government (Behdad 1995: 109). In the 1990s, the NIOC started to attract foreign investors for the Iranian oil and gas industry. The Iranian government also passed a law allowing foreign participation in so-called buy-back oil and gas contracts, under which field developers are compensated with output before the field is returned to NIOC.[22] Revenues from oil and gas export amount to 70 percent of the state's general budget and about 80 percent of all foreign exchange earnings (Karbassian 2000: 629). However, economic dependence on oil revenues is not sustainable in the long run. It needs, even if the government can limit the impact of price fluctuation through the operation of an oil stabilization fund (which was established in 2000),[23] a continuous upward trend in oil prices to continue this economic growth. Thus, only by a structural economic reform could the IRI attain sustainable economic growth. During its Third Development Plan (2000–2005), the Iranian government missed the goal to reduce its dependence on oil revenues to less than US$12 billion. Instead, dependence on oil revenues even increased to more than US$40 billion in 2006 (Nili 2006).

According to Stern (2007), for the IRI the possession of nuclear facilities is not in the first instance connected to its ambition to become a dominant regional power but is rather a source to satisfy increasing domestic energy demands. Between 2000 and 2005, total energy consumption in Iran rose almost 10 percent annually (gasoline 12 percent), which is five times more than the world average. Iran has

[22] See NIOC website: http://www.nioc.ir/brief_history/page9.html.

[23] On the Iranian Oil Stabilization Fund see e.g. Amuzegar, J. 2005. "Iran's Oil Stabilization Fund: A Misnomer." *Middle East Economic Survey* 48(47).

only one percent of the world population but consumes nine percent of global energy production. This is due to subsidized energy prices[24] and the lack of regulations for the construction of private homes, public buildings, and industrial and commercial plants (*Hamshari* 25 November 2004; *Jomshouri Eslami* 25 November 2004). It is estimated that some industrial factories in Iran use 35 percent more energy than the world average; refrigerators built in Iran consume more electricity than imported ones; cars built in Iran use 37 percent more gasoline than foreign cars. Furthermore, Iran's population has doubled in the last 25 years. The country's car industry produces one million cars every year (the highest number in the Middle East). Iran now produces 30 percent less oil than it did in 1979 (Hen-Tov 2006–07: 170).

At the end of 2007 Iran had 138.4 billion barrels of the world's total proven oil reserves, or 11.2 percent of proven oil reserves. The country thus ranks second behind Saudi Arabia in proven oil reserves. In 2007, Iran produced 4.4 million barrels per day (bbl/d), which is a 5.4 percent share of total global oil production. It ranked fourth in oil production behind Saudi Arabia, Russia, and The United States (US). Domestic oil consumption in the same year amounted to 1.62 million bbl/d, which is 0.4 percent of the world's total consumption (British Petroleum 2008: 8, 10, 13). Currently, Iran has 40 producing oil fields of which 27 are onshore and 13 are offshore. Most crude oil reserves are located in the Khuzestan region close to the border with Iraq. In 2006, crude oil and product exports of Iran averaged 2.5 million bbl/d with oil revenues of US$54 billion. The main countries to which oil was exported were Japan, China, India, South Korea, and Italy. With 29 ships, the Iranian oil tanker fleet of the National Iranian Tanker Company is the largest in the Middle East. The main export terminal is Kharg Island. Other important terminals are Kish Island, Abadan and Bandar Mahshar, and Neka. The Strait of Hormuz is an important export point from Iran and other Persian Gulf countries. Every day, 17 million bbl, or about two-fifths of all oil traded by sea, are shipped through the Strait of Hormuz. Iran has a well-developed pipeline infrastructure with five pipelines and various pipeline projections under negotiation and consideration. Iran has nine running refineries with a total capacity of 1.5 million bbl/d, far below domestic demand.

[24] The total domestic energy subsidies amount to US$20–30 billion per year, which is about 15 percent of the country's GDP (Stern op. cit.).

By 2012, about 985,000 bbl/d should be added to the refining capacity through expanding and upgrading the Bandar Abbas, Bushehr, and Abadan refineries. Iran is also in discussion about creating joint ventures with China, Indonesia, Malaysia, and Singapore to expand refining activity (Energy Information Administration October 2007).

At the end of 2007, Iran had 981.75 trillion cubic feet (tcf) of proven gas reserves, or 15.7 percent of the world's total proven gas reserves. The country thus ranks second behind Russia in proven gas reserves. In 2007, Iran produced 111.9 billion cubic meters (bcm) of gas, which is a 3.8 percent share of total production. It ranked fourth in gas production behind Russia, the US, and Canada. Domestic gas consumption in the same year amounted to 111.8 bcm, which is 3.8 percent of the world's total consumption (British Petroleum 2008: 24, 26, 29). About 62 percent of Iran's proven gas reserves have not yet been developed. Major natural gas fields include: South and North Pars, Tabnak, and Kangan-Nar. The South Pars gas field is the world's largest, jointly shared between Iran and Qatar. Nevertheless, due to rising domestic demand it has to be expected that gas exports will remain limited in the coming years. The domestic gas pipeline network has been extended in recent years. There are further pipeline connections to Turkey (a 745 mile pipeline with a capacity of 1.4 billion cubic feet per day [bcf/d]), and to Turkmenistan, whence Iran imports 800 million cubic feet per day [mcf/d] Armenia. This US$195-million pipeline is the first in the Caspian region that bypasses Russia. The 87-mile pipeline was completed in March 2007, transporting 200mcf/d to Armenia in exchange for electricity. Iran might also participate in the Nabucco gas pipeline project (discussed below). The most controversial pipeline proposal is the Iran-Pakistan-India (IPI) pipeline (discussed below) (Energy Information Administration October 2007).

The Iranian energy infrastructure must be modernized extensively. The country needs about US160$ billion for the next 25 years to carry out this job, an amount which it does not have and for which it must rely on foreign investments (Leverett 20 June 2006). The sanctions imposed upon Iran by the US and an insecure investment climate have until now posed an obstacle to large-scale foreign investment. In 1992, the US Administration passed the Iran Non-Proliferation Act, followed by the Iran-Libya Sanctions Act (ILSA) in 1995,[25] that prohibited even

[25] The ILSA imposed sanctions on non-United States companies investing more than US$40 million annually in the Iranian and Libyan oil and gas sectors. The

non-US companies[26] from investing in the Iranian and Libyan oil and gas sector (Karbassian 2000: 632). President George W. Bush extended the law in 2001 until 2006, punishing those oil companies that would spend US$20 million or more in Iran or Libya (Lorenzetti 1 April 2002). On September 30, 2006, President George W. Bush signed a further extension of the law until December 31, 2011. Since 2006 it has changed its name to the Iran Sanctions Act (ISA) (Katzman 25 January 2007). The sanctions imposed on Iran by the United Nations (UN) regarding its nuclear program, in 2006, 2007, and 2008, and the possibility of further sanctions in the future, put additional pressure on both potential investors (Leverett 20 June 2006) and the EU.

6. Iranian Oil and Gas and European Union Energy Supply Security

Whereas relations between the IRI and the US were hostile from the beginning of the IRI's existence, the IRI and the EU (and some of its member countries) were suspicious of each other, but nevertheless willing to negotiate, especially in the economic sector (Rakel 2009: ch. 3). According to Tarock (1999: 44), there are four main reasons for the IRI's differing attitudes towards the main western actors: (1) although European countries, especially (West) Germany, Italy, France, and Britain had good economic relations with Iran before the Islamic revolution, they had not been so deeply involved in Iran's political and military affairs like the US had—except for Britain. (2) European countries did not view the revolution from a Cold War zero-sum perspective as the US did. That means they did not consider the success of the revolution as an indication of the west's 'loss' of Iran. (3) During the

amount dropped to US$20 million one year after the sanction's implementation for those countries that did not undertake measures against Iran's actions in supporting international terrorism and pursuit of weapons of mass destruction, such as the imposition of sanctions (on Iran?—need to be clear here) for a minimum of 2 years.

[26] In November 1997, the EU handed in a formal complaint on the US law at the WTO, based on two grounds: firstly, the law runs counter to the principle of free trade on which the WTO is built; secondly, any punitive action under it would be a violation of international law. The two parties agreed during a meeting in London in May 1998, that the EU would continue its support of the US in combating international terrorism and that the US would grant a presidential waiver to Total, as well as to other European oil companies that intended to invest in the Iranian oil and gas industry. As Tarock notes, in the history of the IRI this was truly the strongest position the EU had taken in favor of the IRI against the US (Tarock 1999: 50–51).

months before Khomeini returned to Iran, France had offered him a temporary home after he had been forced out of Iraq.[27] (4) The political elite of the IRI realized from the beginning that for security, political, and economic reasons, it could not afford to confront both Europe and the US at the same time.

Since the Islamic revolution it can be assumed that Iran needs the EU because of its need of foreign direct investment, technology, and know-how, and its continuing difficult relationship (economic sanctions) with the US. For the EU, Iran is a potential supplier of oil and gas, as well as an important factor for stability in the Middle East and, therewith, in the EU's own backyard. The EU's interests in Iran's energy sector differ regarding the oil and the gas sectors. The oil sector is of relevance from a global supply point of view, while the gas sector is more important to the EU and also politically more complex (Moradi 2006: 182). Oil can be transported by pipeline and by ship, whereas gas has to be exported via pipeline. The construction and usage of pipelines is a delicate issue as often the conflicting interests of producer, consumer, and transit countries are involved. The import of oil and gas are vital for the economies of the EU member countries. The EU therefore needs a long-term strategy for the security of supply at reasonable prices. These goals can only be reached if different producers compete on the European market. At the present time Iran does not play an important role as a supplier of oil and gas to the EU, but it could become an important supplier in the future (Moradi 2006: 182–183).

Regional security problems and US activities in the Middle East have not been central to Iran-EU relations, but they have influenced policies on both sides, particularly since the end of the Cold War and again since the attacks of 9/11, the war in Afghanistan since 2001, and the Iraq War since 2003. Another issue that has more recently impacted on the relations between Iran and the EU is Iran's nuclear program (Rakel 2009: chs. 6 and 7). At the same time, increased diplomatic collaboration between Iran and the EU reflects the desire in Iran, and at least to some extent in the EU, to object to US goals of

[27] During the conflict between the Shah and Ayatollah Khomeini, France declared its neutrality. It even hosted Ayatollah Khomeini during a four months stay in Paris. Khomeini thanked President Valery Giscard d'Estaign for his hospitality: "I am very grateful to my French friends who gave me the opportunity to send my messages from Paris to the Iranian people […]" (cited in Bozorgmehr 1997: 39).

changing political order in the Middle East and domination of the region. Moshaver (Moshaver 2003: 283–284) describes the relations between Iran and the EU as "functional accommodation." He argues that increasing cooperation between Iran and the EU is a "byproduct" of Iran's overall internal and international situation rather than a reflection of a fundamental "political/strategic" change. The Iran-EU functional accommodation has no long-term strategic, political, or security objective as had Iran's relations with European countries during the Mohammad Reza Shah period. The relations between the IRI and the EU are functional in the sense that they focus on mutual economic interests in light of continuing sanctions on Iran by the US.

Still, obstacles to greater economic cooperation in terms of trade and investment have to do with Iran's domestic investment environment, and is hampered by factors such as: a lack of, or unclear, regulations in Iran, lack of transparency in laws and regulations, an unclear tax system, an insecure investment climate and repatriation of investment profits (European Commission 7 February 2001).

6.1 *The Energy Sector in the European Union*

During the oil crisis in 1973/74 the EU put the energy issue on its agenda for the first time. However, it was only in recent years that the EU seriously considered defining a common energy policy. The urge to develop a common energy policy has been driven by several developments: declining energy production in Europe, concerns about Russia as a reliable energy exporter to Europe, increasing energy prices, and the necessity to fight climate change (Closson 2008: 1). The EU Green Papers (European Commission 2006, November 2001) on security of supply express concern that the oil and gas supply will depend in the near future on a limited number of oil and gas producing countries, and that import dependence requires an improvement in economic relations with key producer countries. Thus, the Green Paper forms the basis on which the shape and direction of a future energy policy could be developed. Europe's energy mix is dominated by fossil fuels. Oil accounts for 40 percent of total EU energy consumption, and gas for 24 percent. The consumption of gas has been increasing at the expense of coal (Energy Information Administration January 2006). The World Energy Outlook 2005 of the International Energy Agency states that gas and renewables will increase their share in the energy mix at the expense of nuclear power, coal and oil. Between 2003 and

2030, the share of gas in the EU's energy consumption will rise from 23 percent to 32 percent. In 2003, 6.5 percent of the energy consumed was renewables. By 2030, this amount will have risen to 12 percent. The share of oil in the energy mix in the same period will decline from 38 percent to 36 percent, that of coal from 18 to 13 percent, and that of nuclear energy from 15 to seven percent (International Energy Agency 2005).

The EU's dependence on fossil fuels is a reflection of what is a general trend in energy usage at the global level. The International Energy Agency (IEA) expects fossil fuels to dominate energy use in the future as well. Oil will remain the main fuel used in the primary energy mix, though there will be a slight decrease in its share in overall energy usage. At the same time, the consumption of gas will grow more rapidly than any other energy source (International Energy Agency 2004: 31). In the transportation sector, oil still plays a dominant role. At present, about half of the energy is consumed in the EU is produced at home, while the other half is imported. The EU is becoming increasingly dependent on the import of energy as domestic production capacity is limited. The EU member countries (EU 25) together have about 0.6 percent of the world's proven oil reserves, 2.0 percent of natural gas reserves, 4.0 percent of proven coal reserves, and 18.0 percent of electric generating capacity (Energy Information Administration January 2006). The EU imports its oil from Russia, the Middle East, North Africa, and Norway and most of its gas from Russia and Algeria. Other countries with which the EU is negotiating deals on importing gas are Libya, Egypt, Qatar, Iran, and Azerbaijan. The gas sector has been growing rapidly in recent years. Gas is cleaner and more environmentally friendly than oil or coal. Given the fact that indigenous energy sources in Europe are declining to maintain the high standard of living, the EU will depend even more on foreign energy resources in the future. The EU will therefore establish or consolidate energy partnerships with Russia, Kazakhstan, Azerbaijan, Turkmenistan, and the Middle East (Bahgat 2006: 967–968).

During its summit in March 2007 the EU adopted an *Energy Policy for Europe* (European Commission 2007). However, the formulation and implementation of a common foreign policy in general and a common energy policy in particular has proven to be difficult. The member countries of the EU sometimes have very different ideas on what should be the foreign policy of the EU, and what policy it should follow with regard to Iran. Most cooperation takes place on a bilateral basis and not within an EU context. That means the EU member countries pursue varying, and sometimes contradictory, interests when it comes to Iran (Rakel 2009: ch. 7).

6.2 *Iran-European Union Cooperation in the Oil and Gas Sector*

Iran is not yet a major energy supplier to the EU. Although Iran stands fourth among the main oil exporters in the world, and oil accounts for about 80 percent of the country's exports, oil imports from Iran account for only 5 percent of the EU's consumption (Moradi 2006). However, it has to be expected that Iran will become one of Europe's most important sources of energy and a transit country as, in addition to its huge oil and gas resources, it is the shortest and most economical transit link between the oil-rich Caspian Sea region and Europe (Amineh 2003). In May 1999, the Working Group on Energy was established between Iran and the EU in Tehran. The Working Group meets once a year either in Tehran or in Brussels. Iran also became an observer of the Commission-funded INOGATE[28] (Inter-State Oil and Gas Transport to Europe) with the possibility of becoming a full member. For Iran, due to the sanctions imposed by the US, Europe is the second best option to deliver FDI, modern technology, and know-how to Iran. Despite these sanctions, Iran and European countries have signed several energy contracts in the last years.

The main buyback contracts between Iran and European oil firms include the following. (1) The Cheshmeh-Khosh field, which was awarded to Spain's Cepsa for $300 million. In January 2004, it was re-awarded to state-owned Central Iranian Oil Fields Company (CIOFC). Cepsa withdrew from negotiations in December 2003 when no agreement could be reached on development costs and buyback terms. Nevertheless, Cepsa could still be involved in the project in some way, (2) A one billion dollar, five-and-a-half-year buyback contract was signed by ENI in late June 2001 to develop the Darkhovin field; TotalFinaElf, Shell, Eni, and BP are bidding to develop the Ab Teymour field. (3) In 2001, a contract was signed with the Italian AJIB (an affiliate

[28] For more information on the INOGATE Program see its website http://www.inogate.org.

of ANI) to develop the Darkhoen oil field, with a total investment of 540 million dollars and a buyback of 920 million dollars. (4) In 2000, a contract was made with a consortium consisting of AJIB and the Iranian Petropars to develop phases four and five of South Pars gas field, worth two billion dollars. The total buyback is 3.8 billion dollars. (5) A buyback contract to develop the offshore Balal field was signed in April 1999 between Iran and Elf. The field was turned back over to NIOC in January 2003 after the field had reached its contract-designated level of 40,000 bbl/d. (6) In March 1999 a contract was signed with a consortium consisting of French Elf Aquitaine and AJIB, with a total investment of 540 million dollars and a buyback of one billion dollars. (7) A deal of two billion dollars was made between Total and NIOC on September 29, 1997 to explore the South Pars gas field. NIOC estimates that South Pars has a gas production potential of up to eight billion cubic feet per day (Bcf/d). (8) In 1995, a contract between French Total and the NIOC was signed to develop the Sirri A and E oil fields. It has an estimated investment of 610 million dollars with a buyback of 500 million dollars. (9) In 1991, a contract was signed with Shell to develop the Soroush and Norouz oil fields with a total investment of 799 million dollars and a buyback of 1.6 billion dollars (Hasan-Beygi 2002: 79–106; Energy Information Administration August 2006; Energy Information Administration February 2000).

The Working Group on Energy held its third meeting on October 19, 2002 in Tehran. It followed two earlier meetings, on May 18–19, 1999 and on March 28, 2001, held in Tehran and in Brussels, respectively. Director General of Energy and Transport for the European Commission, Francois Lamoureux, said the committee meeting was important to strengthen ties between Iran and the EU:

> "[T]he EU is ready to cooperate in transfer of technical knowledge to Iran in various areas, notably the renewable energy resources, more efficient fuel consumption, and in the transportation sector" (Cited in Pyvand's Iran News 19 October 2002).

During the meeting, the EU and Iran signed two Memoranda of Understanding (MoU) on energy affairs, one of which was on the third joint session of the Iran-EU expert committee and the possibilities of expanding bilateral ties. The other agreement concerned the creation of an Energy Cooperation Center (ECC). The ECC was opened on October 21, 2002. The EU paid Euro 1.7 million to finance the center's budget. The ECC has several objectives: to investigate the possibilities for further energy cooperation, to exchange ideas on experiences with non-nuclear energy technologies, training, the identification of priority projects, and technical assistance for introducing non-nuclear technologies in Iran. The ECC has a staff of 20 people. It is co-managed by the European Commission and Iran (Pyvand's Iran News 19 October 2002).

Further meetings of the EU-Iran Working Group on Energy and Transport were suspended because of Iran's nuclear program. The talks were reopened again in December 2005 when the Working Group held its fourth meeting in Brussels. During the meeting the Iranian Deputy Oil Minister for International Affairs, Mohammad Hadi Nejad-Hosseinian, met Ria Kemper, Secretary General of the Energy Charter Secretariat (ECS),[29] to discuss developments in the energy sector in the Eurasian region. Kemper stressed that the ECS favors greater energy cooperation with Iran and that it promotes the full membership of Iran to the ECS. Currently Iran only has observer status (IRIB 26 December 2005).

The planned Nabucco pipeline is a more recent indication of strengthening energy ties between the EU and Iran. The Nabucco pipeline[30] is planned to transport gas from the Caspian region, Iran, Iraq, and Egypt via Turkey, Bulgaria, Rumania, and Hungary to Austria. By 2020, the pipeline could transport 25.5 to 31 billion cubic meters of Caspian gas to Europe annually. The Nabucco pipeline should lower EU dependence on Russian gas. However, the future of the pipeline is currently no longer clear, since Russia has signed agreements with Turkmenistan and Kazakhstan on the transportation of their gas. Furthermore, Iraq's future remains very uncertain. This puts into question whether Iraq could be considered a major supplier, and whether it could deliver gas through the Nabucco pipeline. The seven shareholders of the Nabucco pipeline project are OMV (Austria), MOL (Hungary), Transgaz (Romania), BOTAS (Turkey), Gaz de France (France), Total (France), E.ON Ruhrgas (Germany), and RWE (Germany). It is

[29] The Brussels-based ECS is an international forum for energy cooperation, in which 51 European and Asian governments participate. The Energy Charter Treaty was signed in December 1994. Its aim is to strengthen the rule of law with regard to energy issues to reduce the risks of energy investment and trade. For more information, see the website http://www.encharter.org.

[30] For more information on the Nabucco pipeline see http://www.nabucco-pipeline.com

also possible that Russia's Gazprom will be asked to join the Nabucco project (Alexander's Gas and Oil Connection 29 June 2007; EurActiv 19 January 2006).

At the same time, Russia is also pursuing its own pipeline projects that compete with European ones. The Blue Stream Pipeline, inaugurated on November 17, 2005, carries gas from Russia to Turkey via the Black Sea. The pipeline is a joint project of Gazprom and Italy's ENI. The same two companies are now constructing the South Stream Pipeline supposed to carry gas from Russia to Romania, Bulgaria, and Greece and from there by a southwesterly route into southern Italy and on a northwesterly route into Serbia and Hungary, with further connections to Austria or northern Italy. The pipeline is expected to carry 30 billion cubic meters of gas annually. If the project is carried out, the South Stream Pipeline could partly take over the planned extension of the Blue Stream Pipeline from Turkey through Bulgaria, Romania, Hungary, and Austria. Both pipelines will compete with the Nabucco Pipeline. The South Stream Pipeline will also compete with the proposed extension of the Baku-Tbilisi-Erzurum pipeline, which could be integrated with the Nabucco Pipeline or be constructed from Turkey to Greece and Italy (Closson 25 June 2007).

The diversification of European gas imports is confronted with different challenges: the EU stands in competition with Russia, China, and Southeast Asia for Caspian gas. The dispute around the legal regime of the Caspian Sea has not yet been resolved,[31] resulting in negative effects on exploration investments and transportation of Caspian gas to consumers. The nuclear issue in Iran poses an obstacle to large-scale investments (Closson 25 June 2007: 3). However, Iran remains an interesting source for diversification. Therefore, a moderate tone towards Iran, and more constructive efforts, could benefit the EU in establishing a greater economic presence in Iran, or connect its resources to the Baku-Tblisi-Ceyhan pipeline, and perhaps the Nabucco pipeline. The EU could make its own gas contract with Iran, although this could be contrary to Russia's economic and geopolitical interests in Europe (Umbach 2006). The activities of China in Iran with regard to energy could be a challenge both to the EU and Russia.

[31] On the Caspian legal regime dispute, see Amineh, M. P. 2003. *Globalization, Geopolitics and Energy Security in the Caspian Region*. Den Haag: CIEP.

7. Iranian Oil and Gas and China's Energy Security of Supply

Historical ties between Iran and China date back as far as 139 BCE when the Hans and the Parthians established diplomatic and trade relations with one another. Trade was carried out on the so-called "Silk Road" which linked China to Central Asia and the Middle East. When the Mongols conquered both China and the Persian Empire in the thirteenth century, these contacts were further increased (Garver 2006: 14). Even today, Chinese and Iranian leaders refer to these historical relations to legitimize their recent cooperation. The two countries have no history of war and conflict with one another, which helps to strengthen their mutual relationship (Dorraj and Curier 2008: 67).

With the rise of Deng Xiaoping to power in 1976, the implementation of economic liberalization policies in China, and economic growth, China's demand for additional energy sources has been on the rise. Iran's huge oil and gas resources and rising energy import dependency in China has been the most important aspect of the relationship between these two countries (Dorraj and Curier 2008: 70–71).

But this is not all. Several international developments also helped to bring Iran and China closer together in the 1970s: firstly, the withdrawal of the British military from the Persian Gulf region gave Iran the opportunity to fill this vacuum. Iran was supported by China in pursuing this goal. Secondly, the increasing involvement of the Soviet Union in Asia starting in the late 1960s led to a militarization of border conflicts between the Soviet Union and China. Thirdly, both Iran and China placed themselves in opposition to India. Both Iran and China were concerned about India's growing role in South Asia and its attempt to weaken Pakistan. Iran and China had similar interests concerning Pakistan, namely a relatively trouble-free Iranian border to the east, allowing Iran to concentrate on its role in the Persian Gulf region. Finally, rapprochement between China and the US made improving relations between Iran and China easier. For China, a friendship with the US could be used to bring Soviet expansionism to a halt. For Iran, it was a relief that its improved relations with China would not be at the expense of its relations with the US (Garver 2006: 39–45).

Fearing that the Islamic revolution could counter the relationship between the two countries, China recognized the IRI very soon after its establishment in 1979 (Huwaidin 2002: 160). The Iran-Iraq war starting in 1980 brought the two countries even closer to each other. Officially, China's position in the Iran-Iraq war was that of neutrality, mainly because it did not want to ruin its relations with the Arab countries, most of which supported Iraq in the war with Iran (Garver 2006: 69–70). Still, during the 1980s the main feature of Iran-China relations was Chinese arms sales to Iran. In 1982, China and North Korea together accounted for 40 percent of Iran's arms imports (Middle East Defense News 21 November 1988: 1–2), rising to 70 percent in 1987 (Facts on File 8 February 1987: 420).

Since the end of the Iran-Iraq war in 1988, Chinese and Iranian leaders visit each other at least every two years. The end of the war also coincided with the death of the charismatic leader Ayatollah Khomeini in 1989, the election of Hashemi Rafsanjani to president in the same year, and the disintegration of the Soviet Union in 1992. Together with the necessity to attract FDI to reconstruct the war-torn Iranian economy, these developments made a reorientation of Iran's foreign policy possible. From the late 1980s, according to Ramazani, the slogan "Neither East nor West" was replaced by "Both North and South" (Ramazani 1992: 393), or a so-called "de-Arabization" of Iran's foreign policy (Marschall 2003). Iranian policy-makers stated that Iran should no longer focus on Persian Gulf countries, if the latter were not willing to give up their American orientation. Iran should rather stress the importance of countries such as India, Pakistan, Afghanistan, CEA, and China, which were more sympathetic to Iran (Marschall 2003: 119). The so-called "de-Arabization" of Iran's foreign policy was a reaction to US policy in the Persian Gulf and the Arab-Israeli peace process that started in October 1991 (Marschall 2003: 118). Some individuals among the Iranian political elite have played a pivotal rule in establishing and maintaining this relationship with China. The most prominent of these individuals is Hashemi Rafsanjani, who traveled to Beijing in 1985 when he was still Speaker of Parliament. After his visit, Iran and China signed their first bilateral nuclear cooperation agreement. During his presidency, Rafsanjani became involved with Chinese companies in infrastructure construction projects (Calabrese 2006: 5).

The disintegration of the Soviet Union in 1991 was of great geopolitical importance for Iran. While roads to CEA and Europe had been totally blocked during the Soviet era, in 1991 the door towards Europe was reopened (Nahavandi 1996: 2). Iran recognized the independence of CEA countries in 1991, hoping it could profit economically by re-establishing good relations. President Rafsanjani repeatedly declared that with the independence of CEA states, a new "economic trade center" had emerged. Similarly, Iran is a major link for CEA countries to international markets. In addition to bilateral and multilateral transport agreements between Iran and CEA countries, the Economic Cooperation Organization (ECO)[32] is a forum for regional cooperation. The US opposition to a more active involvement of Iran in CEA has hampered the strengthening of ties between Iran and the region. Another important obstacle is the not yet settled dispute over the legal regime of the Caspian Sea.

Another interesting development since the disintegration of the Soviet Union is the Shanghai Cooperation Organization (SCO),[33] which has developed into an important global political, economic, and security organization. Established on June 14, 2001, its six permanent members are China, Russia, Kazakhstan, Kyrgyzstan, Tajikistan, and Uzbekistan. India, Iran, Mongolia, and Pakistan have observer status in the organization. SCO was first established as a security organization to fight terrorism, but more recently has also set economic goals such as the establishment of a free trade area among its member countries and cooperation in the energy sector. At the SCO summit of June 2006 in Shanghai, it was expected that Iran would become a full member of the organization. But concerns about Iran's nuclear program by the US and Europe have put Iran's inclusion as a full member temporarily on hold. However, the fact that the SCO has included energy as one of its priorities makes Iran, due to its huge oil and gas reserves, an attractive potential member. The inclusion of Iran into the organization could have an important impact on the global energy arena (Brummer 2007:185–186). It would also be a clear political statement on the part of Russia and China siding with Iran in its conflict with

[32] ECO was first established in 1977 by Iran, Turkey, and Pakistan as Regional Cooperation and Development (RCD). The organization survived until the Iranian Islamic revolution in 1979. In 1985, the organization was re-established as ECO. ECO's breakthrough took place in 1992 at the Tehran Summit, which paved the way for the expansion of the organization from three to ten members, including Azerbaijan, Kazakhstan, Kyrgyzstan, Tajikistan, Turkmenistan, Uzbekistan, and Afghanistan. ECO is a large economic cooperation organization. Its member states together have a population of 300 million and cover an area of seven million square kilometers. See also the organization's website, http://www.ecosecretariat.org.

[33] The SCO emerged out of the Shanghai 5 organization created on April 26, 1996 to deepen military cooperation between China, Kazakhstan, Kyrgyzstan, Tajikistan, and Russia. In 2001 Uzbekistan was admitted to the Shanghai 5, turning it into the Shanghai 6, and in the same year it transformed into the SCO. See also the SCO website, http://www.sectsco.org.

the west. Furthermore, China and Russia fear that in its attempt at regime change, the US could strengthen US dominance in the Persian Gulf region and establish a political regime in Iran friendly to it, to expand their regional domination. Both Russia and China want to prevent such a development in order not to be excluded from access to the country's oil and gas reserves. The inclusion of Iran into the SCO would undermine US dominance in the Persian Gulf and increase SCO influence in the region and the whole Middle East and CEA at large (Brummer 2007: 190). On the part of Iran, membership in the SCO would expand its international political and economic possibilities. It would grant access to SCO projects and, thereby, to technology, investment, trade, and infrastructure development. In other words, the sanctions on Iran imposed by the US could be undermined by Iran-SCO cooperation (Brummer 2007: 192) and also by Iran-China cooperation on a bilateral level.

7.1 *Iran-China Cooperation in the Oil and Gas Sector*

It is expected that China's energy demand will double between 2005 and 2030. That means that by 2010 the country will probably have overtaken the US as the main energy consumer in the world. Coal will remain the main energy source, accounting for 63 percent of total energy consumption in 2030 (International Energy Agency 2007). China is the second largest oil consumer in the world behind the US. Although the country is trying to increase domestic production, oil imports will comprise almost 70 percent of the country's oil consumption by 2025 (Amineh July 2005: 6). Though gas has not been a major energy source in Iran its share in the total energy mix is increasing. According to the Energy Information Administration (August 2006), China's gas consumption almost doubled between 1999 and 2004 to 1.3 trillion cubic feet, thus making up about 3 percent of the total energy consumption in China. Though natural gas production could significantly be increased due to several natural gas discoveries in recent years, China is not only focusing on domestic production but also considering the construction of transnational pipelines as well as the import of LNG from Iran.

Cooperation in the energy sector between Iran and China started in the 1970s. In 1977 China imported 300,000 tons of oil from Iran. By 1982 the amount had already risen to one million tons, by 1989/90 to two million tons (40,000 bbl/d) (Huwaidin 2002: 165-172), by 2000, to seven million tons, and by 2002, to 11 million tons (Lai 2007: 525). In 2006, Iran was China's third largest oil supplier behind Saudi Arabia and Angola, providing China with 11 percent of its oil imports.

The ILSA imposed on Iran in 1996 was an advantage for China and its aspirations to invest in the energy sector in Iran. China offered Iran to help reconstruct energy facilities that had been damaged during the war with Iraq, as well as offering to participate in exploration activities and the development of new oil and gas fields, to secure long-term access for itself to Iran's energy sector. In 1997 Iran and China signed an agreement for cooperation in prospecting and exploration. One year later Shengli Oil—a subsidiary of Sinopec—transferred to Iran a complete set of oil equipment of 1990s standards (Garver 2006: 268). In 2003, SINOPEC signed a deal with Iran with an estimated value of US$70 billion to import a further 250 million tons of liquefied natural gas (LNG) from Iran's Yadavaran oil field over the next 25 years. The Yadavaran field will, furthermore, provide China with 150,000 bbl/d of crude oil over the same period (China Daily 31 October 2004; Energy Bulletin 30 October 2004). In 2004, the Chinese state owned oil-trading company, Zhuhai Zhenrong, signed a 25 year contract to import 110 million tons of LNG from Iran) (China Daily 18 March 2004). In the same year, Iran and China also signed a US$20 billion agreement on the export of 2.5 metric tons of LNG annually for 25 years from 2008. To date, this is the greatest deal on the export of natural gas in the world (Howard 2007: 95; Lai 2007: 525). Only a couple of months later, SINOPEC and NIOC signed another contract for the purchase of 250 million tons of LNG to be sent to China over the next 30 years. The deal is worth US$70–100 billion (Garver 2006: 271). Part of the deal is also the development of the Yadavaran field from which Iran will export 150,000 bbl/d of oil for 25 years (China Daily 31 October 2004).

It has to be expected that relations between these two countries will intensify immensely, primarily because of China's energy needs and Iran's increasing hunger for consumer goods.

Though China has the financial means to become involved in the Iranian energy sector, it lacks the technological capabilities to modernize the Iranian oil infrastructure. This technology is mainly in the hands of American and European companies (Dorraj and Curier, 2008: 74). To strengthen ties with oil producing countries in the Middle East, China supports them on issues that are of great importance to them as long as this support has no negative effects on China's own strategic interests. For example, China sides with Arab countries on the Israel-Palestine conflict, but it has also acted along US interests. In 2002, China voted in favor of the 1441 resolution against Iraq and its non-compliance with its disarmament obligations, opening the way for the US invasion of Iraq in 2003. In early 2006, China backed the US and EU initiative to refer the nuclear issue in Iran to the UN Security Council, and agreed that Iran should not develop nuclear weapons (Lai 2007: 530). China receives around 50 percent of its oil imports from the Middle East, but is concerned about political instability in the region and the increasing US involvement there. Therefore China also searches for alternative sources in Africa, especially Angola, Sudan, the Democratic Republic of Congo, and Gabon (Lai 2007: 525).

8. Iran-India Cooperation in the Energy Sector

In terms of Iran, China competes with India for the import of energy. As Mudiam (Mudiam 2007: 411–412) argues, Iran and India have similar political interests, strategic outlook, and economic objectives. This has been the case particularly since the end of the Cold War. Both are suspicious of the increasing US military involvement in the Middle East and CEA, and want to ensure that no single power dominates the region. Iran, as a supplier, and India, as a customer, also have complementary interests in the energy sector. They are both key players in Asian energy security and would benefit from a long-term partnership in the energy sector. In January 2005, India signed a US$40 billion deal with the National Iranian Oil Company (NIOC). Iran will ship 5 million tons of LNG annually to India for a period of 25 years. India's ONGC Videsh Ltd (OVL) has a 20 percent share in the development of the Yadavaran onshore oil field (Bhadrakumar 11 January 2005).

To continue its economic development plans, India will need to import more energy and Iran could be a secure source of oil and gas. Being confronted with economic sanctions imposed by the US, Iran needs other economic partners to keep its access to the energy market. The planned Iran-Pakistan-India (IPI) pipeline is an important step in this direction (Behera 2005). The plans for the IPI were started in 1994. The pipeline is supposed to have a length of 1,700 miles and transport 2.8 Bcf/d of natural gas from the South Pars fields in Iran to Gujarat in India. Though both countries have great interests at stake in the pipeline, its construction has been delayed. India demands security guarantees from Islamabad for any pipeline crossing Pakistan. In December 2006, the three countries (Iran, Pakistan, and India) could not agree on the price of gas. Iran asked for US$8 per million Btu (British thermal unit),[34] while India and Pakistan do not want to pay more than US$4.25 per million Btu (Energy Information Administration January 2007). India is also under pressure from the US, which is against the IPI pipeline, and India and Pakistan are in disagreement about how much India will pay Pakistan for the pipeline crossing its territory (Calamur 31 January 2008). China has already shown its interest in participating in the project, should India decide not to take part (Alexander's Gas and Oil Connection 11 February 2008). However, China is carrying out its own pipeline projects. In December 2005, China and Kazakhstan completed the Kazakhstan-China pipeline.[35] Another pipeline, the Turkmenistan-China pipeline, is expected to have opened in January 2009. Through this pipeline, Turkmenistan will deliver 40 billion cubic meters (bcm) of natural gas per year for 30 years to China (Reuters 29 August 2008).

In September 2000, India, Iran, and Russia signed the North-South Corridor Agreement (NSCA). This agreement will provide traders with a shorter Asian-European trade option than through the Suez Canal, with land and sea routes connecting Europe and India via Russia and Iran. The North-South Corridor's land routes could also be linked to the Trans-Asia Railway Network (TAR)[36] that now uses the Caspian Sea to connect Iran with Russia. The TAR agreement was signed on November 10, 2006 to connect the Caucasus with South Korea by

[34] A unit of energy used in the US. One Btu is equivalent to 1054–1060 joules.

[35] The pipeline was officially inaugurated in July 2006. It connects Atasu in northern Kazakhstan with Alashakou in Xinjiang and has a length of 620 miles In July 2006, China began receiving crude oil imports from its first transnational oil pipeline. The pipeline was constructed by the Chinese National Petroleum Company and Kazakhstan's KazTransOil. The pipeline has the capacity to transport 200,000 bbl/d of crude oil, and possibly 400,000 bbl/d by 2010. Of the imported oil, 50 percent comes from Russia and 50 percent from Kazakhstan (EIA August 2006).

[36] Its members are: Armenia, Azerbaijan, Bangladesh, Cambodia, China, Democratic People's Republic of Korea, Georgia, India, Indonesia, Iran, Kazakhstan, Kyrgyzstan, Laos, Malaysia, Mongolia, Myanmar, Nepal, Pakistan, Republic of Korea, Russia, Singapore, Sri Lanka, Tajikistan, Thailand, Turkey, Turkmenistan, Uzbekistan, and Vietnam. Ten of these countries did not sign the agreement on November 10, 2006 but will participate due to its economic prospects.

railway. The TAR would not only be of benefit for CEA's economic development, but as Peimani states (November 15, 2006):

> it could also lift up their regional and international significance by turning them into a hub for intercontinental cargo transportation of Asia, the largest continent rich in mineral and energy resources housing the fastest growing world economies.

There might also be possibilities in the future in the area of arms sales between India and Iran (Calabrese 2002). To prevent any problems with the US, India is cautious in its relations with Iran. However, in regards to the IPI gas pipeline in particular, it also makes clear that it would not give in to US pressure and step out of the project. The country sees the pipeline as an important means of alleviating poverty in its country (The Hindu 14 January 2006, 22 January 2006). India voted for the IAEA resolutions on Iran's nuclear program, but also made it clear that they believed that Iran's nuclear program was not for peaceful purposes (The Hindu 16 January 2006). Despite their competition to secure energy supply, China and India have also started to cooperate. For example, India acquired stakes in the Yadavaran oil field. The development of the Yadavaran field is now a Sino-Indian-Iranian joint project, with China owning a share of 50 percent, Iran 30 percent, and India 20 percent (Bajpaee 17 March 2005). Despite these developments the Iranian nuclear issue impacts Iran-China as well as Iran-EU relations.

9. Impact of the Iranian Nuclear Issue on Relations with the EU and China

In Iran, the nuclear debate is part and parcel of the overall debate on the country's role in world politics since the Cold War. The discussions among the Iranian political elite seem to cut across different political factions. Some Conservatives are against the possession of WMD, while some Reformists argue that the development of nuclear weapons is Iran's right and a national security imperative (Ehteshami 2006). According to Chubin (2006: 28), the primary motivation for Iran to develop nuclear technology is to legitimize the political regime. The nuclear debate is central to the general debate on where Iran is heading in the future, and how it should interact with other countries. It also reflects Iran's quest to be treated with respect regionally and internationally.

Iran began its first nuclear power program in 1957, with the signing of the Atoms for Peace Program between Iran and the US (Bowen and Kidd 2004: 263). In 1967, the first nuclear facility was established at Tehran University. The research reactor came from the US and West Germany. In 1968, Iran signed the Non-Proliferation Treaty (NPT), according to which Iran had the right to develop research, production, and use of nuclear energy for peaceful purposes, and to have access to equipment, materials, and scientific and technological information. Construction of the Bushehr power plant began in 1974 by the West German Siemens Company and its subsidiary Kraftwerke Union (Hibbs 1991). In the 1970s, Iran bought reactors from Framatone (France), Kraftwerke Union (Germany), and via the Atomic Energy Commission (US). Iran also signed an agreement with South Africa to exchange uranium for the financing of a uranium-enrichment plant. The Iranian political elite under the Shah envisioned that 20 nuclear plants would be built by the beginning of the 1990s (Cottrell 1978: 428). At the moment of the Islamic revolution, two nuclear reactors, the one at Bushehr and one on the Persian Gulf, were nearly complete. Had the Shah's regime not been overthrown by the revolution, Iran would probably have become one of the states that now have nuclear weapons, according to Tarock (2006: 652). After the revolution and during the Iran-Iraq War, the Iranian nuclear program was brought to a halt. Germany refused to complete the power plant, as it feared that Iran would try to develop nuclear weapons (Hibbs 1991). After the end of the Iran-Iraq War in 1988, the nuclear program was restarted with Russian and Pakistani assistance (Bowen and Kidd 2004: 263). In 1995, Russia and Iran signed an agreement worth US$800 million to complete construction of the Bushehr nuclear power plant (Cirincione et al. 2002: 257–60). The US demanded that Russia abandon the Bushehr project (US Department of Defense, January 2001). Although Russia argued that the reactor was not a proliferation risk, it partially gave in to the US when it dropped a plan to supply a uranium enrichment facility to Iran (Cirincione et al. 2002).

China is another important factor in the Iranian nuclear energy program. According to an April 1996 US Department of Defense report, in 1991 China supplied Iran with 1,000 kilograms of uranium hexafluoride, 400 kilograms of uranium tetra fluoride, and 400 kilograms of uranium dioxide. The report concludes that at that time, China was Iran's main source of nuclear assistance (US Department of Defense January 2001).

Since the transfer of the Iranian nuclear dossier to the UN Security Council in February 2006, the issue has become internationalized. Not only is the US now openly involved, Russia and China are as well. A possible solution might be the Russian proposal, first suggested in August/September 2005, to enrich Iranian uranium in Russia and ship it back to Iran.[37] The Iranian government has no other choice than to change its policy. What the US and the EU can do is facilitate this change by developing a common more "nuanced" policy approach towards Iran that also includes Russia, China, and India besides the US and the EU.

10. Conclusion and Outlook

For both the EU and China Iran could become an important oil and gas supplier in the future. Iran needs the EU to develop its energy industry and the EU and China to export its oil and gas. Increased cooperation in the energy sector with both the EU and China could strengthen Iran's political and economic position in global politics and in the Persian Gulf and Caspian regions. However, to deepen the mutual relationship with both the EU and China several obstacles need to be overcome. The first is Iran's hostile political and economic relationship with the US. Though EU and China policies towards Iran are not dependent on the Iran-US relationship, a shift from confrontation to dialogue in the mutual relations of the latter two will make especially Iran-EU cooperation and to a lesser extent, Iran-China cooperation, easier. The second obstacle is an unsafe trade and investment climate both for companies and governments in Iran. These structural problems need to be overcome if Iran's aim is to secure foreign direct investments and international trade relations in the long-term. The third obstacle is the nuclear issue, which does not stand by itself, but is related to Iran's overall relationship with the rest of the world. While President Ahmadinejad's confrontational style of rule cannot be sustained in the long-term, it should also be acknowledged that Iran cannot be pressured to suspend its nuclear enrichment and reprocessing activities. A dialogue should be comprehensive and not limited to

[37] Franco-Russian Joint Statement," on the occasion of the visit to Moscow of Mr. de Villepin, French Prime Minister, (14 February 2006), http://www.diplomatie.gouv.fr.

the nuclear issue. A common approach should, besides the US and the EU, also include Russia, China, and India. Threatening Iran with military intervention only weakens the reformist forces in Iran and strengthens the Conservative ones.

CHAPTER FIVE

CROUCHING TIGER, HIDDEN DRAGON: INDIA, CHINA, AND THE DYNAMICS OF ENERGY SECURITY

S. Philip Sen

Abstract: *It is already well known that China's demand for energy imports is increasing rapidly, but less attention is paid to another rising Asian giant: India. India's energy needs are also soaring, and since many of its fossil fuel sources are identical to China's, some might say that competition and even conflict looms.*

However, India's geopolitical considerations are somewhat different from China's. Using an analytical framework derived from Doran's power cycle theory, and with reference to critical geopolitics, the chapter considers India and China's energy requirements, the domestic considerations that drive them, and their relations with energy suppliers—particularly Iran. It looks to India's role and foreign policy in the context of Chinese and US interests in the regional environment, and examines ways in which India's energy security strategy conflicts with or complements China's—hypothesizing that although India's energy demands pose a headache for China, it is one that Beijing can deal with.[1]

Keywords: India, China, Energy Security, Challenges

1. Introduction

Much has been written about the apparent economic miracle in Asia,[2] but—global economic turmoil notwithstanding—its continued success is by no means assured. If the industrial boom is to last and the new middle classes are to remain satisfied, more and more fossil fuels must burn. So, as China 'rises' and India 'shines,' their demand for imported energy will only increase.

[1] Please note that the analysis and views in this chapter are those of the author alone, and not those of the UN, UNV or any associated departments and agencies.

[2] Popular journals such as *Time* and *Newsweek* regularly devote special editions to the theme: see, for example, cover stories by Fareed Zakaria ["Does the Future belong to China?," *Newsweek*, May 9, 2005 and "India Rising," *Newsweek*, March 6, 2006] or Michael Elliot ["India Awakens," *Time*, June 18, 2006 and "China Takes on the World," *Time*, January 11, 2006].

Though their main concern is internal development, since the founding of the modern Republic of India and People's Republic of China (PRC) both have been engaged in a long-running security dilemma fuelled by Cold War politics, regional alliances, and territorial disputes. Both also have nationalist aspirations drawn from historical ideas of their cultural and political predominance over their neighbors (Garver 2001: 3–15). Even before we factor in potential disputes over energy, we can observe a slow-burning conflict of interests. Fearing 'strategic encirclement,' India wishes to reduce Chinese influence over its neighbors (such as Pakistan) while China is dead-set against Indian aspirations to regional hegemony (Garver 2001: 16–18, 31).

There can be little doubt that the rapid industrialization and economic ascendancy of India and China are already having significant effects on the international system, too, and that they will continue to do so. Both are still growing fast and their shares in the system are rapidly increasing (Doran 2004: 21). In this context, the foreign policy roles of China and India, within both the Asian region and the international system, are certainly affected by each other.

At present, both obtain high proportions of their imported fossil fuels from the Persian Gulf. Both look like they will depend on Gulf energy for the foreseeable future and must therefore firm up their foreign policies towards these suppliers, while also looking to develop secure new routes, perhaps through third-party countries.

As demand for the same limited pool of Gulf resources increases from both China and India, logic dictates that sooner or later there won't be enough for both of them, let alone other consumers that rely on the same reserves. Iran's major oil and gas reserves, its geographical location, and its prickly relations with an America that, despite a decline, still has major influence on international affairs make it a particularly interesting supplier. India certainly seeks to take its share of Iran's resource pool, particularly with regard to its growing need for natural gas—a situation that rings alarm bells in Washington and perhaps Beijing too. This entails an evolution of each country's foreign policy position and role within the international system. How can the nascent Asian powers ensure their supply security, and what geopolitical challenges might India's energy needs pose to China? Will the dynamic tend to cooperation or competition, or a more complex interaction of these factors with complementarity and competitiveness, as described by Charles F. Doran?

2. Theoretical Framework

In the globalized 21st century, state and non-state actors, the flows of finances, commodities, and even information, are all interlinked in what we might call an 'international system.' When considering the part that energy security has to play in the interactions between the major actors in this analysis—India, China, Iran, and the US—we must make a distinction between 'role' and 'power' as defined by Charles F. Doran: "Role exists only if legitimized through systemic acceptance, whereas power expresses itself through unilateral action and as control" (Doran 2004: 13–49).

Relative power (as China could currently boast) does not mean a concomitant role in the system. Thus power-role gaps exist that can create disequilibrium and conflict. "Were China's rise to be slowed by faster growth of India [and Russia]," says Doran, "both the regional and global equilibrium would suddenly change" (Doran 2004: 36, 45). A surefire way to slow China's growth would be by India (perhaps allied to the US) practicing a foreign policy that led it to claim a greater share of the energy pool than China would like. Conditions thus appear ripe for a geopolitical competition for resources.

On the other hand, Doran posits that peaceful change *is* possible and that competitiveness (as opposed to *competition*) can even be beneficial. Doran says,

> Competitiveness is a process in which governments clash behaviorally or attitudinally in such a way that something is negated or eliminated—something perceived to be inefficient or unjust, unhelpful or redundant…Competitiveness makes use of the opposing strengths of another actor to eliminate the problematic nature of one's own institutions, laws, or behaviors. Competition, on the other hand, is merely "destructive rivalry".

Thus competitiveness can mean that governments interact in such a way that inefficiencies are eliminated, just as complementarity—"a process in which governments interact, each contributing something the other does not possess, or does not possess in sufficient quantity or with sufficient fit"—means that different governments each contribute something to the international system the other can't (a good example would be China's current role as 'the world's workshop' compared to India's as 'the world's back office').

Competitiveness and complementarity thus help produce an evolution of the system. But mere cooperation and competition that fail to eliminate inefficiency is destructive (Doran 2004: 19–20).

All of the above, however, is based in state-centric realism and it is thus necessary to look above the horizon a little, too. Critical geopolitics seeks to explode the myth that conventional geopolitics promises definitive insight into what is actually the impossibly complex and mutable international system. In essence, it means observing both the big picture and the details. A state's self-contained identity is a social force that impacts its foreign policy behavior, as do domestic actors and private interests. If anything, it is better to avoid closed thinking about states alone, and instead consider less sharply-defined state-society complexes consisting of the relations between state, business, and perhaps the military (Amineh and Houweling 2005b: 7–8).

Guided by their self-perception of their place in the world order, in order to protect their interests (such as energy security), these complexes project their power across national borders. And by doing so, they attempt both to shape the places over which that power is projected, and set conditions for outsiders with relation to those places—for example, access to energy resources (Amineh and Houweling 2005b: 11–15). Just as with conventional geopolitics, our players seek influence in a chess game with their rivals and partners. We simply need to better understand the nature of the players and the environment in which they act.

Therefore, with this critical geopolitical outlook in mind, it is first necessary to examine India's energy demand, what internal and external factors drive it, and how this compares with China's. This will enable us to consider India's energy security strategy and foreign policy outlook with respect to external actors (particularly suppliers such as Iran, potential transit countries such as Pakistan, and also with regional and extra-regional powers such as China and the US). We can then look into how China and India might manage these challenges through the lenses of Doran's ideas of cooperation, complementarity, competition, and competitiveness.

All in all, India's growing energy demand is certainly another geopolitical headache for Beijing, but one it can probably manage via astute foreign policy in the face of India's discomfort with its role within the system.

3. Energy Security: The General Picture

Energy security is defined by the United Nations Development Programme (UNDP) as "the availability of energy at all times in various forms, in sufficient quantities, and at affordable prices, without unacceptable or irreversible impact on the environment" (UNDP 2004: 42). Essentially, a balance between supply and demand defines the situation, and energy security is about overseeing that balance when scarcity arises.

'Supply-induced scarcity' comes down to the inevitable dwindling of reserves. Such depletion can be expected to "provoke a process of competitive power projection" over the territory where the stock is located (Amineh and Houweling 2005a: 80–81). 'Demand-induced scarcity' occurs due to three factors that lead to increased consumption: population growth, rising per capita income (which brings with it a call for both energy and goods that themselves require energy to produce and run), and technological change and industrialization (Amineh and Houweling 2005a: 80). These consumption-boosting factors are apt to the current 'rise of Chindia,'[3] as will be discussed in the next section.

Linked to both supply- and demand-induced scarcity is price. Everyone loves cheap energy, which is fine when it is available in abundance. But the higher the demand, the higher the rate, which in turn reduces a country's ability to fulfill its needs and drive its economy. Energy security is thus as much about market stability and managing the cost as it is about geopolitical considerations. This is especially frightening for China and India, countries that were once committed to *avoiding* global market fluctuations (Verma 2007: 3281). But in 2008, it became abundantly clear that the era of stable oil prices was over.

Beyond the supply-and-demand equation there is a third factor: structural scarcity. Cartels can create headaches by restricting the amounts of fuel on the market, as OPEC (Organization of the Petrol Exporting Countries) did during the 1973 oil crisis (Klare 2004: 9). An individual government may decide to cut off supplies during a dispute over price or payments. Importers thus need to keep a close eye on their sources, the geopolitical space through which energy imports

[3] A neologism attributed to Jairam Ramesh. See Ramesh 2005.

travel, and also upon those actors who might compete, complement, or cooperate with them.

Even though new energy sources are emerging or being exploited in South America, Africa, and the former Soviet states, the Persian Gulf remains the biggest provider by far. By the end of 2007, six Persian Gulf states (Iran, Iraq, Qatar, Kuwait, Saudi Arabia, and the UAE) held 60 percent of the world's proved oil reserves and 40 percent of its natural gas. When it comes to individual producers, Iran is the 'ace in the deck,' speaking for 11.2 percent of world oil and 15.7 percent of gas—this puts it just behind Russia in terms of the proportion of world gas and oil in million tons of oil equivalent (Mtoe) terms (see Table 5.1). The Caspian states of Kazakhstan, Turkmenistan, and Azerbaijan (plus Uzbekistan in Central Asia) are certainly important, but still pale in comparison compared to the Gulf and Russia.

It is also important to remember that the Gulf countries are heavy consumers themselves, especially Iran and Saudi Arabia, both of which rely on their oil exports at the possible detriment of potential gas exports. Though Saudi Arabia is expected to produce 155 billion cubic meters (Bcm) of natural gas per year by 2030, it may well consume all of that domestically (IEA 2005b). The International Energy Agency (IEA) projects that Iran's primary energy demand will double by 2030: it will produce 240 Bcm of gas but export only 24 percent of it (57 Bcm) (IEA 2005a). Should those projections be off, perhaps Iran will be unable to export gas at all.

Geopolitically, therefore, it stands to reason that there will be an element of competition (or *competitiveness*) for these limited resources by major consumers such as China and India, especially considering their geographic proximity to each other. Their roles within the regional and global system are thus likely to direct their foreign policy attitudes and strategies towards their suppliers and transit countries as well as each other. However, as per our critical geopolitical outlook, before looking into this it is necessary to examine the internal situations of the countries under analysis to shed light upon their advantages, problems, and self-perceptions, and how these shape their roles. We will look at India and China's situations in turn.

Table 5.1 Proved reserves of oil and natural gas at the end of 2007.

	Oil: Proved Reserves at end 2007			**Natural Gas: Proved Reserves at end 2007**				**Combination—Oil and Gas**	
	Thousand Million Tonnes	Share of total	R/P ratio	*Trillion cubic metres*	Thousand Million Tonnes Oil Equivalent (Mtoe)	Share of total	R/P ratio	Thousand Million Tonnes Oil Equivalent (Mtoe)	Share of total
USA	3.6	2.4%	11.7	*6.0*	5.4	3.4%	10.9	9.0	2.7%
Total North America	**9.5**	**5.6%**	**13.9**	***8.0***	**7.2**	**4.5%**	**10.3**	**16.7**	**5.1%**
Total S. & Cent. America	**15.9**	**9.0%**	**45.9**	***7.7***	**7.0**	**4.4%**	**51.2**	**22.8**	**7.0%**
Kazakhstan	5.3	3.2%	73.2	*1.9*	1.7	1.1%	69.8	7.0	2.1%
Russian Federation	10.9	6.4%	21.8	*44.6*	40.2	25.2%	73.5	51.1	15.6%
Turkmenistan	0.1	w	8.3	*2.7*	2.4	1.5%	39.6	2.5	0.8%
Uzbekistan	0.1	w	14.3	*1.7*	1.6	1.0%	29.8	1.7	0.5%
Other Eurasia	3.1	2.0%		*8.4*	7.6	4.8%		10.7	3.2%
Total Europe & Eurasia	**19.4**	**11.6%**	**22.1**	***59.4***	**53.5**	**33.5%**	**55.2**	**72.9**	**22.2%**
Bahrain		n/a	n/a	*0.1*	0.1	w	7.4	0.1	w
Iran	19.0	11.2%	86.2	*27.8*	25.0	15.7%	*	44.0	13.4%
Iraq	15.5	9.3%	*	*3.2*	2.9	1.8%	*	18.4	5.6%
Kuwait	14.0	8.2%	*	*1.8*	1.6	1.0%	*	15.6	4.7%
Oman	0.8	0.5%	21.3	*0.7*	0.6	0.4%	28.6	1.4	0.4%
Qatar	3.6	2.2%	62.8	*25.6*	23.0	14.4%	*	26.6	8.1%
Saudi Arabia	36.3	21.3%	69.5	*7.2*	6.4	4.0%	94.4	42.7	13.0%
Syria	0.3	0.2%	17.4	*0.3*	0.3	0.2%	54.7	0.6	0.2%
United Arab Emirates	13.0	7.9%	91.9	*6.1*	5.5	3.4%	*	18.5	5.6%
Yemen	0.4	0.2%	22.7	*0.5*	0.4	0.3%	–	0.8	0.2%

Table 5.1 (*cont.*)

	Oil: Proved Reserves at end 2007			**Natural Gas: Proved Reserves at end 2007**				**Combination—Oil and Gas**	
	Thousand Million Tonnes	Share of total	R/P ratio	*Trillion cubic metres*	Thousand Million Tonnes Oil Equivalent (Mtoe)	Share of total	R/P ratio	Thousand Million Tonnes Oil Equivalent (Mtoe)	Share of total
Other Middle East	+	w	10.9	*0.0*	+	w	18.5	+	w
Total Middle East	**102.9**	**61.0%**	**82.2**	***73.2***	**65.9**	**41.3%**	*	**168.7**	**51.4%**
Total Africa	**15.6**	**9.5%**	**31.2**	***14.6***	**13.1**	**8.2%**	**76.6**	**28.7**	**8.7%**
China	2.1	1.3%	11.3	*1.9*	1.7	1.1%	27.2	3.8	1.2%
India	0.7	0.4%	18.7	*1.1*	0.9	0.6%	35.0	1.7	0.5%
Pakistan	n/a	n/a	n/a	*0.9*	0.8	0.5%	27.6	0.8	0.2%
Myanmar	n/a	n/a	n/a	*0.6*	0.5	0.3%	40.8	0.5	0.2%
Bangladesh	n/a	n/a	n/a	*0.4*	0.4	0.2%	24.0	0.4	0.1%
Other Asia Pacific	2.6	1.6%		*9.7*	8.7	5.5%		11.3	3.4%
Total Asia Pacific	**5.4**	**3.3%**	**14.2**	***14.5***	**13.0**	**8.2%**	**36.9**	**18.4**	**5.6%**
TOTAL WORLD	**168.6**	**100.0%**	**41.6**	***177.4***	**159.6**	**100.0%**	**60.3**	**328.2**	**100.0%**
Of which: OECD	11.9	7.1%	12.6	*15.8*	14.2	8.9%	14.4	26.1	7.9%
OPEC	127.6	75.5%	72.7						
Former Soviet Union	17.4	10.4%	27.4	*53.5*	48.2	30.2%	67.7	65.6	20.0%
European Union 25	0.9	0.5%	7.8	*2.8*	2.6	1.6%	14.8	3.5	1.1%

* More than 100 years + Less than 0.05 w Less than 0.05%

Source: BP 2008: 6, 22 and author's own calculations.

4. 'Shining India': An Economic Boom Fraught with Difficulty

Post-independence India has flirted with authoritarianism and enforced birth control, notably during the 'Emergency' from 1975 and 1977 (Ali 2004: 185–201). But without existing measures like China's 'one child policy,' India's population continues to expand rapidly. By 2007 it stood at 1.13 billion compared to China's 1.32 billion: within a couple of decades it could well overtake China. Meanwhile, though the economy is growing, in terms of both Gross Domestic Product (GDP) and cash-per-capita, India is still well behind its neighbor, and things look set to continue that way. The International Monetary Fund's (IMF's) 2007 figures put India's GDP at $1.1 bn (China's was $3.2 bn): at $964, GDP per capita is less than half of China's (also true when adjusted for purchasing power parity).

4.1 *India's Internal Dynamics*

India's growth comes from the services and information technology industries as well as large-scale industrial production, with services representing 52.8 percent of the GDP in 2007. This is not to say that India is not industrializing: conglomerates such as Tata and Reliance Industries are manufacturing giants within the subcontinent and confidence is high. On the other hand, nearly two-thirds of the population works in agriculture, but contributes only 17.8 percent of the GDP (CIA 2008).

The implications for what might be labeled the world's largest democracy are complex, and as per critical geopolitics it is necessary to understand some of the internal issues that relate to energy as well as energy consumption in and of itself. Rising energy consumption is of course related to economic growth, which is naturally intertwined with India's internal politics and major foreign relationships.

India's recent economic development can in part be attributed to the current ruling Congress Party's liberal agenda, first implemented by current Prime Minister Manmohan Singh back in the 1990s during his tenure as finance minister. The IMF put India's growth at 7.9 percent in 2007 (IMF 2008), down from 2006 but still a similar rate to China's and three times the unimpressive 'Hindu rate of growth' experienced by India prior to the reforms. Growth in 2009 was estimated at just over 6 percent.

However, such rapid growth inevitably heightens the schisms within Indian society and India's rulers must assuage the anxieties of both the middle classes and the deprived rural population to stay in power. What is more, there are communal tensions between Muslims and Hindus (plus other minorities such as Sikhs) and the deep problems of the divisive caste system. No wonder V. S. Naipaul described India as "A Million Mutinies Now" (Naipaul 1990).

Keeping a grip on such a difficult balancing act implies maintaining economic progress, despite the global downturn. A general election during 2009 was a litmus test of public confidence in Singh. His main opposition came from the Bharatiya Janitya Party (BJP) with its agenda of Hindu nationalism. Also in the picture were the socialist parties with which the governing Congress Party had to maintain an uneasy alliance until a vote of confidence in July 2008 (Raman 2008). Singh's Congress retained power comfortably, but the public will expect more of the same from over the next five years.

Finally, just as the PRC must deal with its nemesis Taiwan, there is a Hobbesian security dilemma in South Asia that affects all parties and actors whether they like it or not. Kashmir lies at the heart of both India and Pakistan's national identities: India's due to the fear that an end to the status quo could trigger a fragmentation of the country as other ethnic and interest groups (such as the Sikhs) redouble their efforts for autonomy and a bloody anti-Muslim backlash begins; Pakistan's due to the belief that its secession will never be complete unless all of Kashmir is subsumed. And Kashmir is one issue that fuels the flames of Islamic militancy more than any other.

Despite several horrific incidents such as the Mumbai train bombing in July 2006 and the Mumbai massacres in November 2008 (both possibly perpetrated by Islamist militants sponsored by elements within Pakistan), the India-Pakistan peace process appears to have remained on track. On the other hand, restless Kashmir looked to be on the boil again in 2008, after some years of calm. Both India and Pakistan use Kashmir to justify high military spending, and the issue has helped entrench the military's role in Pakistan's government; whether its post-Pervez Musharraf democracy can survive in the face of increasing Islamic militancy and US incursions stemming from the 'War on Terror' in Afghanistan is one of the key questions in South Asia today (ICG 2004). A huge Indian military and the potential need for it also means more demand for oil and gas, and by extension, energy security. Pakistan also becomes relevant when exploring India's options for transiting Iranian energy, as will be explored in later sections.

In summary, all of the above signpost ever-increasing energy security demands. The dangers of social unrest, famine, electoral defeat, and military rivalry with Pakistan are ever-present. With oil prices at record highs at one point in 2008, the knock-on effects on the costs of food and goods were serious cause for concern. Strikes and protests broke out in June 2008 when the Indian government slashed fuel subsidies (BBC 2008b). Indian inflation in 2008 hit a record 12 percent (BBC 2008c), compared to China's 8.7 percent, itself a record affected by high fuel and food prices (BBC 2008a). Though oil prices tanked in late 2008, the global recession continues to loom. Without energy security to shore up economic growth and the military machine, India and its élites are doomed.

4.2 *India's Energy Picture*

Per capita energy consumption in India remains low; it is currently only a fifth of the world average (EIA 2008). Thus primary energy demand is not on quite the same scale as in China—it is perhaps only a quarter as large—but this is not to say that it is insignificant.

One problem is inefficiency, and if India's high-tech boom is to continue it will have to do something about the parlous state of its power transmission infrastructure and reduce cross-subsidies to Indian farmers. Power infrastructure aside, biomass and waste may account for as much as 50 percent of the total energy supply, mainly in the residential sector (Vikas and Ellsworth 2006). (R. K. Batra of The Energy and Resources Institute [TERI] in New Delhi estimates that these traditional sources represent about 80 Mtoe per year.)[4] This is often forgotten, and is not reflected in BP's figures on India's energy mix in Figure 5.1 below. However, you can't run a major world economy by burning firewood and cow dung. The average Indian may use twenty-five times less oil than the average American, and twelve times less gas, but according to EIA estimates in 2006, India was still the fifth largest consumer of oil in the world (EIA 2007). In terms of the year-on-year increase in oil and gas consumption, India is not far behind China, with its oil consumption set to increase by 60.9 percent by 2020. Use

[4] Interviewed in New Delhi, May 11, 2007.

of coal and nuclear power are also rising, but most significant is the projected increase in natural gas consumption—120 percent.

India's energy mix is more gas-based than China's. Hydropower is limited mainly to the mountainous northern regions, and though coal still represents half of India's power-generation needs, the amount of gas-fired electricity generation is rising by 7.5 percent per year. About 70 percent of India's increase in gas demand—which went up 38.7 percent between 1995 and 2000, and another 36.1 percent from 2000 to 2005—is due to the power-generation sector (Vikas and Ellsworth 2006). By 2025, the plan is for gas to comprise 20 percent of primary consumption (Pandian 2005a: 310). In addition, Compressed Natural Gas (CNG) is taking a share of the transport market, with all of New Delhi's public vehicles running on it as an anti-pollution measure (Narain and Krupnick 2007: 2).

Just like China, India is not blessed with enormous proved fossil fuel reserves of its own. India's oil reserves will run out by 2030: more worryingly, its gas reserves have only about thirty-six years to run, compared to China's forty-seven. By 2006, India's oil and gas imports totted up to $50bn—a third of its total imports (Kumaraswamy 2008).

The economic implications of these imports are indeed grave. Oil imports still represent about half of India's trade deficit (Pandian 2005a: 309). Gas matters, too. Now that India's gas demand has begun to outstrip its own production, it will have to find more sources of gas in particular—in 2004, it imported 2.6 Bcm of Liquefied Natural Gas (LNG) from Qatar, and it will need more and more as demand increases (EIA 2007). The oil ministry has set the target of acquiring 60 Mtoe of overseas oil and gas assets by 2025, and as of February 2007 the national Oil and Natural Gas Company (ONGC) was active in twenty-five projects in sixteen countries (Kumaraswamy 2008).

All of the above inevitably affects India's foreign policy. Figure 5.2 illustrates the extent to which India depends on Saudi and Iranian oil imports in particular: 35 percent compared to China's 31 percent. India enjoys an advantage over other consumers in that it is located closer to the Persian Gulf, and thus transit costs are lower (Vikas and Ellsworth 2006). But a sudden price rise due to a supply- or scarcity-induced shortage of Gulf energy would have a heavy impact.

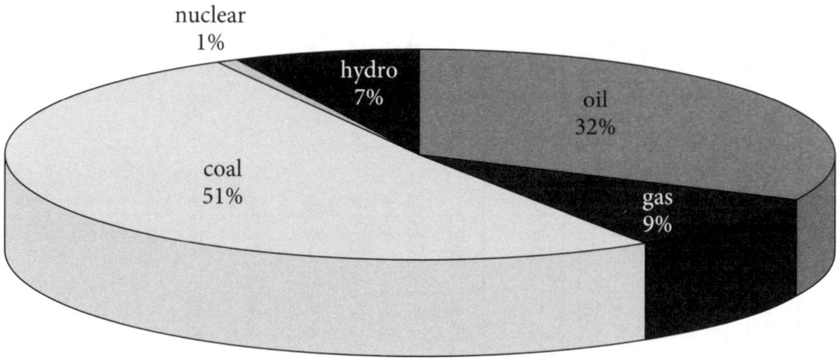

Source: BP 2008, 41.

Figure 5.1 India's Energy Mix 2007 (consumption by fuel).

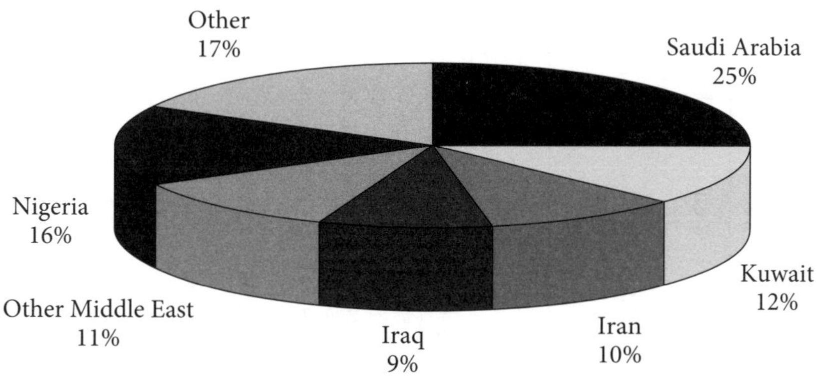

Source: Madan 2006.

Figure 5.2 India's Top Five Oil Suppliers (2004).

5. 'Rising China': A Growing Thirst for Imported Fuels

As the world's most populous country, China's economic rise over the past three decades has been spectacular. IMF statistics show that since the turn of the century, the PRC has experienced double-digit economic growth. Meanwhile, it appears that China's one-child policy is paying dividends, with the expansion of population coming to more manageable levels.

Table 5.2 India's energy consumption 1997–2007.

	1997	1998	1999	2000	2001	2002	2003	2004	2005	2006	2007
Oil (Mtoe)	86.5	92.5	100.3	106.1	107.0	111.3	113.1	120.2	119.6	120.4	**128.5**
Nat Gas (Mtoe)	20.1	22.0	22.6	23.7	23.8	24.8	26.6	28.7	32.1	33.6	**36.2**
Coal (Mtoe)	135.9	136.1	135.8	144.2	145.2	151.8	156.8	172.3	184.4	195.1	**208.0**
Nuclear (Mtoe)	2.3	2.6	2.9	3.6	4.3	4.4	4.1	3.8	4.0	4.0	**4.0**
Hydroelectric (Mtoe)	15.9	18.9	18.6	17.4	16.3	15.5	15.7	19.0	22.0	25.4	**27.7**
Subtotal Non-Oil/Gas [Other] (Mtoe)	*154.1*	*157.6*	*157.2*	*165.2*	*165.8*	*171.7*	*176.6*	*195.0*	*210.5*	*224.5*	***239.7***
Total (Mtoe)	**260.6**	**272.1**	**280.1**	**295.1**	**296.5**	**307.8**	**316.2**	**343.9**	**362.2**	**378.5**	**404.4**
Oil (% of total)	33.2%	34.0%	35.8%	36.0%	36.1%	36.1%	35.8%	34.9%	33.0%	31.8%	**31.8%**
Nat Gas (% of total)	7.7%	8.1%	8.1%	8.0%	8.0%	8.1%	8.4%	8.3%	8.9%	8.9%	**8.9%**
Other (% of total)	59.1%	50.0%	48.5%	48.9%	49.0%	49.3%	49.6%	50.1%	50.9%	51.5%	**51.4%**
Oil (% change on previous year)		7.0%	8.4%	5.8%	0.8%	4.0%	1.6%	6.3%	–0.5%	0.7%	**6.7%**
Nat Gas (% change on previous year)		9.7%	2.5%	5.1%	0.3%	4.4%	7.1%	7.9%	11.9%	4.7%	**7.6%**
Other (% change on previous year)		2.3%	–0.2%	5.1%	0.3%	3.6%	2.8%	10.4%	7.9%	6.6%	**6.8%**

Table 5.2 (*cont.*)

Oil	**1997**	**1998**	**1999**	**2000**	**2001**	**2002**	**2003**	**2004**	**2005**	**2006**	**2007**
Oil—Production (Million tonnes)	75.7	74.2	68.6	71.5	68.0	63.0	57.7	55.2	53.0	49.9	**47.4**
% change on previous year		−2.0%	−7.5%	4.2%	−4.9%	−7.3%	−8.4%	−4.4%	−4.0%	−5.9%	**−4.9%**
Oil—Consumption (Million tonnes)	86.5	92.5	100.3	106.1	107.0	111.3	113.1	120.2	119.6	120.4	**128.5**
% change on previous year		7.0%	8.4%	5.8%	0.8%	4.0%	1.6%	6.3%	−0.5%	0.7%	**6.7%**
Production minus Consumption	−11	−18	−32	−35	−39	−48	−55	−65	−67	−71	**−81**
% change on previous year		69.8%	73.0%	9.3%	12.5%	23.8%	14.6%	17.4%	2.5%	5.9%	**15.0%**
Natural Gas	**1997**	**1998**	**1999**	**2000**	**2001**	**2002**	**2003**	**2004**	**2005**	**2006**	**2007**
Nat Gas—Production (Mtoe)	20.1	22.0	22.6	23.7	23.8	24.8	26.6	26.3	26.7	26.4	**27.2**
% change on previous year		9.7%	2.5%	5.1%	0.3%	4.4%	7.1%	−1.0%	1.3%	−0.9%	**2.8%**
Nat Gas—Consumption (Mtoe)	20.1	22.0	22.6	23.7	23.8	24.8	26.6	28.7	32.1	33.6	**36.2**
% change on previous year		9.7%	2.5%	5.1%	0.3%	4.4%	7.1%	7.9%	11.9%	4.7%	**7.6%**
Production minus Consumption	0.0	0.0	0.0	0.0	0.0	0.0	0.0	−2.4	−5.4	−7.2	**−9.0**
% change on previous year									129.7%	32.5%	**25.0%**

Source: BP 2008: 8-11, 24-27, 41 and author's own calculations.

5.1 *China's Internal Dynamics*

This section will not go into the same depth on China's internal issues as the previous one on India, but some contrasts and parallels will be drawn. The first contrast is between China's one-party system and India's democracy. The governing Chinese Communist Party's (CCP's) top concern is preserving its own power. It cannot allow China's stunning development to derail; such a disaster would not only dislodge the Party but could throw the country into disorder. Therefore, industrialization must continue at all costs (Wang 2005: 676–677). The global slowdown was already beginning to affect employment in late 2008, however, and fears of unrest were growing (Xinhua 2008).

Nevertheless, like India, China has some complex issues of national unity to deal with, notably regarding Tibet, the restive western province of Xinjiang, and Taiwan, which is claimed as an inalienable part of Chinese territory despite going its own way since 1949. Other important considerations include the shocking state of China's environment, and increasing incidences of civil unrest linked to relentless industrialization and the land grabs that often accompany it.

5.2 *China's Energy Picture*

In order for growth to continue at all, especially given the slowdown precipitated by the 2008 global economic crisis, the PRC's manufacturing-based economy needs fuel, and lots of it. At present, coal still represents three-quarters of its needs, but though coal should be in plentiful supply for the next fifty years or so (BP 2008: 32), with 20 of the world's 30 most polluted cities being in China, its use needs to be drawn down considerably (World Bank 2007). In 2005, a law was passed committing to turning 10 percent of the energy mix to renewable fuels by 2020 (People's Daily 2005); more recently, Beijing announced a commitment to acquiring foreign nuclear reactor technology and developing its own (Lin 2007: 1–2). China's commitment to hydropower is represented by huge projects such as the 18.2 Giga-Watt Three Gorges Dam (EIA 2006).

Beijing is working on a series of policies in order to reduce demand for imported oil, such as increasing the supply of natural gas (just like India) and driving down consumption through efficiency measures and regulations. The Party would like to cut back on consumption in general: in the 11th Five-Year Plan presented in 2006, the CCP stated its aim to reduce energy consumption per unit of GDP by 20 percent in five years (Government of China 2006). Whether or not that is possible remains to be seen.

But oil imports are still necessary. Table 5.3 shows that from 2002 to 2007, oil consumption increased by nearly 50 percent. China became a net oil importer in 1993, and—while new reserves continue to be found—BP's reserves-to-production (R/P) calculation (BP 2008: 6) indicates that China's own oil reserves could be out by 2020. It thus relies on an array of suppliers, depicted in Figure 5.4: the Gulf States of Saudi Arabia, Iran, and Oman are clearly a major part of this, with African nations and Russia following up behind.

In any case, with consumption two times higher than indigenous production, China's oil imports are only going to go up. Though Angola became China's single largest supplier in 2006 (EIA 2006), at present, 60 percent of China's crude oil imports still originate in the Middle East, and by 2015 this could be 75 percent (Storey 2006: 4). Thus, despite efforts to diversify, China is still exposed to risk, especially where the Gulf is concerned (Iranian and Saudi oil add up to 31 percent of China's imports).

Table 5.3 shows gas use tripling from 1999 to 2007. In 2007 China was still producing enough gas for its needs (CIA 2008), but in the likely event that consumption outstrips production—just as it may in India—it will soon become a net importer. Moreover, if the CCP moves away from coal to gas for its power generation needs, as is its stated intention (Jiang 2006: 2–3), the role of gas may also increase beyond what is projected.

In order to maintain its growing levels of imports, both oil and gas, China will have to do more and more to ensure the security of its supply, lest the predicted shortfall in world energy supply adversely affects its phenomenal economic growth. Common sense tells us that China will still face increasing difficulties ensuring its energy security.

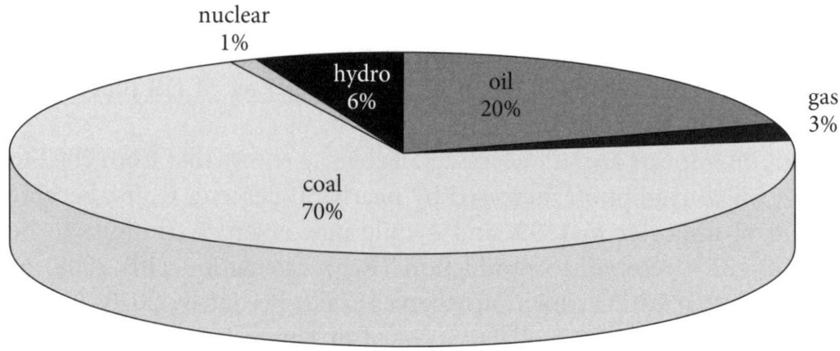

Source: BP 2008: 41.

Figure 5.3 China's Energy Mix 2007 (consumption by fuel).

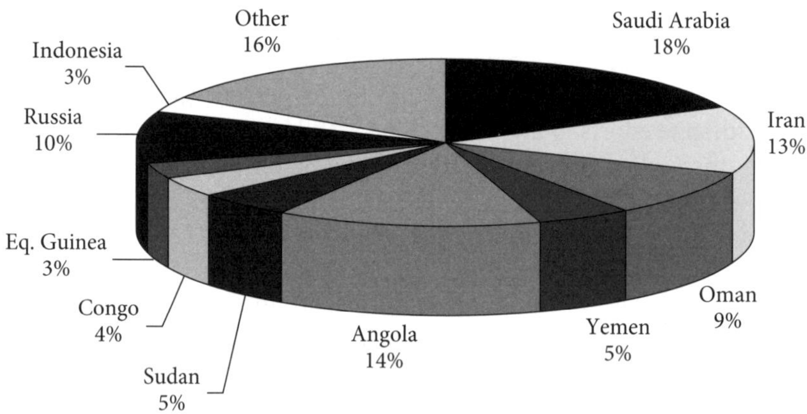

Source: Downs 2006.

Figure 5.4 China's Top Ten Oil Suppliers (2005).

6. Geopolitics, and the Complex of Competition and Complementarity

Following this overview of India and China's domestic situations, we turn to geopolitics. The analysis will be framed in terms of Doran's concepts of positive complementarity and competitiveness as opposed to negative cooperation and competition, with reference also to the 'roles' of China and India within the regional environment, and how these impact their foreign policies.

Table 5.3 China's energy consumption 1997–2007.

	1997	1998	1999	2000	2001	2002	2003	2004	2005	2006	2007
Oil (Mtoe)	196.0	197.0	209.6	223.6	227.9	247.4	271.7	318.9	327.8	353.3	**368.0**
Nat Gas (Mtoe)	17.6	18.2	19.3	22.1	24.7	26.3	30.5	35.7	42.1	50.5	**60.6**
Coal (Mtoe)	700.2	651.9	656.2	667.4	681.3	713.8	853.1	983.0	1088.8	1215.0	**1311.4**
Nuclear (Mtoe)	3.3	3.2	3.4	3.8	4.0	5.7	9.8	11.4	12.0	12.4	**14.2**
Hydroelectric (Mtoe)	44.4	47.1	46.1	50.3	62.8	65.2	64.2	80.0	89.8	98.6	**109.3**
Subtotal Non-Oil/Gas [Other] (Mtoe)	*747.8*	*702.2*	*705.7*	*721.6*	*748.0*	*784.6*	*927.1*	*1,074.5*	*1,190.7*	*1,326.0*	**1,434.9**
Total (Mtoe)	**961.4**	**917.4**	**934.7**	**967.3**	**1,000.6**	**1,058.3**	**1,229.3**	**1,429.0**	**1,560.5**	**1,729.8**	**1,863.4**
Oil (% of total)	20.4%	21.5%	22.4%	23.1%	22.8%	23.4%	22.1%	22.3%	21.0%	20.4%	**19.7%**
Nat Gas (% of total)	1.8%	2.0%	2.1%	2.3%	2.5%	2.5%	2.5%	2.5%	2.7%	2.9%	**3.3%**
Other (% of total)	77.8%	76.5%	75.5%	74.6%	74.8%	74.1%	75.4%	75.2%	76.3%	76.7%	**77.0%**
Oil (% change on previous year)		0.5%	6.4%	6.7%	1.9%	8.6%	9.8%	8.1%	8.3%	8.1%	**8.3%**
Nat Gas (% change on previous year)		3.6%	6.1%	14.0%	11.9%	6.4%	16.2%	17.0%	17.9%	20.1%	**19.9%**
Other (% change on previous year)		–6.1%	0.5%	2.2%	3.7%	4.9%	18.2%	15.9%	10.8%	11.4%	**8.2%**

Table 5.3 (*cont.*)

Oil	**1997**	**1998**	**1999**	**2000**	**2001**	**2002**	**2003**	**2004**	**2005**	**2006**	**2007**
Oil—Production (Million tonnes)	160.1	160.2	160.2	162.6	164.8	166.9	169.6	174.1	180.8	183.7	**186.7**
% change on previous year		0.0%	0.0%	1.5%	1.4%	1.2%	1.6%	2.6%	3.9%	1.6%	**1.6%**
Oil—Consumption (Million tonnes)	196.0	197.0	209.6	223.6	227.9	247.4	271.7	318.9	327.8	353.3	**368.0**
% change on previous year		0.5%	6.4%	6.7%	1.9%	8.6%	9.8%	17.4%	2.8%	7.8%	**4.1%**
Production minus Consumption	–36	–37	–49	–61	–63	–81	–102	–145	–147	–170	**–181**
% change on previous year		2.6%	34.0%	23.5%	3.4%	27.7%	26.8%	41.8%	1.5%	15.4%	**6.9%**
Natural Gas	**1997**	**1998**	**1999**	**2000**	**2001**	**2002**	**2003**	**2004**	**2005**	**2006**	**2007**
Nat Gas—Production (Mtoe)	22.7	23.3	25.2	27.2	30.3	32.7	35.0	41.5	49.3	58.6	**69.3**
% change on previous year		2.5%	8.2%	7.9%	11.5%	7.7%	7.2%	18.4%	19.0%	18.7%	**18.4%**
Nat Gas—Consumption (Mtoe)	17.6	18.2	19.3	22.1	24.7	26.3	30.5	35.7	42.1	50.5	**60.6**
% change on previous year		3.6%	6.1%	14.0%	11.9%	6.4%	16.2%	17.0%	17.9%	20.1%	**19.9%**
Production minus Consumption	5.1	5.0	5.9	5.1	5.6	6.4	4.5	5.8	7.2	8.0	**8.7**
% change on previous year		–1.3%	16.0%	–12.1%	9.6%	13.4%	–29.7%	28.0%	25.7%	11.0%	**8.9%**

Source: BP 2008: 8–11, 24–27, 41 and the author's own calculations.

India and China's growing economies and internal political issues mean that imports of oil, and especially gas, are going to grow. Projections issued by the IEA confirm this, even if all bets on economic growth are off at the moment due to the global financial crisis. But since reserves are limited, there surely must come a point when both countries, not to mention other major consumers such as the US, the European Union, and Japan, are going to have to enter a phase of competition or cooperation—and possibly competitiveness or complementarity—as per Charles F. Doran.

In the context of a multipolar international system, the potential competition for access to fossil fuels may be evolving into what some writers call the New Great Game:

> The battle for the vast oil and gas riches of landlocked Central Asia... [and] the intense competition between the regional states and Western oil companies as to who would build the lucrative pipelines which are needed to transport the energy to markets in Europe and Asia (Rashid 1997).

Thus the policies, relationships, and roles of each player deserve close scrutiny. In this analysis, the actors are of course our foci of interest, India and China, plus other major powers such as Russia, the US, and finally supplier and transit countries.

During the Cold War, India's foreign policy veered towards 'non-alignment,' though in reality it leaned towards the Soviet Union and rejected western liberal markets and capitalism. Today, in keeping with Singh's reformist agenda, India's foreign policy is more pragmatic than ideological, focusing on economic development and also shoring up its military strength in the quest to find a role as a major power within the international system (Ganguly 2004: 41–43).

On the other hand, lacking a sympathetic superpower in the shape of the USSR, New Delhi now appears to be closing ranks with Washington. One reason is a need to dilute America's influence over Pakistan due to the 'War on Terror.' Another is anxiety over China's evolving role in the region, especially considering that India still smarts from a disastrous war in 1962 over border issues (Ganguly 2004: 44–45). The third, of course, is energy. India has struggled for years to acquire US nuclear technology that might give it the power transmission capacity it so desperately needs. There are also oil and gas to be considered, as will be discussed in the remainder of this paper.

China's trump card in any 'New Great Game' could be the way it reinvents the rules. According to the author of 'The Beijing Consensus,' Joshua Cooper Ramo: "China's rise is already reshaping the international order by introducing a new physics of development and power."[5] This "new physics" is essentially a realignment of the assumptions and policies of the 'Washington Consensus,' whereby the developed West (mainly the hegemonic US) instructs the rest of the world on how to run their governments and economies via institutions such as the IMF and World Bank.

The credit crunch and recession the world now faces may diminish the Washington Consensus further, and state intervention in national economies is back in a big way. Whether or not this will mitigate the appeal of the Beijing Consensus remains to be seen, but on the other hand, more and more developing countries may want the no-strings-attached economic assistance that China seems to offer.

Energy relations are particularly critical to this continuation of geopolitics by other means. Chinese analysts (such as Guo Xuetang of Shanghai Tongji University) say Beijing's oil strategy is driven by economic benefits based on the principle of mutual trust and equality (Guo 2006: 136). Though the other countries involved may not necessarily see it that way, this is the line that Beijing likes to push—and it is a far more comforting one than the so-called 'Bush Doctrine' of interventionism in the Gulf and Central Asia.

---

[5] "It replaces the widely-discredited Washington Consensus, an economic theory made famous in the 1990s for its prescriptive, Washington-knows-best approach to telling other nations how to run themselves. The Washington Consensus was a hallmark of end of-history arrogance; it left a trail of destroyed economies and bad feelings around the globe. China's new development approach is driven by a desire to have equitable, peaceful high-quality growth, critically speaking, it turns traditional ideas like privatization and free trade on their heads. It is flexible enough that it is barely classifiable as a doctrine. It does not believe in uniform solutions for every situation. It is defined by a ruthless willingness to innovate and experiment, by a lively defense of national borders and interests, and by the increasingly thoughtful accumulation of tools of asymmetric power projection. It is pragmatic and ideological at the same time, a reflection of an ancient Chinese philosophical outlook that makes little distinction between theory and practice. Though it is decidedly post-Deng Xiaoping in structure, the 'Beijing Consensus' still holds tightly to his pragmatic idea that the best path for modernization is one of 'groping for stones to cross the river,' instead of trying to make one big, shock-therapy leap. Most important, it is both the product of and defined by a society that is changing so fast that few people, even those inside China, can keep up with it." See Ramo 2004: 4.

There is a variety of ways in which China and India seek to address their energy security and concomitant foreign policy challenges. The first, turning back to critical geopolitics, is through the actions of what we might term state-business complexes.

6.1 *Commercial Considerations: Complementarity and Competition outside the Gulf*

In this world of globalization led by economic and business interests rather than military expansion, the main arms in the quest for secure fossil fuel resources must be transnational oil companies (TNOCs) that operate beyond state borders.

Though TNOCs are not directly controlled by their states, most governments and international institutions still have an obligation to monitor and even veto their activities. China's internationally operating companies have relative freedom to roam at will, seeking out the best deals and opportunities. Beijing is already pursuing an aggressive policy of acquiring equity stakes in foreign producers via National Oil Companies (NOCs) such as the China National Petroleum Corporation (CNPC), the China Petroleum and Chemical Corporation (Sinopec), and the China National Offshore Oil Corporation (CNOOC) (EIA 2006).

By contrast, the political hand in India's NOCs remains firmer than the economic hand. While Indian companies need cabinet approval to buy stakes in overseas energy interests, their Chinese counterparts have more flexibility. This renders the options for cooperative efforts more limited.

This is not to say that they do not exist, and the need for energy allows TNOCs to broaden their remit beyond their government's geopolitical considerations. Despite their rivalries, China's CNPC and India's ONGC have a broad cooperation agreement in which they jointly acquire international assets. In 2006, for example, they cooperated to purchase a 38 percent stake in a Syrian joint venture for \$600m (Dittick 2006: 18). Sinopec has teamed up with ONGC in Colombia; they now co-own a 50 percent share of a major oilfield company, Onimex de Colombia. By agreeing to work together and not to compete, China and India are also able to keep down the equity prices of overseas companies (Eurasia Group 2006: 19, 23). It is a more or less a win-win situation—what Doran would call complementarity.

But this is not the case everywhere. Just like China, India continues to look for energy from the Caspian region. ONGC holds a $2.8 billion 20 percent stake in Russia's Sakhalin-I project, for example, and other investments in countries such as Qatar (Vikas and Ellsworth 2006). India's political and financial clout there is lacking. China's CNPC outbid India's ONGC for both PetroKazakhstan and a stake in Iran's Yadaravan field, for example (Guo 2006: 123; Verma 2007: 3286).

Beijing's strategy is more defined than India's. China's economic heavyweight status means that it can often simply outbid India for energy stakes. Each of China's TNOCs has a clear set of targets, and commercial and political decisions are fast. India's are slow and cumbersome, and lack the necessary capital to buy up equity stakes in the face of ruthless competition. Of course, this may put pressure on the liquidity of Chinese TNOCs and on Beijing's cash reserves, but in the long term China's leaders seem ready to live with that. Finally, China has the political capital that permanent membership in the UN Security Council can buy (Nair 2007: 17).

6.2 *Potential Complementarity through the SCO*

Of possible relevance in mitigating potential Sino-Indian conflict within the 'New Great Game' is the Shanghai Cooperation Organization (SCO). The institution was originally conceived as a security cooperative but is now turning to economic partnerships, too; India, Pakistan, Iran and Mongolia are already observers. The latter three have applied for full membership though India continues to dither (Blank 2006).

As a regional organization, the SCO presents a number of advantages. In the absence of the US there is no danger of Washington taking it over for its own interests (as it did with the Asia-Pacific Economic Cooperation forum, APEC) (Buzan 1998: 68–87). Russia provides a useful counterbalance to China's power within the SCO, ensuring that no single big player can assume a dominant role. The forum is concerned with security cooperation as its main focus, and energy security is a part of that (Allison 2004: 463–483). The presence of the Central Asian Republics or 'CARs' (though not Turkmenistan) means that producers are represented as well as consumers.

If India could be persuaded to join, and is accepted, then the institution could bring India, Pakistan, and China together in a formalized context. Bilateral agreements and disagreements could be resolved more effectively in the quest to promote stability. From a Russian point of view, Indian involvement in Central Asia could present a useful way to counter US influence in the region (with the added benefit of drawing India away from the US). The CARs would welcome India, since it would dilute Russian and Chinese influence (Akbardzadeh 2003: 224–226). The Chinese might also like to exploit the SCO to tease India away from the American sphere as well as bolstering its own efforts to counter Washington's sway over the CARs (Guo 2006: 120–123).

Thus, the SCO could become a useful countermeasure to competition for influence over Central Asian energy suppliers, providing a basis for complementarity and mutual self-help. On the other hand, the SCO is still very much in its infancy and should not be viewed yet as some kind of regional version of NATO.

All this being said, both China and India will continue to seek diverse and reliable sources, and as has already been stated, the key suppliers are still in the Persian Gulf. Thus it is upon this area that we should focus most when asking whether the Asian giants will be drawn into competition or if they can exist with complementary roles within the overall international system.

7. Competition, Cooperation, Complementarity, and Competitiveness in the Persian Gulf

In US military doctrine, the Middle East remains an "arc of instability" (Joint Chiefs of Staff 2004: 5). Moreover, it includes one of the most important energy 'chokepoints,' the Strait of Hormuz, through which 16.5–17.0 million barrels per day (Mb/d) flows. Almost all—90 percent—of Gulf exports travel through the Strait, representing 40 percent of world traded oil, and this deeply affects energy flows to both India and China. Another such chokepoint is the Malacca Straits, which sees 11.7 Mb/d (EIA 2007b).

India also knows full well that conflicts in the Gulf mean disruptions and high energy prices for itself and others (Pandian 2005a: 311). However, the Indian Ocean is the Indian Navy's (IN's) main field of operations (Berlin 2004: 248–251). For the time being, lacking a complete 'blue water' capability, the People's Liberation Army Navy (PLAN) presence there is weak, whereas, on paper at least, India has a superior fleet with aircraft carriers to boot (GlobalSecurity.org 2007).

In energy transportation terms, then, China faces a triple problem, since once its tanker traffic passes through the Strait of Hormuz and then India's strategic domain in the Indian Ocean, it still has to get through the Malacca Straits. About 80 percent of China's oil passes through this route (Ni 2006; Bajpaee 2006). With India holding a rare 'hard power' advantage in the Indian Ocean region—especially if allied with US naval power—conditions are ripe for competition or competitiveness in the 'soft power' arena of foreign policy.

7.1 *Interactions with Saudi Arabia*

Saudi Arabia is India's biggest supplier of oil, while India is its fourth biggest customer behind the US, China, and Japan. That makes India's Saudi policy critical. However, though Indian companies such as Reliance are investing in Saudi refineries, relations between New Delhi and Riyadh could be better.

The reason for this is political Islam. Second only to Indonesia, India has a huge Muslim population. Though the Indian and Saudi governments have signed agreements on energy, trade, and counterterrorism (such as the 2006 'Delhi Declaration'), fears remain in New Delhi that Saudi interests are bankrolling terrorist groups such as Lashkar-e-Toiba, which operate in India, Pakistan, and disputed Kashmir. Moreover, it is often Saudi money that funds radical Salafi madrassas and mosques in India. Riyadh, for its part, is skeptical of India's growing ties with Israel and Iran: it subsidizes oil exports to India's own nemesis, Pakistan, and may have been involved in funding Islamabad's nuclear and missile technology. There are many Indian voices that would therefore like to see political reform in Saudi Arabia (Pant 2006).

China has fewer qualms. In 2004, it established a regular dialogue with Riyadh, and state company Sinopec has signed a deal to explore for gas in Rub al-Khali. In 2005, Sino-Saudi trade amounted to $14 bn, and Beijing is offering assistance with diversifying the Saudi economy (Pant 2006).

It would stand to reason, therefore, that in the event of a political disagreement or world oil shock, India and Saudi Arabia are less likely to close ranks, leaving the door wide open for China. With Beijing's policy of non-interference at odds with New Delhi's misgivings over Islamic extremism, Riyadh, too, would rather stick with its friends in the Far East. There is little breathing space for mutually beneficial collaboration or competitiveness here: India's foreign policy gurus must look elsewhere.

7.2 *Interactions with Iran*

We now turn to Iran. Iran is worth particular attention due to its huge reserves of oil and gas, its geographical proximity to China and India, and its interesting role within the international system. Thus, there is much scope for complex relationships of competition and complementarity.

India and Iran are historic partners. Indeed, before the British Raj, Persian was the language of the Moghul courts and today there are 20 million Sh'ia Muslims in India, the world's second biggest population. While Pakistan and Saudi Arabia once supported the Taliban in Afghanistan, both Iran and India sided with the anti-Taliban Northern Alliance. The two have engaged in military cooperation and joint exercises for some years.

India is participating enthusiastically in the modernization of Iran's Chahbahar port on the Makran coast—a port that bypasses the vulnerable Strait of Hormuz. There is a memorandum of understanding whereby Indian goods can be transported across Iran to Central Asia (Vakil 2006: 51–60). Furthermore, Iran is the source of about 10 percent of India's oil, ranking third in 2004 behind Saudi Arabia and Kuwait. More importantly, now that India is moving towards natural gas in its energy mix, *Iran is potentially India's most critical gas supplier, too.*

If India's relationship with Iran is well established, then China's is even more so. Since the late 1970s, when both China and Iran experienced revolutions of a kind—Iran's political and China's economic—the countries have entrenched their cooperation deeper still.

"In spite of multiple difficulties," says John Garver, an analyst of China's foreign relations, "there is a high degree of economic dependence between Iran and China" (Garver 2006a: 280). In 2003, Iran exported $3bn of goods to China—80 percent in the form of crude oil—and China's trade amounted to $2.5bn, mainly manufactured goods. China also provides Iran with know-how, for example technical expertise in building the Tehran metro system, and is helping Iran to diversify its economy via capacity-building assistance in vehicle manufacturing and power generation. Significantly, it is heavily involved in Iran's oil and gas development. In 2004, for example, several major gas deals were signed to the tune of $100bn over a thirty-year period, and China is also investing $750m in Iran's Yadaravan gas field (Garver 2006a: 271–276).

More than this, China is often a useful political supporter, providing desperately needed help to Iran during its 1980s conflict with Iraq. There is undoubtedly something of a balancing act between this closeness of Beijing to Tehran set against the CCP's desire not to displease Washington too badly. China has bowed to pressure from the US, cutting off the supply of nuclear and ballistic missile technology to Iran during the 1990s (Garver 2006a: 281–282). But for the most part it remains a very stable partner, something of a lifeline for Iran, just as Iran is one of China's biggest energy suppliers.

Garver also theorizes that the Sino-Iranian relationship may have a strategic value in the event of a war over Taiwan. Should the US Navy attempt to block China's access to Gulf hydrocarbons by cutting off the vital Strait of Hormuz chokepoint, Iran's cooperation will be essential for keeping some fuel coming into China, perhaps bypassing the Strait via pipelines to the Gwadar port discussed later. Counting out Japan, India, and Russia, Garver postulates that in a multipolar international structure a resurgent Iran would be China's key partner as a counterpoint to America and India (Garver 2006a: 288–300).

Iran, in turn, knows full well that cultivating relationships with both China and India provide it with the economic and political coverage the west as yet will not provide, and that such mutually supportive trade and energy relationships bolster its regional role. The next sections will look at two initiatives of particular relevance: the Iran-Pakistan-India pipeline and the Gwadar port facility.

7.2.1 *The Iran-Pakistan-India (IPI) Pipeline*

A potentially definitive aspect of India's diplomatic and energy relationship with Iran is still under discussion. India's most obvious overland access route to Gulf energy becomes obvious with a glance at the map of South Asia, and indeed for several years there has been talk of an Iran-Pakistan-India (IPI) natural gas pipeline. By February 2006, the pipeline was estimated to cost $4.5bn and be around 2,775 km long. Its capacity was to be 150 million cubic meters per day (PGJ Staff 2006: 14).

The situation moved on in April 2007 and the project was revalued at $7bn. Talks between the then Pakistan Prime Minister Shaukat Aziz and Murli Deora, the Indian Minister of Oil and Natural Gas, led to a tentative new formula whereby Pakistan would sell gas directly to India at the border, rather than act as the transit country for Iran-India sales. Pakistan would buy 60 million cubic meters per day (21.9 Bcm per year) from Iran and sell half of this to India (Dutta 2007).[6] Talks were ongoing in 2008, but there were few signs of reaching a settlement.

The first problem is price. Beyond India's actual *demand* for energy, the IPI is also about relative cost savings of natural gas over Liquefied Natural Gas (LNG). India's LNG infrastructure for transport and regasification is not great, and importing the product is estimated to cost about $4–8 per million British Thermal Units (Mmbtu). In 2005, the Iranians were selling natural gas at $2.20–2.50 per MMbtu. It is estimated that the total savings for India at this rate (including skipping some of the LNG infrastructure upgrades it requires) would be about $10bn over twenty-five years (Pandian 2005a: 316–317). In short, India would be financially better off buying plain old Iranian natural gas via the IPI than shipping in LNG; as well as avoiding importing energy via the Strait of Hormuz.

But it doesn't seem that the bargain rate mentioned above is on offer right now. It remains to be seen whether a price acceptable for all parties can be decided; the wrangling has continued into 2009. Iran's only competitor (so far) in supplying gas to India is Qatar, which is still undercutting it dramatically with its LNG, so if it wants the deal to succeed it might be wise to drop the cost. But if Iran were to sell below the market price, it would expect something in return from India.

The second problem is Pakistan. Due to India's suspicions of Pakistan, the Indian government has publicly stated that it must be *Iran's* responsibility to ensure the safe delivery of gas to India rather than Islamabad's. The World Bank and Asian Development Bank (ADB) have been suggested as guarantors (Pandian 2005a: 316). Pakistan, for its part, would probably rather just supply gas to India at its border,

[6] Previous options included: 1) Iranian companies or an Iran-India-Pakistan consortium owned and operated the pipeline, delivering gas at the India-Pakistan border; 2) India and Pakistan bought the gas from Iran but the pipeline was owned and operated by foreign TNCs. See also Verma 2007, 3283.

and would consider this acceptable under America's ISA (Iran Sanctions Act, formerly the Iran-Libya Sanctions Act or ILSA).[7]

The third problem is capacity. In the long term there is a question mark as to whether Iran is even capable of supplying all of the energy it promises. The combination of the American ISA and Iran's prohibition of foreign ownership has resulted in deterioration of its energy infrastructure. Production is impaired, meaning that Iran faces a shortfall in the amount it can export in contrast to its inefficient and highly subsidized domestic consumption. Within Iran, while demand is growing by 9.2 percent per year, usable production is growing by just 3 percent. Much of this gas is also needed for reinjection into the oil production process, making its availability for export highly questionable (Stern 2007: 378–380).

There is a real danger that Iran won't be able to meet demand as it increases. If (as the IEA projects) Iran is set to become a net exporter of only five Bcm of gas by 2010 rising to 30 Bcm by 2020 (IEA 2005a), when is the 21.9 Bcm for the IPI going to come online for Pakistan and India, and how many other markets will Iran be able to supply? Perhaps the Iranians are being optimistic as to their gas export capabilities, and the IEA projections could always be overstated. Surely Tehran wouldn't want to risk two-thirds of its export capacity on just one pipeline for two customers, especially considering that it may soon have to supply gas to another major consumer, China. The success of the IPI may therefore be out of Pakistan and India's hands—and in those of America and China.

7.2.2 *The Geopolitics of the IPI*

Despite its accommodation of nuclear technology, America has big issues regarding the IPI pipeline. Outgoing US Secretary of State Condoleezza Rice certainly voiced concerns that the IPI pipeline may violate the ISA (Rahman 2005). Moreover, Iran's nascent nuclear program meant it continues to be a pariah to western powers. There were remarks about "extending a hand" to such pariah states during the inauguration speech, but in early 2009 the Obama administration's

[7] Syed Hussain, USAID SARI Energy Program. Interviewed in Islamabad (via teleconference) May 9, 2007.

policy on Iran was not yet clear. Inevitably, America's stance will affect the Iran-India relationship.

On the other hand, India's fractious left supports the IPI, precisely because it doesn't require US technology or funding, and the socialist parties have censured Manmohan Singh for backing down on Iran to favor the Bush regime (Kumaraswamy 2008). But despite positive talks held in 2008, Singh's Congress Party realizes that with the acrimonious dispute over pricing, Iran is proving itself an unreliable partner. India needs energy security, not years of unproductive haggling.

Were the IPI to go ahead, there would be Pakistan to consider as well. Whoever has the most influence over Pakistan if it becomes a key energy corridor commands an element of power over their strategic opponent. If China took the upper hand in Pakistan, it would have the option to make India's energy access more difficult, and vice versa.

And the PRC has a clear advantage. It is telling that, despite Pakistan's dealings with Washington, the Sino-Pakistan relationship has arguably been the most stable of all of Beijing's foreign contacts over the last fifty years. According to Garver, China has even threatened to intervene when India is breathing down Pakistan's neck, such as during the 1965 Kashmir conflict. A strong Pakistan is useful for China in countering Indian regional hegemony, and historically, whenever the US has aided India, China has responded by building up Pakistan (Garver 2001: 187–215).

Pakistan ranks fifth as a recipient of US aid but the *top* recipient of Chinese military assistance. According to the Stockholm International Peace Research Institute (SIPRI), between 2001 and 2006, Pakistan received 43.5 percent of China's total arms exports (Iran came second with 21.2 percent). SIPRI does not provide dollar values, basing its statistics on 'Trade Indicator Values' instead; details of actual expenditure tend to be highly classified. Still, the figures indicate that Pakistan received nearly four times more from China than it did from the US during the same period (SIPRI 2007).

This military backing is a clear snub to New Delhi. Since the 1980s, however, in order to reduce Sino-Indian tensions China has attempted to delink its alliance with Pakistan from its dealings with India. But this has not meant a reduction of ties with Islamabad in favor of improving relations with New Delhi. Lacking leverage and fearing the Chinese and US reactions should it argue too hotly, India has tacitly accepted this (Garver 2001: 218–225).

On the Kashmir question, having belligerently supported Pakistan in the past, China is now looking to move to a more neutral position. China's President Hu Jintao has even hinted that China would be willing to assist in cooling down the Indo-Pakistan animosity (BBC 2007). But despite the Sino-Indian rapprochement, Pakistan appears to be confident that China would lend a hand if it became absolutely necessary. And Indian planners have to assume that in the event of war, China's People's Liberation Army (PLA) would intervene (Garver 2001: 228–234, 242).

Beyond China's political and military support, Sino-Pakistani economic relations have also been strong. Though some of the figures are disguised by the way that money flows are routed through third parties, ultimately Beijing is Islamabad's second biggest investor and trading partner (Wang and Engle 2007). This partnership looks set only to grow: during a business forum in 2006, for example, Memorandums of Understanding (MoUs) totalling $555m were signed.

The US is still number one for Pakistan, representing 40 percent of its Foreign Direct Investment (FDI) from 1989 to 2003 (Govt of Pakistan undated). However, compared to the levels China is and will be pouring into Pakistan's infrastructure projects, US investment is not as effective or visible.

The advantage may thus pass to China, and indeed in 2008 there were reports that Beijing had already expressed an interest in the IPI project. Since Islamabad and Tehran were already agreed on the price of Pakistan's share of the gas coming through the pipeline, if New Delhi continued to drag its feet then Beijing would step in and take the Iranian gas for itself (Economic Times 2008). By October 2008, Iran and Pakistan appeared to have reached agreement but were irked by India's intransigence over the IPI, for example a request for a protective 'war clause' in the contract. Open threats to allow China to take on the IPI instead of India were beginning to surface. Thus, without even having to exert itself in prickly negotiations with India, China could make use of its good relations with Pakistan to bypass geopolitical competition or cooperation over Iranian natural gas supplies.

7.2.3 *China's Counterweight in Pakistan: Gwadar*

In fact, China's energy policy regarding Iran is already beginning to impinge on India's interests. While India develops Chahbahar in Iran,

200 kilometers to the east, Chinese transnational companies (TNCs) continue to sponsor a rival project at Gwadar in Pakistan.

The grand opening of Gwadar by Pakistan's President Musharraf (notably accompanied by the Chinese Minister of Communications) in March 2007 (Aziz 2007) heralded only the beginning of the project. Aiming to emulate Rotterdam, Dubai, and China's Special Economic Zone (SEZ) at Shenzhen (Arthur D. Little 2006: 6), the former fishing town is slated to become an internationally competitive port facility. Construction on the port began on March 22, 2002 under the supervision of the China Harbour Engineering Company (CHEC), a TNC headquartered in Beijing. The facility will be linked to Iran and Karachi via the Makran coastal road, for which the ADB is contributing $500m. When fully completed, Gwadar looks like it will boast superior capacity to Iranian ports such as Bandar Abbas and Chahbahar, with which it may compete (Hassan 2002: 16, 23). Gwadar began functioning commercially on December 21, 2008 with the arrival of a cargo of urea.

Though the first phase was sponsored mainly by China, which contributed an estimated $200m, or 80 percent of the $248m costs (Fazl-e-Haider 2007), Gwadar will be operated not by a national company but by the Port of Singapore Authority International (PSAI) (PSA 2007). Phase 2 construction costs alone are estimated at around $600m, and China has already pledged $200m (Garver 2006a: 1–22).

A lot of that investment is down to Gwadar's potential as an 'energy hub'. A management consultants' report highlights Gwadar's potential to attract future investment for crude oil and also LNG terminals, oil refineries, and petrochemical plants, envisioning liquid bulk and containers as the main contributors to the port's success by 2020 (Arthur D. Little 2006: 7–21). In the first three years of operation, the oil refinery is expected to handle about 10.5m tons per year, which would be set to double in due course (Fazl-e-Haider 2007).

7.2.4 *The Geopolitics of Gwadar*

Should China take over the Iran-Pakistan pipeline, it would already have an advantage. Gwadar's geostrategic position is clearly useful for China—otherwise, why pay? It is in a good location to connect to China's disadvantaged western province of Xinjiang, for example. There is already a Sino-Pakistani agreement to upgrade the Karakoram highway between the two countries that China helped built in the 1970s

(Beijing Review 2006), and this may include construction of the Himalayan pipeline discussed below (Siddiqui 2007).

While China emphasizes that Gwadar's value is mainly commercial, this does not mean that the facility can't serve as a useful 'port of call' for the PLAN too. Moreover, it could also provide Pakistan's maritime forces with a better strategic position in the event of another confrontation with India (Masood 2004), or it could double as a listening post to monitor the US Navy. Should Beijing need to protect oil supplies from the Gulf in the event of a confrontation with Washington, New Delhi, or Tokyo, it might even act as a temporary military base (Anwar 2007; Niazi 2005).

The point to note, however, is Gwadar's potential role in regional energy traffic. As a major regional container port and refining facility, Gwadar's final role as an 'energy hub' could be as the meeting point of no less than five oil and gas pipelines.[8]

One ambitious project on the drawing board is the Trans-Himalayan or Karakoram pipeline. Beginning in Gwadar, it would continue northeast through Pakistan to the Khunjerab pass on the border of Pakistani-controlled Kashmir and China, from where it would go on to connect with existing networks in Xinjiang province. A 2006 sales pitch on the project gives some figures: with a length of 3,300km the 30-inch pipeline would deliver approximately 12m tons of crude oil or other products per year; costs of construction are estimated at $4.5–5bn; assuming the feasibility study is positive, the earliest date of completion could be 2012 (SSGC 2006).

Thus India has reasons to desire the downfall of the Gwadar project: if Gwadar failed it would boost Chabahar; it would degrade China's potential to siphon off Iranian and Gulf energy; moreover, it would deny the PLAN a naval outpost to cover the Strait of Hormuz. Senior Indian officers have expressed concerns. In this sense, Gwadar could become a focal point for the destructive competition for energy and transit options.

As one would expect from the leaders of the 'Beijing Consensus', China is downplaying the significance of Gwadar for India. It may

[8] Natural gas pipelines are probably more suitable than LNG imports by sea or pipeline: since Pakistan is close to potential suppliers and LNG pipelines are more economical when they are at least 3:000km long. See Fazl-e-Haider 2007; see also Pandian 2005a: 315.

say that any problem over Gwadar is purely down to Indian geopolitical paranoia, and that in fact it is a perfect opportunity to explore avenues for Indo-Pakistani *detente* via the pipelines and other energy and transport channels that will flow from it. Chinese analysts note that if Beijing wanted to control the port it could, but it has allowed the Singapore Port Authority to control it so as not to ignite tensions with India. China has the influence but it chooses not to use it. It wants to show India that it's not a threat: it's just business, and business is good for everybody.

And there is room for complementarity between India and China's plans. For example, on July 17, 2004, the Government of Pakistan signed an MoU with Crescent Petroleum regarding a plan to build an undersea pipeline to Gwadar from Qatar, the country with the world's third-largest proven gas reserves (behind Russia and Iran). Interestingly, there was an option to extend this line to India (Economic Review 2004: 26).

Once Gwadar is fully up and running, there is also potential for reviving the Turkmenistan-Afghanistan-Pakistan (TAP) pipeline. After the fall of the Taliban in 2001, momentum returned to project, and in 2003, the ADB initiated a feasibility study that envisaged the TAP as "a natural-gas transmission pipeline of about 1,700km to transport about 30 Bcm of gas annually from the Dauletabad gas fields in southeast Turkmenistan to consumers in Afghanistan, Pakistan, and, possibly, India." The cost of the reborn project was estimated at about $3.3 billion (Jung 2003).

The ADB's feasibility study on the TAP pipeline was positive (OGJ 2005: 10), and delegates at a regional conference in November 2006 decided to "accelerate" work (Pradhan 2006), though the security situation in southern Afghanistan obviously needs to be resolved for it to really take off. Still, a MoU was signed in 2006 for Turkmenistan to supply 32.7 Bcm per year of gas to Pakistan over thirty years, and both India and China have expressed interest in extensions (OGJ 2006: 48).

The point of the TAP is to bring previously unavailable Turkmen gas onto the world market. Nevertheless, its economic viability requires two things: first, stability in Afghanistan, and second, access to markets other than just Pakistan (Olcott 2006: 228). Since one market could be India, the line is also known as the 'TAP (I)'—Turkmenistan-Afghanistan-Pakistan-(India). The idea was endorsed by the Indian cabinet in May 2006 (Bhadrakumar 2006).

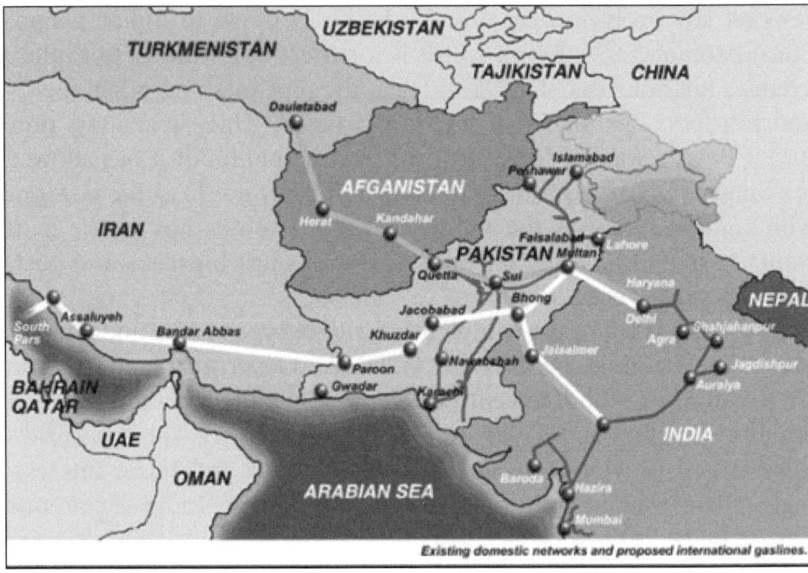

Existing domestic networks and proposed international gaslines.

Source: Millison 2006.

Figure 5.5 Alternative Routes of Pakistan Pipelines.

The impediments are many. Russia's stranglehold on Turkmen energy tightened in May 2007 with an agreement to reconstruct and build a pipeline with a capacity of ten Bcm per year to bring the gas into the Russian infrastructure (BBC 2007). But if both India and China are interested in reviving the TAP pipeline, then all is not lost—and via the development of Gwadar port, it may just have a definitive destination.

8. Conclusions

Of course, concentrating only on oil and gas does distract us from other energy issues that affect the roles and interactions of our countries of concern. The US-India nuclear deal under negotiation during 2007–2008 is a case in point. Seeing the arrangement as another facet of US attempts to contain Chinese power, Beijing insisted that US-Indian nuclear cooperation must conform to the rules of the global non-proliferation regime—though, ironically enough, China itself is partly responsible for the nuclear status of India's rival, Pakistan (Malik 2006).

Outgoing US President George W. Bush's waiving of the rules on civilian nuclear technology shows that the US considers India a potential regional ally. On the other hand, this relationship impinges in particular upon New Delhi's energy relationship with Tehran, making defining its foreign policy outlook a complex balancing act. The deal was finally approved by the US Senate on October 2, 2008 and the long-term implications remain to be seen (though the first outcry came from Pakistan, of course, which demanded a nuclear deal of its own) (BBC 2008d).

Moving on to the security of oil and gas supplies, it is not possible to characterize all of the foreign policy interactions between India and China in Doran's simple terms of 'competition,' 'cooperation,' 'competitiveness,' or 'complementarity'. China has something of a hands-off leadership role characterized by economic investment and respect for sovereignty; India, on the other hand, remains unsure of its role (especially in the face of China's relative political and economic power) and may already be seeing a decline in its regional influence.

If India remains more concerned about China than it does about its energy security, and the IPI remains stalled, it may wish to subdue China's energy relationship with Pakistan via covert means (such as sponsorship of Baloch ethnic nationalists who oppose the Gwadar project, or subversive policies regarding Kashmir) rather than see China make the relative gain. If the IPI goes ahead, however, China's leverage over Pakistan could have implications for India's gas supplies via the pipeline. India is also concerned about Gwadar's double purpose as a naval listening post (Berlin 2004: 244).

On the other hand, in economic terms, India could benefit from Gwadar's refinery capacities and as a potential terminal for the TAP(I). Buying energy via Gwadar would bind India into a complementary commercial relationship with Pakistan and China that would have benefits for all. And China's construction of a conducive environment for Pakistan as an energy hub could even do India a favor by encouraging growth and stability in Pakistan, thus reducing the risk of an explosion over Kashmir and continued cross-border militancy.

In summary, when it comes to routes through Pakistan there is space for complementarity through pipelines such as the TAP(I) to Gwadar, even if India does not see this. On the other hand, India's issues over the IPI—driven by considerations such as its enmity with

Pakistan and burgeoning alliance with America—reveal its discomfort within the regional and international system.

However, though Doran says that competition is a mutually destructive force, this is less worrying if you think you can win the competition. China, simply put, should be able to rise above the bickering in South Asia and take advantage of Iranian resources for itself. It also holds the cards when it comes to Saudi Arabia and other sources such as Myanmar. Though it has shown that it can constructively cooperate with India when it chooses, in other situations it uses its economic and political advantage to out-compete. It also remains ascendant in terms of foreign policy role. Interestingly, while asserting that America can't influence its Iran policy, India has displayed a high degree of deference to China's views on foreign affairs in recent times. It did little to criticize Myanmar during the crackdown on the monks' protests in late 2007, and in April 2008 it effectively suppressed potential pro-Tibet protests during the passage of the Beijing 2008 Olympic torch through New Delhi.

This is not necessarily a good thing for China in the long term. Its approach does seem to lean more towards competition rather than competitiveness, and ultimately the more China pays for energy security interests, the higher prices will become. That will have a negative knock-on effect on India and other countries. One could also argue that India's nervousness about China's influence over energy suppliers and its difficult relations with Pakistan mean that it is falling into the arms of an equally fearful America, which is keen to provide nuclear technology (on the condition that New Delhi distance itself from Tehran) (Doran 2004: 21–24).

Such situations are fraught with danger. Prior to 1914, Doran suggests, an ascendant Germany didn't see Russia as a threat until too late, its ensuing panic a factor in triggering World War I (Doran 2004: 28–30). Whether or not Doran's historical judgment is correct, it is possible that, in similar fashion, China may not yet view India as a threat to its energy security and foreign policy aims. But this could change if factors such as global economics, internal political strife, sparring with Pakistan, and general energy scarcity increase the pressure on India. For the system to reach equilibrium, it is often necessary for players to yield role and responsibility to ascendant actors: just as the US is in a period of self-examination over conceding power and influence to China and the EU, so soon could China find itself mulling over a rising India.

China's advantage over India is deeply connected to India's geopolitical concerns, particularly with respect to its rivalry with Pakistan and its growing alliance with America—in the face of the perceived challenge from China. Both of these fit into Doran's destructive categories—competition with Pakistan and cooperation with America. The best confidence-building measures that could alleviate India-Pakistan tensions are economic, and the IPI could be a major aspect of any trend towards building economic linkages. In the words of former Pakistani Prime Minister Shaukat Aziz: "the confidence-building measures and the dialogue we've initiated at all levels is helping to create a conducive atmosphere… [and] the [proposed] pipeline creates linkages and interdependencies and builds trust and interaction" (ICG Staff 2004: 4–6; Pandian 2005b: 308, 318). Sino-Indian collaboration on the IPI, TAP(I), Gwadar, and Chabahar would be of benefit to all, Pakistan and Iran, too—if the US will allow it.

Likewise, Beijing would do well to ease off its aggressive energy policy and cut India some slack. Though ultimately there is not enough oil and gas to go around, ironically it might be in China's interests to promote more complementary energy security efforts with India. In such a way it could avoid the claws of a thirsty tiger that just might lash out in desperation, backed up by its American allies. Though China is well able to manage the headache of India's challenges to its energy security, it should not push it into a corner.

CHAPTER SIX

CHINA'S OIL SUPPLY STRATEGY: THE CASE OF SAUDI-ARABIA AND SUDAN

Chen Mo

Abstract: *This study has two parts.*[1] *The first part is an evaluation of the policies and measures China has adopted to achieve greater oil supply security. This may be seen through two perspectives, namely "easier access to oil" and "keep away from oil"*[2].

"Easier access to oil" refers to enabling China to have greater possibilities in acquiring oil. It includes, in the first instance, diversification of the sources of supply and implementation of energy diplomacy; China should get petroleum through various channels, and develop relationships with oil exporting countries. If one source is suspended, other sources will remain available. In the second instance, this involves "going global": Chinese companies should control oil resources and establish offshore oil bases. In the third instance, this involves the establishment of strategic oil reserves: China should be able to use its own oil reserves in case of supply interruption.

"Keeping away from oil" refers to less dependency on oil. This will reduce the problems encountered in oil import security, and can be achieved by reducing imports, practicing energy-saving methods, increasing energy efficiency, and exploiting alternative sources of energy. In the following, we do not distinguish between the policies of China's three large mainly state-owned oil companies that operate overseas and those of the Chinese government, because in principle, these companies follow policies set by the Chinese state and their parent companies, and they have parallel interests.[3]

The second part is a case study of Saudi Arabia and Sudan, two of most significant oil exporting countries to China. After coal, oil is and will remain the second most important source of energy for China. Over the past twenty years, China's oil import volume outweighed domestic output. In the future the Middle East and Africa are likely to remain the chief sources of China's oil import. It is difficult to examine the prospects of oil supply security for China without conducting research on these major oil exporters to China.

[1] I wish to thank Eduard B. Vermeer for his suggestions and additions to this article.

[2] An Weihua & Qian Xuemei (eds). *Commenting on A New Theory of Gulf Petroleum*. Social Science Documentation Publishing House, 2000.

[3] "Chinese foreign policy is now driven by China's unprecedented need for resources." David Zweig and Bi Jianhai, "China's Global Hunt for Energy," *Foreign Affairs*, Sept–Oct. 2005.

China's oil cooperation with foreign nations will inevitably lead to a restructuring of geo-political relations, especially China's relation with the US. In terms of international oil partnership, the Sino-US relationship has both conflicts of interest and common strategic benefits. Therefore, it is vital for China's international oil cooperation to develop a proper understanding of Sino-US ties.

Keywords: China, Energy Diplomacy, Saudi Arabia, Sudan, Oil Supply Security

1. Policies and Measures to Enhance China's Oil Supply Security

1.1 *Background of China's Strategy of Oil Supply Security*

To ensure the security of oil supply, China should first safeguard easy access to petroleum, then increase its possibilities of acquiring oil. The most convenient way is to rely on domestic supplies. Because of rapid economic growth and optimization of its energy structure, China's demand for oil will increase substantially. As a developing country, China has given its energy-intensive industries an important priority in its overall economic development. Meanwhile, with urbanization and improvement in people's living standards, an ever larger number of residents will become consumers of modern energy resources. The above factors increase the demand for oil. In addition, the past heavy reliance on coal has become a major threat to the environment, as a major contributor to air pollution. Thus, the need to optimize China's energy structure requires more oil and natural gas, too.

China has a large population and rather low oil resources per capita. By the end of 2007, proven oil reserves hit 169 billion tons worldwide, of which China had only 1.2 percent, namely 2.1 billion tons. The world's oil reserve-production ratio is 42 years, yet the figure for China is merely 11.3 years.[4]

Since the 1990s, China has seen the gap between domestic supply and demand widening. Daqing (Northeast China) and eastern China used to account for three-quarters of total output, but their oil fields have entered their late phase and output is stagnating. Western China, as the strategic substitute region for China's oil industry, has contributed to the increase of onshore crude oil output since the 1990s. However,

[4] BP Statistical Review of World Energy June 2008. P. 6.

Western China needs time and huge amounts of investment to increase its production capacity, because of its complicated geological conditions, adverse natural environment, and high production cost. Offshore oil requires high-tech, huge investment and involves high risk. Although some marine spaces in the East China Sea and South China Sea enjoy good prospects of oil and gas resources, finds along the South China coast have been disappointing so far, and unsolved territorial disputes with neighboring countries make it hard to exploit oil and natural gas resources in those marine areas in the short term.

China's rate of self-sufficiency in petroleum plummeted from 102% in 1993 to 67% in 2003. Since it became a net importer in 1996, overall crude oil imports have continuously risen. 1990 imports were a mere three million tons, but by 2006, they had increased to 145 million tons.[5] China's total consumption hit 350 megatons, nine percent of the world total. It could only be met by increased imports.

China relies heavily (45% in 2007) on the Middle East for its crude oil imports, now and in the future. There are two main reasons. First, the Middle East leads the world in oil reserves, production volume and production capacity, and will continue to be the center of global oil supply. Second, China has longstanding good political relationships and a sound foundation for highly promising economic and trade cooperation with countries in the Middle East. These are favorable conditions for oil imports from this region.

However, such heavy reliance on the Middle East for oil supply is a potential threat to energy security. The Middle East has witnessed ethnic and religious conflicts for centuries. Affluent natural resources and crucial geographical location make it a place of strategic importance for the big powers. In addition to potential regional conflicts, political instability of major oil exporters could also impact global oil supply and even cause a dramatic rise of the oil price and an oil crisis. The possibility of oil supply discontinuity from this area is a potential threat to China's energy security. The security level of oil resources and oil transportation channels are other unstable factors for the region.

[5] 2006. China Customs Statistics Yearbook. P. 8.

1.2 *Methods Adopted by China for its Strategy of Oil Import Security*

China has adopted a series of energy policies and approaches to realize a steady oil supply, which may be put under different headings, as below.

1.2.1 *Diversification of Oil Imports*

Greater diversity of oil imports has been an important security strategy of major oil-importing countries since the 1970s. Principal goals include diversity of oil import resources, diversity of oil trade methods, control of overseas oil production through investment ('equity oil'), and control of international oil transportation channels.

Greater diversity of oil imports is important to China, and its efforts to reduce its dependence on the Middle East have made considerable progress in recent years. A promising prospect is in Russia, a country with one of the largest oil reserves in the world. According to the *Russia's Energy Strategy for the Period Until 2020*, adopted by the Russian federal government in 2003, Russia encourages foreign investment in oil and gas exploration and exploitation. Russia also stresses greater diversity of energy export markets. For China, this is a valuable opportunity to realize greater diversity of oil imports. For Russia, large investments and foreign exchange earnings from oil export will help to achieve rapid economic recovery. The complementarity of their economies constitutes a solid foundation for a win-win cooperation in the oil business. The 1996 *Sino-Russian Strategic Cooperative Partnership* has created favorable conditions for mutual benefit and extensive cooperation in every field, including oil. In February 2009, China and Russia signed seven agreements on energy cooperation, including the construction and financing of an oil pipeline and annual exports to China of 15 million tons of crude oil.[6]

Central Asian countries have vast potentials of oil and gas reserves and are likely to become new stars in the global oil market in the next decade. Cooperation between China and Central Asian countries is not only in line with the interests of China, but also of those countries. The Central Asian countries had to rely on Russia's pipeline system for oil exports, and the high expenses charged by Russia made them eager to find other partners to guarantee oil export security. Strife between the

[6] Xinhua News Agency, March 3, 2009.

US and Russia is another factor that makes Central Asian countries welcome the arrival of a third party to keep down tensions. China's increasing annual oil demand creates a huge potential market for them. China will benefit also from more trade with this area, particularly in the implementation of the *Strategy for Development of West China*. Cooperation with China's Central Asian neighbors in oil trade can reduce transportation costs and risks. However, China should not overlook the difficulties in exploitation and transportation. Since Central Asian countries are located inland, it is necessary to construct oil pipelines first. Hence oil trade cooperation between China and those countries is still at the initial stage, but it will be indispensable in the long term. The *Shanghai Cooperation Organization (SCO)* will play an important role in developing a friendly and cooperative relationship between China and central Asia. As for oil cooperation with central Asian countries, China has made considerable progress with Kazakhstan. China and Kazakhstan have established a strategic cooperative partnership and signed a treaty of cooperation and have good relations. Petroleum and petrochemical cooperation between the two countries consequently looks promising.

Latin America is another key oil production area, and cooperation with Latin American countries is another means for China to realize greater diversity of oil imports. For example, Venezuela is also looking forward to greater diversity of oil exports in order to overcome dependence on the US market as its oil production volume grows. In 2007, President Hugo Chavez said in public that China was on the rise and would be their future market.

African oil is playing an increasing role for China. China's oil imports from Africa were 17 million tons in 2000 and surged to 49 million tons in 2007.[7] In recent years, Angola has been the second largest source of crude import for China, after Saudi Arabia.[8] Sudan is another important overseas oil source for China and has received the largest and most comprehensive oil investment of China, over six billion US dollars till 2007. After long years of investment and operation, China made considerable achievements in Sudan in the fields of oil exploration, exploitation, pipelines, refinery and harbor transportation.

[7] BP Statistical Review of World Energy 2007.

[8] In 2008, 36, 30 and 21 million tons, respectively, of crude oil were imported from Saudi Arabia, Angola, and Iran. Dow Jones Newswire, Feb. 9, 2009.

Southeast Asia does not have abundant oil reserves. Indonesia had been the largest crude oil supplier of China in the past, but since 1996 its share in China's overall oil import has dropped continuously, from 6.5 percent in 2000 to zero, because of Indonesia's declining oil supply potential.[9] Russia is another oil supplier of China, but its oil export mainly aims at Europe. The Middle East and Africa abound with oil resources and have superior price and transportation conditions. Therefore, the Middle East, Africa, Russia, and Central Asia will be the main sources for China in its efforts to create a greater diversity of oil imports.

1.2.2 *Energy Diplomacy*

Energy consuming countries are conducting energy diplomacy and various diplomatic activities in order to ensure steady energy supply and guarantee energy resources security. Energy diplomacy mainly features facilitating energy cooperation with other countries and regions through diplomatic action, as well as creating better conditions for domestic energy companies in overseas exploration and exploitation. For China, oil security mainly refers to avoiding oil supply disruption and cushioning the impacts of dramatic oil price fluctuation. China's energy diplomacy is a substantial guarantee for energy security, and indispensable for ensuring oil import security, creating better conditions for domestic companies in the international market, introducing energy saving and environment friendly technologies and carrying out international cooperation on alternative energy resources. Energy diplomacy is a new policy and a new integrant of China's diplomatic efforts since the 1990s.

The Sino-US relationship is becoming increasingly important in global affairs and its strategic significance extends to the international economy, politics, security and other fields. Its influence has expanded beyond bilateral issues and exerts a direct impact on the development of Asian-Pacific and even global situations. China and the US are finding increasingly common interests in maintaining regional and international security. Since both countries are big energy consumers, oil supply and oil price changes have greater impacts on their respective economy. Therefore it is important to work together on the stability

[9] China Customs Statistics Yearbooks, 1991–2001.

of the global oil market and on energy resources. When President Hu Jintao visited the US in April 2008, he discussed with US leaders their further cooperation in international affairs, and both parties agreed to strengthen strategic negotiations on energy resources, deepen their cooperation in research and development of clean energy as well as new and renewable energy resources, and strive to achieve greater stability of global energy supply and prices.

As for the European Union, during the Helsinki ASEM meeting in September 2006, premier Wen Jiabao proposed to avoid the influence of regional political disputes on global energy resources supply, and to forestall turning the energy issue into a political one. He advocated closer communication and cooperation in order to optimize the international energy market mechanism, rationally exploit traditional energy resources, and develop renewable energy resources. He also stressed efforts in advanced energy technology, energy savings, and energy efficiency, and called for an improvement of technology transfer mechanisms to help developing countries.

The relationship with oil exporting countries also plays an important role in China's oil diplomacy. Most are developing countries, with which China has cultivated relations for many years. In 2006, the *Government Report at the 16th CPC National Congress* emphasized that China should strengthen friendship and cooperation with the third world, enhance mutual understanding and trust, increase mutual help and support, and expand cooperation and its effectiveness. China's strategy is based on the 'Five Principles of Peaceful Coexistence,' and stresses equality, mutual benefits and help to developing countries. China needs their support on the Taiwan issue and other major issues concerning national sovereignty and territorial integrity, while developing countries value China's position in global affairs and as a permanent member state of UN Security Council, expecting China will stand up for their interests. They also share common interests in fighting an unreasonable international business order and supremacy.

China needs steady energy supplies and oil dollar investment in the Chinese market. Oil exporters earn from China's growing imports, and hope this flow will help maintain international oil prices at a reasonable level. They also hope to reduce their reliance on the export of primary products and realize greater diversity of their economy. Lacking advanced technology and labor resources, they strive after international support, including from China. China has serious experience and mature technologies in its manufacturing and agricultural industries, and a favorable reputation in overseas construction projects. Thus, there is a favorable environment for mutually beneficial cooperation and development.

As a latecomer, China has been viewed as a threat to prevailing, mainly western interests in the international energy market, but this is unfair. Competition between enterprises is normal, and in fact, companies from developed countries have a dominant power. The Chinese companies cannot rival these long-established leading companies, whether in terms of technology, market share, or investment volume. Most Chinese 'equity oil' comes from projects that the large international oil companies have sold or were unwilling to undertake, so there is no conflict of interest. From the strategic view, China's investment in the international oil industry serves the interests of the US, Japan and all European oil importers, as it helps reduce a shortage of oil on the international market, some of which has been caused by big-power politics and unilateralism. China's extensive oil cooperation with oil providers on the basis of mutual respect of sovereignty will not only help alleviate global oil production capacity, but also lead to the easing of regional political tensions. Oil cooperation should be recognized as a factor that tempers political tensions and helps maintain steady oil supplies and stable prices.

1.2.3 *The "Going Global" Strategy*

The "going global" strategy, initiated by President Jiang Zemin in 1997, underpins China's efforts to participate in economic globalization and compete in international market. It is important for raising China's GNP and expanding the business scope of Chinese enterprises. Since 2000, encouraging domestic enterprises to globalize is a national strategy. China is a large developing country with limited reserves of oil, gas, other key mineral resources, forest, and fishing resources. China must learn from other countries and regions, use the global market, strengthen cooperation in developing overseas resources, and secure raw materials from overseas in order to realize long-term development of its national economy.

Although there is a trend for nationalization of domestic oil production, more and more countries are opening up their oil industry. The high oil price has stimulated the search for oil exploration and exploitation opportunities, also by China. Most developing countries with rich oil reserves do not have the capability for independent oil resources exploitation, and need international capital and technology. As domestic oil consumption of these countries is limited, they have to rely on the fiercely competitive global oil market. Inviting foreign capital under favorable contract terms to exploit domestic resources will help them compete in the global oil market. There are hundreds of oil exploration and exploitation bidding opportunities around the world every year, and numerous opportunities for bilateral negotiation on oil trade and cooperation. Developing countries with rich oil reserves may have great exploration and exploitation potential, with low exploration costs. These new trends indicate that the global oil resources market is expecting new allocation and distribution. This is an opportunity China should grasp, whether for security of oil import diversity or for the development of domestic oil companies.

In terms of cost and technology, Chinese oil companies still have a long way to go before catching up with their well-known multinational rivals, but they have other advantages. After the oil industry restructured in 1998, domestic oil companies have obtained control over both upstream and downstream businesses and become stronger through market competition. China National Petroleum Corporation (CNPC) ranks among the top five global international independent oil exploration and exploitation companies in terms of overseas project exploration and operation costs. China's oil companies have low-cost experts in exploration and exploitation planning, geology, drilling, oil field exploitation, pipeline construction and refinery design. In addition, they have built up professional oil industry construction teams. They have gained some experience in overseas operations, and have shown an ability to retain their position in the global oil industry by making full use of their low-cost advantage. For instance, China's oil projects in Sudan have a lower production cost than comparable international projects.

1.2.4 *Establish Strategic Oil Reserves*

Since the 1990s, strategic oil reserves have helped developed countries stand the test of two global oil crises. It has become a universal obligation for International Energy Authority (IEA) members to establish strategic oil reserves. After years of preparation, China established the National Oil Reserve Center at the end of 2007. It will establish and improve an oil reserves management system with Chinese characteristics, accelerate construction of strategic oil reserves, and regulate their operations. The first batch of four bases (in Zhenhai, Zhoushan,

Dalian and Huangdao) has been filled, and sites have been selected for the second batch (two of which to be located in Lanzhou (Gansu) and Shanshan (Xinjiang). China's strategic oil reserves will reach 12 million tons by 2010. Site selection was based, in part, on the need for development of China's western regions and the fact that Russia and Central Asia will become a major source of oil and gas imports.

1.2.5 *Seek Energy Saving Approaches, Improve Energy Efficiency, Reduce Energy Consumption, and Strengthen Environment Protection*

China must realize energy savings, improve energy efficiency, reduce energy consumption and strengthen environmental protection. China has a higher elasticity coefficient of energy consumption than developed countries. At the same time, due to its energy structure and high energy consumption, China suffers from air pollution, and needs to invest in prevention, treatment and reduction of exhaust emission. Since 2006, the Chinese government has taken a number of measures in this respect. As a result, energy consumption increased considerably less than GDP in 2007 and 2008. Industrial restructuring, control of energy-intensive industries, technological upgrading, revamping traditional industries, faster development of services, and high-tech industries have been major contributing factors. Moreover, advanced technologies are being utilized to realize energy saving in key industries. China welcomes more international cooperation in this field. Regulations on energy savings have been strengthened under the new Energy Conservation Law.

China adheres to the requirements of the UN Framework Convention on Climate Change and the Kyoto Protocol. In the past 15 years, China has reduced the share of coal in its energy consumption and thereby curbed the growth of CO_2 emissions. In the *Outline of the 11th Five-Year Plan for National Economic and Social Development*, China has explicitly set a goal that between 2006 and 2010, its per unit GDP energy consumption will decrease by 20% and its chemical oxygen demand and CO_2 emission will drop by 10%.

The 2006 *Renewable Energy Law* brought China's renewable energy resources development under legal supervision. Implementation of the law drives the exploration and utilization of renewable energy as well as development of the industry, and plays an important role in energy resource restructuring, environmental protection as well as ensuring energy resources security. Various renewable energy resources have seen rapid growth. Taking hydropower for instance, annual installed capacity exceeded 10 GW for the first time, while total installed capacity reached 125 GW, accounting for one quarter of China's developable potential. For wind power, total installed capacity hit 1.3 GW at the end of 2006, higher than the total of the previous 20 years, and a growth of 270% over 2005. Annual renewable energy consumption volume added up to 200 megatons of standard coal (excluding traditional biomass energy), about 8 percent of overall energy consumption. Hydropower contributed the equivalent of 150 million tons of coal, and solar, wind and biomass energy another 50 million tons.[10] Renewable energy has contributed to optimizing the energy structure, improving the ecological environment, and building an energy-saving and environmentally- friendly society.

2. Energy Cooperation between China and Saudi Arabia

2.1 *Background*

China and Saudi Arabia are strategic partners who share some common interests. Saudi Arabia plays an important role in the Middle East. It supports the one-China policy on the Taiwan issue, and seeks China's backing on the stability and security of the Middle East and the Gulf Region. It hopes China can play the role of a major power, and take a fair and just stance in Arab-Israeli issues. It also expects China to exert its influence in solving the Iran nuclear problem and address the threat of weapons of mass destruction.

In January 2006, the Saudi King Abdullah visited China, his first official visit to another country. In-depth talks were held in the field of energy cooperation. With Chinese President Hu Jintao, he signed a treaty of cooperation in fields of, amongst others, oil, gas and minerals. The visit had a significant impact on oil cooperation. Three months later, President Hu Jintao visited Saudi Arabia for the first time. The exchange of high-profile visits within one year had never happened before in the history of relations between China and the Arab world, and was a sign that both sides attached great importance to bilateral relations and energy cooperation.

[10] Cui Minxuan (ed.). "The Energy Development Report of China." 272, *Social Science Documentation Publishing House.* 2008.

China needs oil from Saudi Arabia, while the latter wishes to invest in China, as an export market with great potential, which is relatively safe for oil dollars. China could become an important market for petrochemicals. The positive quality record maintained by Chinese construction contractors is another basis for fruitful cooperation.

2.2 *Overview of Oil and Gas Cooperation*

Transnational joint ventures have seen increased development since China and Saudi Arabia established diplomatic ties. While China increased its oil imports from Saudi Arabia, Saudi oil companies are making their way into the Chinese market as well. Bilateral trade has grown significantly, from 4.1 billion US dollars in 2001 to 25 billion US dollars in 2007. Among the latter, exports to Saudi Arabia were 7.8 billion US dollars and imports from there 17.6 billion US dollars. The primary imports were oil and oil-related products.[11]

Large-scale Sino-Saudi cooperation in the energy field kicked off in 2003, when bilateral investment grew rapidly. An oil refinery and ethylene factory with a capacity of 240,000 barrels per day was jointly built in Quanzhou (Fujian) by Sinopec, Saudi Aramco, and ExxonMobil. It may be expanded with another 800,000 tonnes-per-year ethylene cracker and associated chemical units.[12] Sinopec Group won an international bid in the second half of 2003, which involved 1.9 billion US dollars Chinese investment in the tapping of a large gas field in southern Saudi Arabia's. In April, 2004, Sinopec and Aramco set up a joint (80%-20%) company to that purpose.[13] Sinopec Shanghai Engineering Company, Sinopec Second Construction Company, and Netherlands AK jointly won a bid in the construction of production facilities with an annual capacity of 400,000 tons of polyethylene and 400,000 tons of polypropylene offered by Sabic, a Saudi company. The project, under a contract worth around 700 million US dollars, was built in Yenbo on the Red Sea. The production line was put into use in March 2008.[14] The project, aimed at downstream petrochemicals, is the first of its kind

[11] 2006–2007 China Customs Statistics Yearbook

[12] SABB Notes, *Saudi-China Trade Relations*. Feb. 10, 2009.

[13] Energy cooperation between China and Saudi Arabia (http://news.tom.com/2006/01/21). In the light of Sinopec's limited experience in gas development, the agreement may have reflected political considerations.

[14] "Sinopec Group in Saudi Arabia" http://sa.mofcom.gov.cn/aarticle/zxhz/tzwl/200509/20050900458773.html.

which Sinopec acquired with a foreign company. Sinopec is responsible for half of the total investment. The success of the project will enhance the competitive edge of China's petrochemical industries in the international market in terms of facilities designing and installation, and yield experience in cooperation with international counterparts. So far, in Saudi Arabia Chinese investments in aluminum, cement, a high-speed railway, and other construction projects have been more important than those in oil or gas, because of the Saudi preference to use consulting companies such as Haliburton, instead of foreign oil companies.

Saudi Arabia has invested significantly in refinery projects in Japan and South Korea and now it is also turning to China for that end, for example, in Qingdao. Such investments connect import and export markets, and skip the profits that intermediaries and transnational companies usually take. Thus, losses to both countries can be avoided. The increasingly open Saudi policies towards downstream oil production provide an opportunity for Chinese oil industries to "go global."

Major headway has been made in Sino-Saudi oil cooperation, and there is further potential. Mutual investment will strengthen bilateral ties, ensure China's oil trade with the region and create a win-win situation. China's economic growth has boosted the export of industrial products and Saudi oil has found increasing demand from the Chinese market. Thus, oil dollars can be recycled and export of Chinese products driven by oil trade. For a long time, major oil exporting countries in the Gulf region have been expanding their market shares by investing in oil refining, storage, and retail sales in countries with great market potentials. China faces the problem that its existing refineries are unsuitable for the Middle East high sulfur crude oil, and needs capital to revamp its refinery facilities. The interest of both countries in cooperation is obvious. The long-term oil supply agreement concluded between Saudi Aramco and Sinopec in 2008 to deliver 1.5 million barrels per day of its heavier crude by 2015[15] promises relative stability in the amount, category, origin, and timing of oil trade. For a country like China, which is heavily dependent on foreign oil, such long-term oil supply contracts are a major means of maintaining secure crude oil imports.

---

[15] SABB Notes, *Saudi-China Trade Relations*. Feb. 10, 2009.

2.3 *The US Factor in the Sino-Saudi Energy Relationship*

Saudi Arabia and the United States are known for their longstanding and close relations, through which US companies are gaining great benefits in the Saudi oil industry. The Sino-Saudi relationship does not pose any threat to the US. Politically, America tends to take sides with Israel on Middle East issues and Saudi Arabia wishes China could play a more important role in the region, but China's influence is quite limited when it comes to Arab-Israeli issues. China is excluded from the "four-party mechanism" and can only offer its mediation. It is China's long-standing policy not to interfere in other nation's internal affairs. In contrast, the US has attempted to enforce democratic reform in the Middle East. Its apparent lack of success indicates that the western political modality does not fit Middle East countries. The US government is aware of the importance in maintaining a close relationship with its Middle East allies so as to assert its traditional influence in the region. Economically, Saudi Arabia has not yet opened the upstream sectors of its oil industry and the Arabian American Oil Company is technologically supported by US companies. The biggest overseas investors in the Saudi upstream gas industry, which has been opened up, are still American companies. Even in the construction contracting market, expensive contracts are still obtained by contractors from the United States and other developed nations, leaving China with contracts cheap in value and small in scale. The emergence of Chinese companies in the Saudi Arabia is not yet a threat to American businesses.

3. Energy Partnership between China and Sudan

3.1 *Background*

China and Sudan have been on friendly terms for a long time, and cooperation between the two countries has been constantly promoted. In November 2006, Sudanese President Al-Bashir visited China for the Beijing summit of the China-Africa Cooperation Forum. In February 2007, China's President Hu Jintao visited Sudan and exchanged views with his Sudanese counterpart on friendly relations and reached a general consensus on cooperation in fields of energy, trade, finance, agriculture, and human resource training. President Hu also joined President Al-Bashir in resolving the Darfur issue. Economically, China and Sudan share a win-win basis. Sudan is an underdeveloped country with rich resources, especially oil. Due to the US and UN sanctions, those resources remained untapped and the Sudanese people were living under chronic poverty. Sudan needs Chinese oil companies to exploit its oil resources and also agricultural technology transfer, as ways to improve the people's livelihood and promote its economy. The Sudanese government aims to resolve the country's ethnic conflict, realize its agricultural potentials and achieve a diversified economy. For its part, China needs Sudan's abundant oil supply in order to diversify its imports. As oil investment markets are largely dominated by developed countries, Sudan presents itself as a potential investing target. In addition, Sudan has great agricultural potentials, which can be an important field for cooperation. Thus, China and Sudan share grounds for mutual benefit, both politically and economically.

In fact, oil cooperation between China and Sudan is mutually beneficial. Chinese investments in Sudan, which totalled over six billion US dollars by 2007, enable a joint development of Sudanese oil resources. Moreover, China has extended five preferential loans totalling 184 million US dollars.[16] At the request of the Sudanese government, China provides capital, technology and human resources to build a complete oil industry from upstream to downstream sectors. Sudan, once the most heavily poverty-stricken country in the world ten years ago, has achieved 760 US dollars per capita GDP and a 8% economic growth, transforming itself into one of the fastest growing economies in both Arab and African regions. Under the 2005 Peace Agreement between the North (GONU) and South (GOSS), oil revenues from Southern Sudan (where the main producing blocks are located) should be divided equally between GONU and GOSS, with 2 percent going to the local states. The demarcation line passes right through major oil fields. Unfortunately, development of the oil fields in Southern Sudan has often been preceded by violent army actions against and displacement of the local population, villagers, or migrants, who were perceived as a potential rebel threat. Also, the environment for traditional farming was affected.[17]

---

[16] Sudan Tribune, Nov. 6, 2007.

[17] European Coalition on Oil in Sudan, *Oil Development in northern Upper Nile, Sudan*. May 2006. www.ecosonline.org.

China has been criticized about its purchase of oil from and sales of small arms to the Sudanese government by the US and others, as undercutting international sanctions. Rebel leaders in Darfur have threatened Chinese and other foreign oil companies, and recently four Chinese workers were killed.[18] However, the Sudanese government could turn to many countries for its trading needs. In any case, the war in Darfur is fought far away from the oil fields and oil is not seen as a driving factor in this ethnic conflict. Actually, China has played a positive role by bringing development and providing income to the Sudanese.[19] Moreover, in 2006 it has pressed the Sudanese government to accept the UN peace-keeping force and its UN Ambassador Wang Guangya contributed to the political compromise reached in Addis Abeba. North and South will have to find a way to cooperate, even if the South would opt for secession in 2011, as the main source of revenue, oil, has to be transported north to the Red Sea by pipeline to Khartoum and the Red Sea. Unfortunately, according to some foreign observers the Sudanese oil industry is poorly supervised and highly politicized, and a source of strife and division.

3.2 *Overview of Energy Cooperation between China and Sudan*

Since the establishment of diplomatic ties, the two countries have experienced smooth development of bilateral trade. In recent years, bilateral trade increased significantly, from 1.2 billion US dollars in 2001 to 5.7 billion US dollars in 2007, of which export to China amounted to 4.1 billion US dollars. Energy cooperation, in particular, has increased sharply. In 2007, Sudan exported 10.3 million tons of crude oil to China, more than double the year before.[20]

The oil cooperation between China and Sudan was initiated by China National Petroleum Corporation (CNPC) in 1995. The first oil field began to produce oil by the end of 1999. So far, CNPC owns 40% in the now oil producing blocks 1/2,4 and 41% (and Sinopec another 6%) in 3/7, and 95% in block 6, and also 40% and 35% shares in the (so far little productive) blocks 13 and 15 near the Red Sea, under a

[18] "Darfur rebel leader plans attacks on Chinese oil firms," Sudan Tribune, Dec. 8, 2007, and "Darfur rebel JEM wants Western oil Cos to replace Chinese," Sudan Tribune, Apr. 16, 2008.

[19] He Wenping, "The Darfur issue and China's role," Symposium on Chinese-Sudanese Relations, Beijing July 26, 2007.

[20] 2007. China Customs Statistics Yearbook.

general contract term of 25 years. In 2002, the world-class Palouge oil field was discovered, a break-through in the prospecting section 7. Subsequently, more discoveries were made in sections 3/7, which started producing the paraffinic Dar Blend oil in August 2006. The typically split of profit oil (after cost recovery) is government 80%, companies 20%. Oil companies do not pay taxes in Sudan, but they face pipeline transport costs of 4 to 6 US$ per barrel and Dar Blend is traded at a deep discount because of its very high acidity and arsenic content, which makes it hard to sell and refine.

In addition, China Petroleum has invested in the Khartoum Refinery, Khartoum Chemical Plant and Petrochemical Trading Company. China Petroleum and the Sudan Ministry of Energy and Minerals each hold 50% of the Khartoum refinery (China Petroleum's share will be reduced to 10% after 2015). Its first phase construction was completed in 2000, with an annual processing capacity of 2.5 million tons. The second phase was completed and put into production on June 30, 2006. Both projects were designed by China Petroleum and mostly equipped with China-made materials. So far, it is Sudan's only large refinery, as the agreed 2 billion US$ Petronas refinery in Port Sudan (also designed for Dar Blend) has been delayed because its cost estimates have more than doubled.[21] By placing Sudanese employees at every job level in the refinery, China provides Sudan with core technologies in the refining industry, including catalyst-related technologies. In fact, there are twice as many Sudanese employees in the refinery as there are Chinese employees. A joint venture agreement was sealed with Khartoum Refinery, giving 95% share to China Petroleum and 5% to Sudan Ministry of Energy and Minerals. Additionally, a 1,540-kilometer-long oil pipeline, the second longest in Africa, was built in 1997–1999 by the consortium Greater Nile Petroleum Operating Company (GNPOC) founded by China, Malaysia and Sudan after it took over blocks 1, 2, and 4 originally exploited by Chevron and Shell.

The oil cooperation between China and Sudan covers areas including prospecting, exploitation, production, oil pipelines, refining, polypropylene, plastic processing and refined oil selling. The oil industry has become a major driving force behind Sudan's economic growth, putting momentum into the development of transportation, manufacture and construction. Under the principle of "mutual benefit and

[21] Sudan Tribune, May 28, 2008 and Jan. 14, 2009.

development," China Petroleum, while investing in the oil industry in Sudan, offers free support of funds for local agriculture, education, culture, health care, and construction of roads and bridges, which is widely welcomed by the Sudanese people.

China Petroleum also stresses the use of environmental technologies and facilities in Sudan, striving to improve local environment. In recent years, GNPOC (1/2/4 area project) has built an experimental project for sewage treatment that covers four square kilometers. The project, located in Heglig, is aimed at treating sewage by using biotechnology and vegetation. At present, the level of oily sewage disposal is below the international standard, occasionally reaching zero discharge.

China Petroleum has built Khartoum Refinery Hospital, Fula Friendship Hospital and Palouge Friendship Hospital in Sudan and donated medical equipment to Melovue Hospital, Al Zariba Hospital, Hassan Hospital and Jilie Health Center. Meanwhile, the China Petroleum Sudan Project gives out medicine to residents living on the periphery of the southern oil fields and dispatches medical staff to provide civilians and soldiers with vaccines against malaria. Thousands of people infected with malaria have been cured by using Chinese Artemisinin.

Currently, China's Sudanese oil projects employ 11,000 Sudanese employees and workers. They emphasize training, and send many Sudanese employees to China and other countries for further education, thus improving their management and technological skills. China Petroleum also invests heavily in infrastructure and transportation. In 2002, it donated Babila Airport, 1,800 meters long and 45 wide, fit for 5-ton cargo planes and passenger planes. It provided great convenience for the operation of oil field and the transportation of residents and officials travelling in and out of Khartoum. Section 6 raised funds to build a 580-kilometer paved highway and ten bridges, a stimulus to local economic growth. These measures have benefited 1.5 million local people already.

The oil industry has put Sudan on the track of economic growth and industrialization, improved infrastructure in Sudan, and boosted employment for the Sudanese people. The resulting increased wealth has furthered Sudan's political reconciliation process. Over the past few years, the Sudanese government has signed peace deals with some anti-government organizations. The Sudanese president Al-Bashir has said that it was China that had helped them discover oil, the development of which had made peace between north and south of the country possible.

3.3 *The US Factor in the Energy Relationship between China and Sudan*

The US-Sudan relation has been dominated by hostility for a long time. Dissatisfied with Sudan's Islamic regime, the United State blacklisted Sudan as a terrorist country in 1993. In 1997, the United States declared unilateral sanctions against Sudan, imposing a ban on US companies making investment in the Sudanese oil sectors. After the bombing of US embassies in Tanzania and Kenya in 1998, America condemned Sudan for its involvement and destroyed the Al Shifa Pharmaceutical Plant in Khartoum, claiming it produced chemical weapons. Meanwhile, America pulled out its oil companies in Sudan. In 2004, it defined the tribal warfare in Darfur as "genocide" and accusing the Sudanese government of ethnic massacre. The statement was an attempt to overturn Sudan's Islamic regime by instigating UN sanctions. For this reason, America views China-Sudan relations and China's investment in the Sudanese oil industry as a challenge to American interests.

Invoking the international principles of mutual respect for sovereignty and non-interference, China has advocated political solutions to the Darfur issue, while making tangible contributions to Sudan's peace process and economic growth. Peace mediation became effective after President Hu Jintao visited Sudan in 2006. With Chinese persuasion, the Sudanese government accepted the three-phase solution proposed by UN Secretary Kofi Annan, allowed in UN peacekeepers and reached a cease-fire agreement with some anti-government organizations. 300 Chinese peacekeepers have already been stationed in Darfur, and more are to come. With China's efforts, the Darfur issue has been settled gradually, the result of peaceful solutions rather than sanctions. China's achievement in Darfur has been recognized by the international community, including the United States. Where peaceful settlement prevails, cries for sanction abate. The difference between China and the United States on the Darfur issue is diminishing, as greater consensus is reached.

4. Conclusion

The oil supplies in China function as a complicated system that includes both domestic and international strategies. Domestic political strategies include energy-saving measures, energy efficiency improvement, use of alternative energy, and establishment of a strategic oil reserve.

International strategies include "going global," diversification of oil supply sources, and developing relationships with oil producers and major powers. As a new major oil importer, China has used its previous experience as an oil exporter. As a developing nation, China also learnt lessons from those countries. The result has been an oil import security policy with Chinese characteristics. Its main feature is the cultivation of good relations with oil exporting countries, based on mutual benefit and development. This is in the interest of both oil exporting and importing countries, and represents a new concept of energy security.

Appendix

Table 6.1 Oil reserves and production of Saudi Arabia and Sudan.

	Reserves (billion tons)			Production (million tons)		
	2005	2006	2007	2005	2006	2007
Saudi Arabia	36.3	36.3	36.3	526.2	514.6	493.1
Sudan	0.9	0.9	0.9	18.1	19.6	22.5

Table 6.2 China's oil imports, total and from selected regions/countries (million tons).

	Total	Saudi Arabia	Sudan	Middle East	Africa
2005	126.8	22.2	6.6	70.6	35.3
2006	145.2	23.9	4.8	77.6	42.3
2007	163.2	26.3	10.3	83.4	58.5
2008	178.8	36.4	10.5		

Sources for Table 6.1 and Table 6.2: BP Statistical Review of World Energy 2006–2008; China Customs Statistics Yearbook 2006–2008; Statistical Communiqué of the PRC on the 2008 Economic and Social Development, Feb. 26, 2009; www.china5e.com accessed March 9, 2009.

PART TWO

RENEWABLE ENERGY & SUSTAINABLE DEVELOPMENT

CHAPTER SEVEN

CHINA'S RENEWABLE ENERGY DEVELOPMENT TARGETS AND IMPLEMENTATION EFFECT ANALYSIS

Shi Dan

Abstract: *Renewable energy development is an important means for China to solve the problem of 'energy poverty' in rural areas, enhance its energy structure and increase energy supply. In order to facilitate the development of renewable energy resources, China has formulated corresponding legal measures, priority policy measures and development plans. This article summarizes China's legal and policy measures on renewable energy, and on the basis of four renewable energy development plans issued by the government it describes the government's main development targets: to eliminate energy poverty in rural areas, optimize China's energy structure and establish a renewable energy industry with independent intellectual property rights. This article also analyzes China's achievements in attaining its planned targets, problems in the implementation process and the practical contributions of renewable energy development to realizing the above-mentioned three development goals. In the conclusion, some remedial measures and suggestions are put forward.*

Keywords: renewable energy, plan, target, effect

1. Introduction

China is experiencing accelerating industrialization and urbanization, with its energy demand ever increasing. China's primary energy consumption in terms of standard coal equivalents doubled between 2000 and 2007 to 2.7 billion tons of standard coal. Coal constituted over 70% of China's total primary energy consumption while oil took up 23%. 50% of China's oil consumption relied on imports. The continuous growth of energy demand has exerted great pressure on China's energy resources, environment and energy security. According to forecasts by IEA and other institutions, by 2020 China's oil shortage will amount to approximately 8 million barrels per day, and its dependence rate on oil import is likely to exceed 70%.[1] Today China ranks first in the world

[1] Source: 2006, The Brookings Institution, Brookings Foreign Policy Studies Energy Security Series: China, p. 9.

in SO_2 and CO_2 discharge, with more than 40% of its territory suffering from acid rain. The situation will become even worse unless fossil energy production and consumption are properly controlled. In addition, a considerable proportion of China's rural population is plagued with a shortage of energy, particularly electricity, and this 'energy poverty' hinders social and economic development in rural areas.

China regards renewable energy development as an important channel to reduce energy poverty, increase energy supply and improve the energy structure and has given priority to renewable energy in its energy development plans, also by legislative means. So far, renewable energy is responsible for only a minor part of China's electricity supply: in 2007 21.6% of total electrical power generating capacity and 14.8% of electricity output. Almost all of this is hydropower. As large hydropower is able to compete in terms of price with conventional power, it is not included in preferential policies for renewable energy, but small hydropower stations (below 25kW or 50kW) are.

In order to stimulate the development of renewable energy resources, China has formulated a series of renewable energy development plans and priority policy measures. In the following, we will sort out the targets, focal points and priority policy measures of China's renewable energy development, analyze the achievability of its targets and look into some problems that have emerged during implementation. Finally, some remedial measures and suggestions for the development of renewable energy resources will be put forward. The article has four parts: 1) a summary of targets and priority policy measures of China's renewable energy development; 2) China's achievements in attaining its planned targets, and problems that emerged during implementation; 3) actual results of China's renewable energy development; and 4) conclusion and suggestions.

2. China's Renewable Energy Development Targets and Priority Policy Measures

2.1 *Development Targets and Focal Points*

To advance the development of renewable energy, China started to formulate its first renewable energy development plan in 1996, namely An Outline of New Energy and Renewable Energy Development (1996–2010). To date China has issued four plans on new and renewable energy development, including: An Outline of New Energy and

Renewable Energy Development (1995), the 10th Five-Year Plan of New Energy and Renewable Energy Development (2001), the 11th Five-Year Plan of Renewable Energy Development (2006) and A Medium and Long Term Plan of Renewable Energy Development (2003).

When the first Outline of New Energy and Renewable Energy Development was formulated, China's economic development was experiencing an extreme shortage of energy resources in its rural areas. 120 million villagers were living without electric power and more than 900 million relied on straw and firewood to meet the energy demands of their daily life. This situation led to the denudation of forests, deterioration of the vegetation cover and a degraded eco-environment, all of which seriously hindered the social and economic development of rural areas. Therefore this outline focused on solving the energy supply problem for rural areas. The guideline featured "Develop and promote the use of clean energy that suits local conditions, such as solar energy, wind energy, geothermal energy, tide energy and bio-energy". The specific development targets included planting area of firewood, bio-energy-generated methane, small hydropower stations, solar energy utilization, development and utilization of wind energy and wind powered generators, geothermal energy development.

As environmental problems and other issues become increasingly prominent after the year 2000, the Medium and Long Term Plan of Renewable Energy Development and the 11th Five-Year Plan of Renewable Energy Development gave first priority to the substitution of fossil energy with renewable energy and to an optimization of the energy structure. Wind power, solar energy and bio-energy received great attention, with the first two being used for on-grid electricity *generation*, and the third for bio-power generation and the production of biogas, biomass solid fuel, and liquid biofuel. The concrete development goal is to have renewable energy make up ten percent of the country's energy consumption by 2010 and fifteen percent by 2020. By 2020, China should have 300GW of hydropower and 30GW of wind power and biomass power each. This means that China's renewable energy development should follow the path of large-scale and industrialized development, a completely different one from previous stages. Therefore, the target of developing a renewable energy industry was put forward in the 11th five-year plan (2005–10). It aimed to drive industrialized development of new technology for renewable energy industry through large-scale investment and construction. The planned targets of technological standards and capacities of equipment

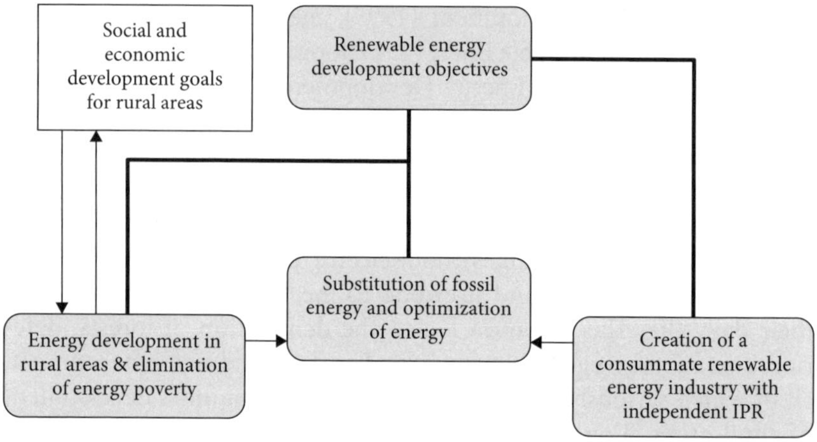

Figure 7.1 Components of China's renewable energy development objectives and correlations.

for the renewable energy industry may be summarized as follows: for the current stage, China will focus on enhancement of technological levels and establishment of a complete industrial system. The existing mature technologies should realize large-scale and modern production, and form strong production and service systems. For the decade 2010–2020, China will strengthen research and development of renewable energy technologies, increase the volume of domestically-made equipment, establish a complete renewable energy industry system and reduce the cost of renewable energy development and utilization by a wide margin.

Based on the above-mentioned four development plans, China's renewable energy development targets cover three aspects (see Figure 7.1): first, elimination of energy poverty and protection of the ecological environment; second, substitution of fossil energy and optimization of the energy structure; and third, development and improvement of the renewable energy industry with independent design capacity. Among these, elimination of energy poverty was the earliest target put forward for China's renewable energy development. This target that corresponds with China's specific situation is a fundamental component of the country's social and economic development goal for rural areas. The targets of optimization of the energy structure and creation of a renewable energy industry were brought forward more recently.

Table 7.1 Planned targets of China's renewable energy development from 2000 to 2020.

Renewable energy	2000 target	2005 target	2010 target	2020 target
Equivalent standard coal (million tons per annum)	(298)	[Traditional utilization of small hydropower and bio-energy excluded in the 10th Five-Year Plan: 13]	(390) 300, {300}	{600}
Installed capacity of hydropower generators (100 million kilowatts)			1.9 {1.9}	{3.0}
Installed capacity of small hydropower generators (10 thousand kilowatts)	(1985)		(2788) {5000}	{7500}
Installed capacity of biomass power generators (10 thousand kilowatts)	(5)		(30) 550 {550}	{3000}
Installed capacity of wind power generators (10 thousand kilowatts)	(30–40)	[On-grid 120]	(100–110) On-grid 1000, off-grid 7.5 {500}	{3000}
Installed capacity of solar power generators (10 thousand kilowatts)	Establish 9 independent PV power stations in Tibetan counties where electricity was unavailable	Cumulative possession reaches 53 million watts	30 {30}	{180}
Annual utilization of solid bio-fuel (million tons)			1 {1}	{50}

Table 7.1 (*cont.*)

Renewable energy	2000 target	2005 target	2010 target	2020 target
Annual utilization of marsh gas (100 million cubic meters)	(22.6)	[Utilization of large-sized marsh gas tanks reaches 2 billion cubic meters in the 10th Five-Year Plan]	(40) 190, of which 40 is utilization of large-sized marsh gas tanks {190}	{440}
Annual utilization of bio-diesel (million tons)			0.2 {0.2}	{2}
Annual utilization of fuel alcohol (million tons)			3 {Annual utilization of non-grain fuel alcohol 2}	{10}
Solar energy collector area (100 million square meters)	Equivalent to 1.23 million tons of standard coal	[10th Five-Year Plan: possession to reach 0.64]	1.5 {1.5 }	{3 }
Solar energy collector area in rural areas (100 million square meters)			{0.5}	{1}
Terrestrial heat utilization	(880 thousand tons of standard coal)	(Utilization of terrestrial heat reaches 20 million square meters by 2005)	30 million square meters of heat supply, 600 thousand households of hot water supply (1.51 million tons of standard coal) {4 million tons of standard coal}	{12 million tons of standard coal}

Table 7.1 (*cont.*)

Renewable energy	2000 target	2005 target	2010 target	2020 target
Hydrogen energy			(5000 cubic meters daily)	
Commercial fuel forest (10 thousand hectares)	(640)		(1340)	
Installed capacity of tidal power generators (10 thousand kilowatts)				{10}

Source: Outlines of New Energy and Renewable Energy Development (1995), 10th Five-Year Plan of New Energy and Renewable Energy Development (2001), 11th Five-Year Plan of Renewable Energy Development (2006), Medium and Long-term Plan of Renewable Energy Development (2003).

Note: Data in round brackets are from the Outlines of New Energy and Renewable Energy Development (1995), data in square brackets are from the 10th Five-Year Plan of New Energy and Renewable Energy Development (2001), data in curly brackets are from the Medium and Long-term Plan of Renewable Energy Development (2003), data with no bracket are from the 11th Five-Year Plan of Renewable Energy Development (2006).

The substitution of fossil energy and optimization of the energy structure are common targets shared by all countries and regions concerned with renewable energy development. Establishing a renewable energy industry with independent IPR is seen by China as an inevitable choice, if it wishes to protect its national interests amid the global competition in renewable energy development. Moreover, it serves as a fundamental condition to realize the other two targets. Their relationship may be illustrated as in the figure below. (An arrow denotes supportive relations.)

2.2 *Planned Targets and Adjustment*

The scale of renewable energy development, namely the specific quantitative targets, has been defined in all plans according to the development status and conditions in each phase. See table 6.1 for reference. Generally, medium and long-term plans provide guidance for the five-year development. The succeeding plan adjusts the targets set in the previous one. In table 7.1, the targets and their quantities demonstrate that China's renewable energy development is increasingly focusing on hydropower, wind power, solar energy and bio-energy, among which wind power and solar energy aim at higher targets because of their low base and faster development pace. However, the total quantitative target for 2010 has been reduced from 390 million tons of standard coal equivalents to 300 million tons.

2.3 *Legal and Policy Measures*

Compared with regular fossil energy, renewable energy has great external benefits, but higher investment cost. Some types of renewable energy, such as wind power and solar power, are poor in stability and uncompetitive in cost and price. Under the present economic and technological conditions, renewable energy cannot develop at a favorably high speed without support from government policy, which is the key factor in furthering renewable energy development.

China's rural energy development mainly depends on special projects and composes a major part of the economic development in rural areas. Funds are mainly drawn from central and local government finance. In recent years, the central government has allocated increasing amounts from its budget. Since the 1980s, China has constantly promoted the construction of rural electrification. Small hydropower stations have been built to solve the problem that power supply was unavailable to some rural areas. Since 2000, the central financial departments have set up special development funds to support rural energy projects, such as the "Brightness" project, "Bringing Electric Power to the Countryside" project, and "Substituting Small Hydropower for Firewood" project. They have also offered great support to renewable energy utilization projects in agricultural and pastoral areas, as well as to the construction of independent systems of renewable energy power generation in remote areas and on islands.

The Renewable Energy Law (referred to as "the Law" hereafter) enacted in 2006 has a great influence on China's renewable energy development. The Law establishes five systems for promoting renewable energy development, viz. the overall target setting system, the compulsory on-grid system, the pricing-by-category system, the nation-wide extra cost distribution system, and the special funding system. The Law is a brief umbrella document, which provides the provincial governments with a mandate to develop renewable energy feed in tariffs and quotas for the purchase of renewable energy within their locality. It is these detailed provincial government regulations that impact directly on renewable energy project development in China.[2]

In order to thoroughly enforce the Law, the Chinese government has enacted some supporting administrative regulations, such as the Trial Measures for the Administration of Pricing and Distribution of Expenses of Renewable Energy Power Generation (2006),[3] Measures for the Administration of Renewable Energy Power Generation (2006),[4] Opinion on Wind Power Industry Development (2006), Strengthening the Administration of Biomass Fuel Alcohol Projects, and the Development of the Industry (2006). These regulations explain and define the renewable energy administration system, on-grid pricing and distribution of expenses as well as renewable energy development in some detail. For instance, the cost sharing system requires all power consumers to pay a surcharge to support the development of renewable energy (the present surcharge is only 0.001 yuan per kWh), and regulations provide for the compulsory use of solar power in some buildings. Also, the rules on the compulsory grid connection system stipulate that all energy generated from renewable sources must be purchased and that utilities must provide grid-connection services (including construction) and related technical support. However, so far the government has not set mandatory renewable energy targets at the level of utility companies, as this is considered too complicated. Rather, obligations are set on administrative regions and generators. The separate pricing laws for different types of renewable energy use both feed-in tariffs ("government-fixed prices") and competitive tendering ("government-guided prices") to strengthen the economic position of generators of renewable energy.

In addition to the above-mentioned legal and regulatory measures, China also takes some economic stimulus measures, including (1) Tax reduction and exemption. Imports of renewable energy equipments

---

[2] Baker and MacKenzie, *Relaw Assist*: Renewable Energy Law in China—Issues Paper. June 2007.

[3] These set out the principles for renewable energy power pricing and cost sharing. In particular, they identify the level of wind and biomass power pricing and clarify all costs related to renewable energy power that will be covered by the renewable energy surcharge.

[4] These stipulate that utilities are obliged to allow renewable energy facilities to connect to the grid.

enjoy a relatively low tariff. Renewable energy enterprises also enjoy a preferential policy of half value-added tax. Small hydropower enterprises below county level pay a 6% value-added tax for their products according to simplified methods. Domestic enterprises enjoy import tax and import value-added tax rebate for key parts and raw materials they import to develop and manufacture large wind power generators that meet technological specifications. The rebated tax is mainly used for new product development, production and innovation capability construction. Moreover, China's local manufacturing industry for wind turbines is protected by a 70% local content requirement for wind turbine components. The wind turbine generators must be assembled in China, including in ordinary wind farm projects. Nevertheless, foreign manufacturers such as Gamesa and Vestas, and Nordex and GE (in joint-ventures) have been quite active in China. Renewable energy technologies such as wind power generation and PV power generation are covered in the high-tech spectrum to which the country offers great support. Eligible renewable energy enterprises benefit from the preferential policy of income tax reduction and exemption according to the law.

(2) Discount interest loan policy. The central government formulates a policy of special discount interest loans, which are mainly used for large and medium-sized marsh gas projects, solar energy utilization and wind power promotion. Enterprises will be awarded if their demonstration projects meet criteria after being put into production. The central finance arm of the government offers an interest discount of 2% to large and medium-sized renewable energy power generation projects, which are built upon approval from the government with a capacity of over 3,000 kilowatts. It also supports industrialization of large and medium-sized wind power generators. As yet it is unclear on which scale these interest subsidies have been granted. Some local governments also implement preferential policies, such as discount interest, small loans and tax reduction and exemption to rural marsh gas tank construction.

(3) Fiscal subsidy policy. Subsidies are mainly for renewable energy development, demonstration projects and bidding prices of renewable energy power. The first 50 megawatt wind power generators get a subsidy of 600 yuan per kilowatt, half of which goes to the generator manufacturing enterprise and the rest to the key part manufacturing

enterprise. The subsidy is mainly offered for new product research and development. Power grid enterprises enjoy additional subsidies for the prices of renewable energy power generation, including the part of bidding price that is higher than the bidding price of local desulfurized coal power generation, as well as the grid connection expense. Doubling the value, namely an increase of 3 billion yuan in 2008, the subsidies largely fill the gap of on-grid cost of renewable energy power generation. The subsidies are meant to make renewable energy competitive in the power market, and therefore vary with the average cost of generation of each category. In 2008, wind power (costs of which are about 0.50–0.60 yuan per kWh, twice those of thermopower) received a subsidy of 0.23 yuan per kWh. Biofuel received the (legally stipulated) subsidy of 0.25 yuan per kWh (from 2010 this subsidy will be progressively decreased). Solar power received a subsidy of 3 yuan per kWh, based on a cost-plus principle. Over the period of October 2007–June 2008, renewable energy subsidies totaled 20 billion yuan, most of which went to wind power.[5] Also, the newly introduced fossil energy taxes, and less subsidization of coal for power stations have helped to reduce their cost advantage over renewable energy. Of course, the extreme price changes in commercial oil and coal over the past year have increased the difficulties of setting long-term competitive prices in the renewable energy sector.

(4) Direct investment by national finance. The national finance arm of the government sets up special funds for renewable energy development and supports the following programs: sci-tech research on renewable energy development and utilization, standard formulation and demonstration projects; renewable energy utilization projects in rural and pastoral areas; independent renewable energy power generation system and electricity extension in remote areas and on islands; exploration, evaluation and relative database construction of renewable resources; and localized production of equipment for renewable energy development and utilization, resource surveys and the establishment of technical standards. The Ministry of Finance, together with other relevant authorities, is currently formulating regulations to guide the management of the cost sharing fund. As with the special funds themselves, these regulations are specific to particular renewable

[5] www.china5e.com.2008-11-06.

energy technologies and applications. By heavily investing in the expansion and upgrading of national and regional power grids, the central government and power grid companies have considerably improved the access of small renewable energy generators to power grids and thereby contributed to their economic viability.

Based on the sphere of influence, China's policies and measures on renewable energy development can be summarized in four aspects. First are the fundamentals of renewable energy development, e.g., technological research and development, trial projects, resource evaluation and formulation of technological standards. Second is rural energy construction that aims to solve the problem of unavailable energy supply in rural and remote areas. Third, the effort to increase and ensure the share of renewable energy in the current market. Fourth, the promotion of technological progress and building of an industrial system.

3. Achievement of Planned Targets and Existing Problems

In recent years, China has experienced rapid development in renewable energy. In light of the achievements in 2007, which fell somewhat short of planned targets, China must keep up a high-speed development in order to realize the targets planned for 2010 (See table 7.2 below).

China's renewable energy development encounters major problems:

- rapid growth rate of installed capacity, but incomplete utilization
- interested parties receive no compensation for losses incurred from using renewable energy
- blind expansion and inadequate spatial planning
- relatively high production cost
- and irrational competition between state-owned enterprised (SOE) and private companies

3.1 *Rapid Growth Rate of Installed Capacity, but Incomplete Utilization*

Since 2000, China's installed capacity of wind power generation has been increasing at high speed and now its scale ranks 5th in the world. According to statistics from the Chinese Wind Energy Association,

Table 7.2 Actual accomplishments vs. planned targets of China's renewable energy development.

	2007 actual accomplishment	2010 maximum planning target	Actual/Plan	2007–2010 annual average growth (%)
Utilization of renewable energy (Mtce)	193.8	300	64.3%	15.7%
Hydropower (100 million kilowatts)	1.47	1.9	77.4%	8.9%
On-grid wind power generation (10 thousand kilowatts)	605	1000	60.5%	18.2%
Biomass power generation (10 thousand kilowatts)	260	550	47.2%	28.4%
Utilization of marsh gas (100 million cubic meters)	113.8	190	60%	18.6%
Utilization of solar energy (collector area 100 million square meters)	1.1	1.5	73.3%	10.9%
Annual utilization of bio-diesel (million tons)	0.2	0.2	100%	–

China has built more than a hundred wind power plants.[6] Over 350,000 small off-grid wind power generators supply electricity to households in remote areas. However, the growth rate of China's power generation fails to be in step with the growth rate of its installed capacity.

[6] China Industry Research Information Net, 2007–2008 Annual Research Report on China's Wind Power Industry, http://www.36021.cn.

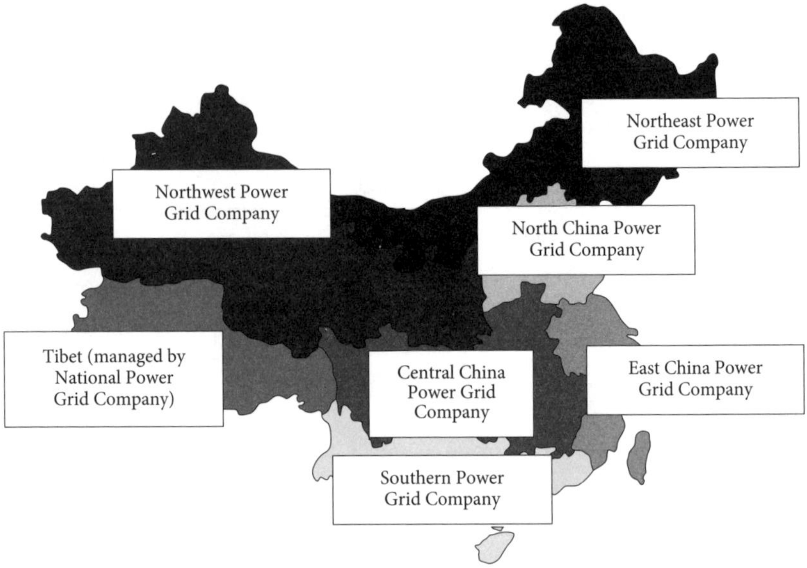

Source: Baker and Mackenzie, 2007.

Figure 7.2 China's seven main power grids after the 2002 break-up of the central government monopoly.

Some wind power generators have not started operation because of the electricity price and the power plant design. In 2006, China's installed capacity of on-grid wind power generation reached 2.5933 million kilowatts, but the on-grid electricity was only about 3.8 billion kilowatt-hour. That means that on average, the wind power generators ran only 1,465 hours, much less than the design requirement of over 2,000 hours per year. Among the 1.26 million kilowatts of installed capacity of wind power generation by the end of 2005, at least 25,000 kilowatts[7] have not started work due to the quality of generators. Small hydropower stations, many of which were constructed decades ago, also encounter such problems. Due to old equipment and lack of follow-up technological services and maintenance, a number of small hydropower plants are unable to continue power generation after running for years.

[7] Wang Zhongying, Li Junfeng, 2007 Annual Report on China's Renewable Energy Industry, Chemical Industry Press, p. 79.

3.2 *Interested Parties Receive No Compensation for Losses Incurred from Using Renewable Energy*

A major problem is that China's renewable energy policies only define legal liabilities of interested parties for renewable energy development. Moreover, the economic stimulus focuses on renewable energy producers, but other interested parties do not receive compensation for their losses from purchasing or utilizing renewable energy. For example, as a kind of intermittent energy, wind power reduces the accuracy of grid load forecast, and consequently negatively affects the dispatching and operating model, frequency control, voltage adjustment, power flow distribution, power quality, failure, stability, and operating cost. Therefore, wind power plants and power grids face conflicts of interest. The Law has established a system of full purchase of on-grid wind power generation, but does not define the entitlement to economic compensation for impacts and losses that power grids may suffer due to the above-mentioned problems. As a result, power grid enterprises show little enthusiasm about purchasing wind power, and they absorb wind power in a limited way, on the condition that it poses no harm to their regular business.

For these reasons, some power grid companies have been unwilling to contribute financially to connecting RE projects to their grid, or make the additional investments in creating a higher peak load capacity needed to accommodate the variable amounts of wind or hydropower. Also, they may charge electricity transport losses to the RE producers and refuse to offer a fair price (or the state-set price) to small hydropower stations (even if government regulations stipulate that RE projects must include construction of feed-in-power lines, the responsibility for constructing them is unclear, as it is shared by government, producer and grid operator). While the large power companies tend to follow the rules, local power grid companies sometimes refuse to conclude proper contracts with suppliers, or put in very unfavorable conditions, for example, concerning the supply of excess electricity. Also, there may be inadequate registration of the electricity supplied.[8]

[8] State Power Regulatory Committee, Inspection Report on the Implementation Situation of the Purchase of Electricity from Renewable Energy and the Electricity Price Policy (in Chinese). January 21, 2009, taken from www.China5e.com.

According to foreign observers, gaming of the tendering process through the submission of artificially low bids has been common, leading to extremely low prices (e.g. 3.8 euro cents per kWh). Low or non-existent projected returns have consequently led to a low contract implementation rate, hampering industry development and reducing economic development through taxation revenue. Where projects have been implemented, low prices have led to the use of low quality equipment during construction, which increases operational risks.[9]

3.3 *Blind Expansion and Inadequate Spatial Planning*

Many local authorities have pushed projects without considering overall plans. Compared with other types of renewable energy, China's biomass power generation grows at a low rate. In 2007, China's installed capacity of biomass power generation reached approximately 2.6 million kW and power generation was 6.24 billion kWh, with the generators running only 2,400 hours in a year. It is primarily due to the fact that some projects have an excess scale and encounter a shortage of raw materials, such as stalks, within a reasonable transport radius, because too many producers are concentrated in one region, as, for example, in Jiangsu. A similar lack of proper spatial planning may be noted in wind power projects, which by their very nature are concentrated in wind-rich regions such as Inner Mongolia and the coast. This has caused safety problems for the regional power grids. That also goes for small hydropower stations in some regions.[10]

In contrast, China's biomass liquid fuel grew by leaps and bounds since it began in 2003 and reached a top in 2005. Subsequently, the industry has plunged (See figure 7.2). The main cause for that was the rise in grain prices. In as far as it depended on grain (particularly maize), biomass liquid fuel contributed to tight grain supplies and price hikes. Thus, China decided that expansion of grain-based biomass fuel production is not suitable. Development of liquid fuel production with wood requires forest areas, which in turn will affect the grain area and production. China's biomass liquid fuel development should find a model that does not impact grain cultivation areas.

[9] Baker and MacKenzie, op. cit.

[10] State Power Regulatory Committee, Inspection Report on the Implementation Situation of the Purchase of Electricity from Renewable Energy and the Electricity Price Policy (in Chinese). January 21, 2009, taken from www.China5e.com.

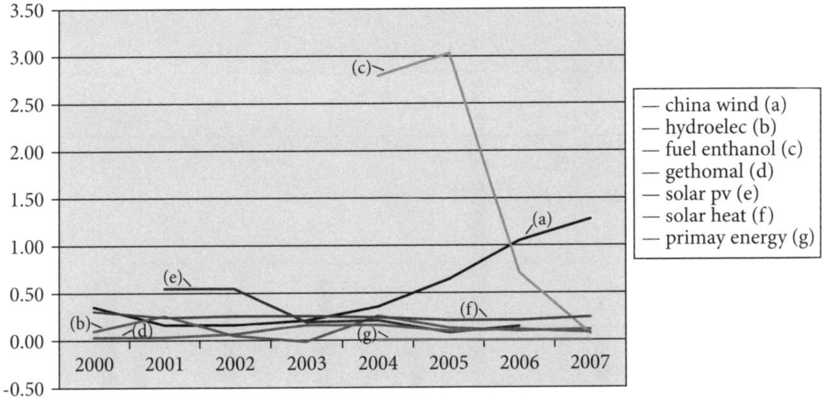

Source, based on: China New Energy Chamber of Commerce '2007 Annual Report on New Energy Industry' and China Industry Research Information Net, 2007–2008 Annual Research Report on China's Wind Power Industry, http://www.36021.cn.

Figure 7.3 Development of different renewable energy categories in China.

However, there is no answer to the question of how much land can be used to plant energy trees, since such land is still waiting for reasonable evaluation and scientific planning. In addition, the technology of bio-diesel production with oil plants is in the research and experimental phase, therefore it cannot be used for mass production until industrial experiments are completed. The technology of fuel alcohol production from cellulose and bio-diesel production that requires relatively abundant resource reserves is also at an early stage, far from industrialization.

3.4 *Relatively High Production Cost*

Although China's solar PV industry has developed at a rapid speed in recent years, the country has yet to master the technology of solar-grade silicon production. As the raw material is in tight supply, the domestic price of solar grade silicon is 2 to 2.5 times of the international level; therefore it is difficult to reduce the cost of solar PV power generation. The current cost of polycrystalline silicon, which leads the solar battery industry chain, accounts for 70% of the total cost of solar power generation. In recent years, taking advantage of its labor force, China has become the world's PV battery processing and assembling base. Its capacity of high-purity silicon, silicon ingots, silicon wafers, battery and components production has reached 25 million watts,

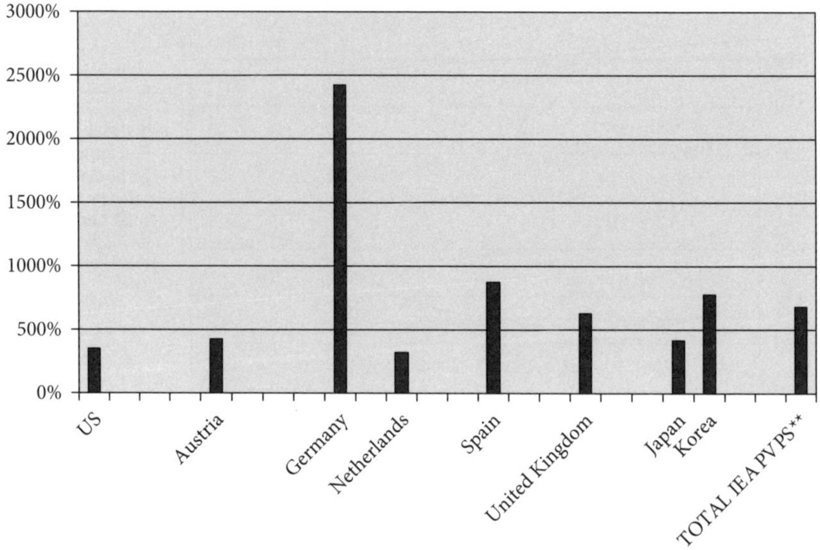

Source: Chinese Renewable Energy Industry Association under China Association of Resource Comprehensive Utilization, Greenpeace, European Photovoltaic Industry Association and World Wide Fund; 2007, "Report on China's Photovoltaic Development".

Figure 7.4 Growth of solar power generation (2000–2007).

580 million watts, 500 million watts, 1.4 billion watts and 1.087 billion watts respectively, ranking 3rd in the world. Nevertheless, as China's domestic demand in PV power lags behind severely, 96% of its PV products are for export. The major market of China's solar PV power industry is not the country itself, but developed nations.

Currently, rural off-grid power generation and telecommunication power generation respectively take up 41% and 34% of the installed capacity of PV power generation. PV power generation in rural areas mainly relies on government subsidies. Such a structure restricts the development of the PV power market to some extent. The growth rate in recent years is lower than the one in some developed countries (see Figure 7.4). On-grid power generation takes up a share of only five percent of the PV power market.

3.5 *Irrational Competition between State-Owned Enterprises (SOE) and Private Companies*

China implements concession bidding for wind power projects. Concession bidding is an admission license and only those who win the bid can take a slice of the wind power pie. As a result, state-owned power grids offer a below-cost bidding price in order to seize the market. SOEs are much more sensitive to scale than income, and defeat non-state-owned enterprises that must be concerned about investment risks. Therefore, SOEs are the major participants in wind power generation. The fact that SOEs lack risk consciousness eventually will affect the efficiency of wind power development, and counteracts cost reduction through technological progress.

4. Actual Effects of China's Renewable Energy Development

4.1 *Effects on Poverty Alleviation in Rural Areas*

Renewable energy development proceeds smoothly in China's rural areas. In particular, it has achieved remarkable success in reducing the number of counties where electric power is unavailable and has also publicized electricity services. By the end of 2000, the number of counties without electricity supply decreased to 3, the number of such townships reduced to 766, the number of such villages declined to 16,509, the number of such households dropped to 7.06 million, and the population with no electric power supply fell to about 30 million. By 2005, electric power was available to all counties and the population without power supply was reduced to 20 million. By the end of 2005, China has built independent PV power plants in 12 counties and more than 700 towns, promoting over 500,000 sets of PV systems for household use. According to the country's renewable energy development plan, the population without power supply will decrease to 11.5 million by 2010 and power supply will be available for all people by 2015.[11]

By the end of 2007, marsh gas tanks for household use numbered 25 million, producing 10 billion cubic meters of marsh gas. Large

[11] CHANGE Serve agriculture, rural areas and farmers, balance urban and rural development, The Economic Daily, Oct. 12, 2007.

and medium-sized marsh gas projects that use livestock and poultry manure and industrial liquid waste numbered over 3,800, producing approximately 1.5 billion cubic meters of marsh gas and providing high-quality living fuel for more than 80 million rural people. According to the achievements of two five-year plans, biomass marsh gas production was 54% higher than the planning target by 2000, and was 100% higher than the target by 2005. By 2007, total marsh gas production accounted for 60% of the target planned for 2010. However, as China has a large rural population, those who are using high-quality household fuel take up less than ten percent of the total population. It is planned that 30% of rural households will have access to clean renewable energy and approximately 40 million households (about 160 million people) will mainly use marsh gas as their household fuel by 2010. Ten years later, the two figures will respectively reach 70% and 80 million (about 300 million people). Based on the actual accomplishment of biomass marsh gas production and the conditions for bio-energy development in rural areas, it is absolutely feasible to achieve the planned targets for 2010 and 2020.

China has achieved great success in alleviating energy poverty, mainly because its rural energy development relies on government investments and subsidies. The government, consumers and energy producers share common interests. Nevertheless, China has a large rural population. Based on an urbanization rate of 60%, clean energy will still be unavailable for over 120 million rural people by 2020. Therefore, China should make greater efforts to speed up rural energy development.

4.2 *Effects on Energy Structure Improvement*

Some developed countries improve their energy structure through increasing consumption of renewable energy and decreasing consumption of fossil energy. In order to improve its energy structure, China has to develop renewable energy at a higher speed because of its rising fossil energy consumption. For example, renewable energy accounted for 8.1% of China's total primary energy consumption in 2006; the figure dropped to 6.4% in 2007, because fossil energy consumption grew at a faster pace. Based on China's renewable energy development plan, the renewable energy consumption will reach 300 million and 600 million tons of standard coal equivalents by 2010 and 2020 respectively, with the total energy consumption approaching 3 billion and 4 billion

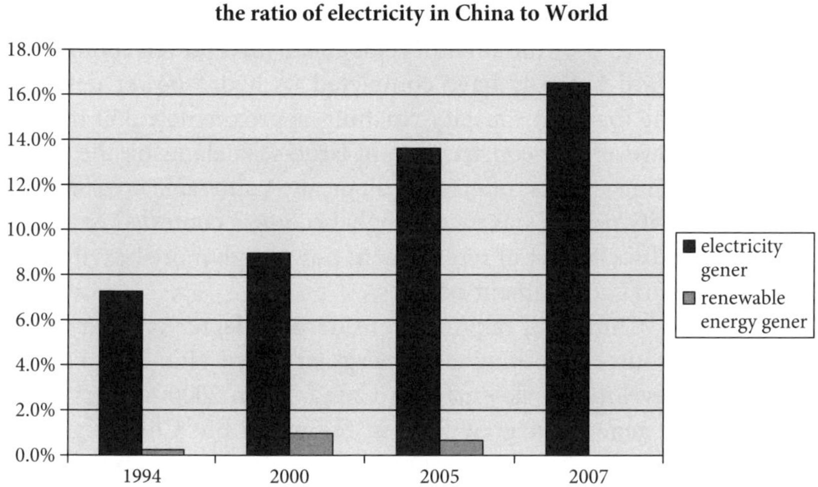

Source: State Power Regulatory Committee, Inspection Report on the Implementation Situation of the Purchase of Electricity from Renewable Energy and the Electricity Price Policy (in Chinese). January 21, 2009, taken from www.china5e.com.

Figure 7.5 Percentage of power generation & non-hydro renewable energy power generation in China to world.

tons of standard coal. Fossil energy consumption will increase by 700 million tons of sce, at a higher rate than renewable energy.

Compared to the rest of the world, China plays a leading role in the renewable energy development scale. Its solar thermal energy utilization, solar PV power generation and wind power generation, for instance, rank 1st, 4th and 5th in the world (see table 6.2). However, fossil energy takes up over 90% of China's total energy consumption and is mainly used for industrial production. Aiming at replacing fossil energy with renewable energy, China must utilize renewable energy in the large-scale industrial production area. This will require on-grid renewable energy power generation (wind power, solar PV power and biomass power generation) and biomass liquid fuel that can replace petroleum.

At present, among all types of renewable energy only hydropower has realized large-scale utilization. In 2007, hydropower generation reached 161 million tons of standard coal, accounting for 83% of total renewable energy. It is predicted that China's installed capacity of hydropower generators will reach 190 million kilowatts by 2010, including 50 million kilowatts of small hydropower. The country's installed capacity of hydropower generators will achieve 300 million kilowatts by 2020, including 75 million kilowatts of small hydropower, accounting for 75% of the total developable hydro-power volume. By 2030, China will basically have completed its hydro-power development, with the installed capacity reaching approximately 350 million kilowatts. However, the construction of large-scale dams on the upper reaches of China's rivers, often in ecologically vulnerable areas inhabited by minority people, has increasingly become a contested issue.The economic and social cost of resettlement now weigh more heavily than before in central government policies.

After 2030, China will rely to a considerable degree on non-hydro renewable resources to optimize its energy structure. Non-hydro power generation develop at a slow pace in China. From 2000 to 2005, non-hydro power generation grew 52% in the world but China's growth rate was only 3.5%. Non-hydro power generation takes up less than 1% of China's total power generation and the percentage was even lower in 2000 than in 2005 (see figure 7.5). In 2005, China's power generation accounted for over 13% of the world's total, but the percentage of non-hydro power generation was below 1%. Renewable energy development plays a limited role in optimizing China's energy structure.

4.3 *Bolstering Effects of Renewable Energy Industry*

The renewable energy industry boosts renewable energy utilization mainly in two aspects. First, it meets the market demand for technological equipment for the production of renewable energy; second, it reduces the cost of renewable energy utilization.

Solar heater industry leads China's renewable energy industry in its scale. By the end of 2007, China's annual production capacity of solar heaters has reached 15 million square meters. The country has established more than 1,000 solar heater enterprises, mainly composed of private firms. The industry has achieved an annual production value of nearly 12 billion yuan and formed a complete system, absorbing more than 200,000 employees. China-made solar heaters have their own intellectual property and production technologies. The industry also develops on a basis of market competition and the cost of products is low, largely stimulating China's solar heat utilization.

China's wind power industry has developed at a fast pace since 2003. Through the end of 2007, the number of wind power generators and parts manufacturers has exceeded 40. China's wind power generators and equipments manufacturing capacity has enhanced considerably. It has established the largest small wind power generator industry and market in the world. It exports a series of mature small wind power generators with the single system capacity ranging from 100 watts to 10 kilowatts to more than 20 countries and regions, with export volume reaching one sixth of the total production. According to statistics from China Wind Energy Association, domestic enterprises took up only 18% of the entire wind power generator market in 2004; 82% was partitioned by foreign-funded companies, most of which were from Europe, such as Denmark, Germany and Spain. In 2006, China's wind power generator market witnessed a big shift, when the market share of domestic enterprises reached 41% and that of foreign-funded enterprises dropped to 55%, while joint ventures held the remaining market share.

China's wind power generator market has experienced changes mainly because the government enforces strict rules on the percentage of domestic-made equipments in wind power plants. On the other hand, China has made great achievements in small wind power generator manufacturing technology. However, it still lags behind in large wind power generator manufacturing technology and lacks core technologies. Key parts for large wind power generators still depend on imports. Currently, China's wind power generators mainly have a capacity of below 850 kilowatts. In 2006, wind power generators with a capacity of below 850 kilowatts accounted for 79% of total. China's wind power generator industry cannot meet the demand for large-scale wind power plant construction.

In recent years, developed countries promote solar PV industry driven by PV power development. China has become the world's PV battery processing and assembling base thanks to its advantage in labor force. However, China has not mastered the solar-grade silicon production technology. As raw materials are in tight supply, the price of China's solar-grade silicon materials is 2 to 2.5 times of the international level. Therefore, it is difficult to reduce the cost of solar PV power generation. Chinese polycrystalline silicon enterprises generally take the improved Siemens model, with which 1 kilowatt solar battery costs over 10 kilograms of polycrystalline silicon and consumes a large amount of 5,800 to 6,000 kilowatt-hours of electricity. The president of Himin Solar Energy Group has stated that "over 90% of raw materials and equipments of the solar energy industry are imported and over 90% of products are for exports. Our job is to weld the silicon wafers. China becomes another processing base in the world. Solar energy is still a low-grade processing industry, similar to leather shoe and tie production." Hence, China is in urgent need to improve the technological contents of solar PV power industry.

5. Conclusion and Suggestions

China's renewable energy policies have achieved a great success in facilitating rural energy construction, increasing electric power supply and alleviating energy poverty. As it has a large population suffering from energy poverty, China needs to increase investment in rural energy construction and speed up renewable energy cleansing and utilization.

China encourages renewable energy development via interest-driven measures, i.e., fueling renewable energy development through preferential pricing, tax reduction and exemption, subsidy and the like. But renewable energy development and utilization involve the interests of various parties. During the policy design process, the interests of consumers, producers and transmitters should be taken into consideration, and compensation should be given for losses.

Poor market affordability has become one of the greatest obstacles facing China's renewable energy utilization. In order to substitute fossil energy on a large scale, China (as other countries) must reduce the cost and price of renewable energy production. Even so, for technological and economic reasons, renewable energy will have only a limited effect on the improvement of China's energy structure in the next two decades. With low-cost coal serving as the main energy source, the replacement of fossil energy with renewable energy faces a big challenge in economic cost. Consequently, China should focus on developing low-cost, commercially viable renewable energy products and industries.

China's renewable energy industry is still at a starting phase with insufficient core technologies and market share. The nation should invest more in researching renewable energy technologies, accelerate technological progress and innovation, and improve the technological content. As soon as possible, it should also establish a reasonable market competition mechanism for those renewable energy enterprises that already meet the conditions of commercial development. This would prevent excessive or inadequate competition from having negative impacts on renewable energy development. Blind development should also be curbed, in order to be able to fully utilize completed projects.

CHAPTER EIGHT

CHINA'S ENERGY SECURITY: INCREASING DEPENDENCE ON FOREIGN OIL AND SOLUTIONS FAVORED BY BEIJING STUDENTS AND RESEARCHERS

Eduard B. Vermeer*

Abstract: *Driven by a rapid increase of energy consumption, the Chinese government favors an all-out approach to energy security, under which as many as possible sources of energy are tapped simultaneously. Choices for particular sources of energy and technologies do not reflect much coordination, and the absence of public debate makes it difficult to understand the processes of decision-making. That goes even more for China's foreign energy policy. The first part of this article presents an analysis of major Chinese achievements and perceived problems. An opinion survey of advanced students and researchers was conducted in Beijing in early 2008, in order to gain more insight into Chinese attitudes towards domestic and foreign energy issues, and particularly energy supply security. We found significant attitudinal differences between men and women, and between the younger and older age group. Reluctantly, the Chinese have become more willing to accept high levels of foreign dependency as inevitable. This conflicts with the general nationalist attitude, even stronger among the populace than in government circles, which feels that China should be self-reliant and foreign countries and markets in general cannot be trusted.*

Keywords: energy security, China's energy policy, opinion survey, China's oil imports, foreign dependency, environmental concerns.

1. China's Energy Problem: Rapid Growth of Energy Consumption, Weak Policy Implementation, and Need for More Coordination

An increasing dependency on foreign oil imports, high and volatile energy prices, concerns about environmental pollution and global warming, and serious disruptions of the electricity network have put energy security high on the agenda of China's policy makers. The production and use of energy have become a hot topic again. This is not

* I wish to thank the University of Turku, Finland, for providing the time needed for this research.

so much because of a domestic supply shortage of electricity this time, as China's generating capacity has been expanded greatly in the past few years and caught up with China's economic growth. This was quite an achievement, as since 2000, growth was faster than expected and driven mainly by energy-intensive industries, resulting in an energy elasticity of above one until the adoption of stricter energy-saving measures in 2006.[1] Confusion about actual energy consumption because of faulty figures on coal output (China's main energy source) lasted until 2003, when the official figures were finally revised.[2]

The present IEA predictions for China's oil consumption in the next decades may be too low.[3] The OPEC secretariat predicts China's demand will grow from 6.5 million barrels a day in 2005 to 16.4 m. b/d in 2030, which means that 30 percent of world demand growth will be China's.[4] Chinese and Western expert predictions for China should not be accepted too quickly, bearing in mind China's still abnormally high rate of economic growth, capacity for state interventionist policies, possibility of new oil and gas finds in China,[5] and the general uncertainty about world oil prices and substitution of coal for oil.

---

[1] This rapid rise in energy consumption was unexpected: in the late nineties faulty predictions on the basis of understated coal production figures and unrealistic assumptions about technological improvement and substitution of coal by oil led China's Ministry of Energy (later abolished) to believe that under a Business-As-Usual scenario with 7.5% annual GDP growth between 1996 and 2005 its coal consumption would remain constant at 1,360 Mt. In reality, it doubled! See: Yanjiu ketizu (Research group), *Zhongguo nengyuan zhanlüe yanjiu (2000–2050) (Study of China's Energy Strategy (2000–2050).* China Power Publishing House, 1997.

[2] The State Statistical Bureau reported an annual decrease of energy production of 3.68% and therefore negative energy elasticity of minus 0.445 for the 1996–2000 period, and an annual increase of energy production of 14.43% and an energy elasticity of 1.76 for the years 2001–2003. Subsequently, production and elasticity figures have been smoothed out.

[3] This conclusion was drawn on the basis of a minimum statistical (lower-bound) annual oil consumption for developed countries of 11 barrels per capita. Willem P. Nel and Christopher J. Cooper, "A critical review of IEA's oil demand forecast for China." *Energy Policy* 2008:1.

[4] *Oil outlook to 2030: OPEC Secretariat background paper*, 11th International Energy Forum, Rome 20–22 April 2008. Table 2. 2008 consumption of crude and refined oil increased by 5.6 and 11.9 percent, respectively.

[5] China's oil reserve-to-production ratio is only 13 per year as against a world average of 40. After a decade of disappointing new oil finds, in 2007 PetroChina made a large discovery in the Bohai Bay, the Nanpu oil fields with total reserves of 1.18 billion tons of oil, "a shot in the arm for both the company and China's energy security." It increases the likelihood of further discoveries. Chinaview.cn 2007–06–28 and 2007–08–15; http://gasandoil.com/goc/news/nts72406.htm. However, only 445 million tons were certified by the ministry as economic reserves. China's Ministry of Land and

Before the global economic crisis, there were no signs that the energy intensity of China was reduced much, in spite of official policies that China should do its utmost to achieve a target of 20% energy reduction per unit GDP between 2006 and 2010. In 2006 and 2007, energy consumption increased by 9.6% and 7.8%, respectively, and energy intensity dropped slightly. Since September 2008, a shift away from heavy industry, reduction of inventories, and a downward trend in manufacturing all contributed to a fall in China's energy consumption even as its GDP continued to grow. Thus, the target may be reached after all.[6] Most likely, this is a temporary phenomenon. Key drivers have remained: rapid urbanization, rising incomes (resulting a.o., in a fast growth of private car ownership) and China's development into a global manufacturing base. Future energy consumption (the IEA projected an annual 6.4% energy consumption growth for China between 2000 and 2020) has become rather uncertain.

China's energy use is unsustainable. A joint study by the World Bank and the Development Research Centre of China's State Council, completed in June 2006, made a number of recommendations. It advocated institutional strengthening, investment in new coal technologies and energy efficiency, free market prices for electricity and refined oil, and greater use of the tax instrument. So far, some recommendations have been followed, and others have not. That may be attributed in part to their findings that China's energy policy making has been uncoordinated, and regulation is weak, mainly because of understaffing and underfunding of the governments' energy and environmental institutions.[7] Chinese opinions are divided on whether establishing a Ministry of Energy would help. It might help push and coordinate energy savings, but also come into conflict with the powerful Development and Reform Commission (NDRC). Considering the mounting importance of environmental issues, a Ministry of

Resources expects to discover ten oil fields larger than 100 m.t. and 8 to 10 gas fields larger than 100bn. cu.m. in 2008–2010, increasing proven oil reserves by 4.5 to 5 bn. tons and gas reserves by 25 to 28 bn. cu.m. according to a report presented in April 2008, *Xinhuanet* 2008–04–03.

[6] However, in 2008, crude and refined oil imports were 180 m.t. and 39 m.t., respectively, up 9.6% and 15%, Chinese Customs, Jan. 22, 2009 and in 2009, imports of oil, gas and coal have continued to increase.

[7] Noureddine Berrah et al., *Sustainable Energy in China: The Closing Window of Opportunity*. Washington: The World Bank 2007. State-set prices were moved closer to market prices and an energy tax was introduced on Jan. 1, 2009.

Energy *and* Environment might be needed.[8] The solution adopted in June 2008 has been to raise the status of the Energy Bureau under the NDRC and increase its authority in price determination, supervision of oil reserves, and coordination of energy investments, including those in foreign countries. Its new name is the National Energy Administration.[9]

The State Council issued a White Paper on energy use in December 2007. It declares China's fundamental energy strategy to be based on thrift, reliance on domestic energy sources, diversification, S&T, environmental protection, and international cooperation. It highlights China's hidden potential in coal and hydropower ("only 13% and 20%, respectively, of which have been developed"), but also in oil and gas ("of which only 33% and 14%, respectively, have been discovered"), and sets an ambitious goal for renewables (10% of total primary energy production in 2010, 15% in 2020).[10] However, it does not seriously consider measures to achieve fundamental energy use reductions that would result from changes in lifestyle, a choice for urban mass transit systems rather than automobiles, reduction of energy-intensive industrial production in favour of greater imports, and subsidization of investments in energy-efficient industrial processes, buildings, and heating systems. One should realize that the state no longer has the legal and financial instruments to direct investments, output, and exports of most energy-intensive products.[11] A draft new Energy Law, still under discussion, may provide more regulatory powers.

However, under the 2006 Renewable Energy Law the Chinese government has started to provide support for wind, solar, hydro, and biomass energy in the form of tax breaks, loan subsidies, and special

---

[8] Han Xiaoping, *Shouxi zixun guan* 2008–03–16, www.china5e.com/dissertation/zonghe/20080316093631.html. The present State Environmental Protection Agency (SEPA) lacks teeth.

[9] Zhang Libin and Jason Lee, "Untangling China's Energy Policy." *China Security* 2008 (3)58–62.

[10] Guowuyuan (State Council), "Zhongguodi nengyuan qingkuang yu zhengce." (China's energy situation and policy) *Shuili xinxiwang (Water Conservancy News Web)* 2007–12–28.

[11] By far the most energy-consuming export is steel. In spite of promises by the Iron and Steel Association, closure of some outdated less-efficient plants, abolition of export VAT rebate and imposition of an export tax on steel, in 2007 steel exports jumped by 46% to 63 million tons. *Xinhua* March 6, 2008. In 2008, iron ore imports increased by 16% to 440 m.t., while imports of steel dropped by 9% to 15 m.t. Steel exports dropped 5% to 59 m.t. *Chinese Customs data (in Chinese)*, released on January 22, 2009.

funding.[12] Half a year earlier, a NDRC paper on global warming mitigation measures had summed up China's priorities in power sector technologies: cleaner coal, hydro-power, and nuclear.[13] Earlier experiences have shown that Chinese policy planning for renewable energy tends to be overly ambitious, and that actual implementation has lagged far behind.[14] (As a matter of fact, the same could be said for many other countries.) The present contribution of alternative sources of energy, such as wind and solar, to China's energy security is minimal. A negative aspect is that, by dominating public opinion, they may reduce political willingness to invest in improvement of conventional coal technologies.

Thus, foreign and Chinese energy analysts are confronted with a complex Chinese situation.[15] Future demand is highly uncertain, because of abnormally high rates of growth. Policy messages are confused and even contradictory. Different departmental views (China has at least a dozen that are involved in energy issues) and clashes of interest on vital issues such as energy pricing, hydropower development, support for cleaner coal technologies, and environmental constraints are passed over by central government without making firm choices.[16] Let us give four examples. 1) Some experiments in coal liquefaction have received central government approval, but others have not.[17] There are many doubts about 'oil from coal:' the demand on

[12] See *The China Business Review* July–August 2006, p. 40 and William Brent, "Cleantech Boom…or Bust?" *The China Business Review.* July–August 2007, pp. 32–35, 54.

[13] Priority was given to technologies of: high-efficiency, low-pollution coal-burning power generation, large hydropower generation units, new generation nuclear plants, renewable energy, building energy conservation, clean fuel vehicle, hybrid vehicles, urban rail-based traffic, fuel cell, and hydrogen. NDRC, *China's National Climate Change Programme*, June 2007, pp. 60–61. China operates 11 nuclear power stations. According to the China National Nuclear Corporation, newly added proven uranium reserves in China are larger than demand. Sinohydro signed a US$140 million uranium mining contract in Niger. *Chinaview.cn* 2008–02–27.

[14] Zhang Zhengming, Dr. Jan Hamrin et al., *Renewable Energy Development in China: The Potential and the Challenges*. Center for Resource Solutions, 2000, p. 4.

[15] The best recent study of China's oil policies and policymaking is Erica Downs, *China*, in the Energy Security Series of the Brookings Institution. December 2006.

[16] The NDRC, State Energy Leading Group, Electricity Regulatory Commission, Water Conservancy Department, State Land Resources Commission, Ministry of S&T, State Environmental Protection Administration, and Ministries of Railways, Agriculture, Communications, and Commerce are most involved.

[17] China's largest coal company Shenhua is building a risky 1.5 billion US$ experimental plant in Erdos that will produce 20,000 barrels of synthetic oil per day. It is based on direct liquefaction that bypasses the syngas step (the Bergius process), a

scarce local water supply, risks in production technology, market risks, the huge investment needs (much of which for imports), and its profitability.[18] The NDRC has acted as a brake on provincial and company plans, for good reasons.[19] 2) Coal bed methane (CBM) was planned to reach an utilization rate of 36 percent in 2006, but only 23 percent was achieved because of lack of supporting investment policies, difficult implementation of state regulations on the joint development of coal and CBM (partly because most mining rights to CBM and coal are held by separate companies), and because prices for feeding into the electricity network are very low.[20] 3) In 2002, clear targets and locations were decided for building up a strategic oil reserve,[21] but from 2007 high oil prices upset the purchasing schedule and the role of private enterprises is undecided. 4) There have been a firm commitment and program to equip thermal power plants with desulphurization equipment (emissions of SO_2 should be reduced from 13 to 5 million tons between 2005 and 2010), but many domestically produced installations are hastily assembled, cheaply made, defective, and not

very complicated process Shenhua has developed further itself on a small-scale in Shanghai. www.technologyreview.com/Energy/17963. A number of smaller liquefaction plants proposed by provinces have been forbidden in the past few years. In September 2008, the NDRC ordered all CTL projects except two involving the Shenhua Group to be stopped, on the grounds of business risks (as coal prices remained high while oil prices went down). Another problem is the emission of CO_2, *China Daily* Sept. 19, 2008.

[18] Gong Guojun, "Five Views on Oil from Coal." *Zhongguo shiyou jingji (China Oil Economy)* 2007(7):20–21.

[19] Liquefaction of coal to produce oil produces 6 tons of CO_2 and requires 7 to 13 tons of water per ton of oil produced. Not considering these costs, at a coal price of 520 Yuan/ton liquefaction becomes economical only at an oil price of 74 US$ per barrel. A NDRC analyst concluded that liquefaction does not fit with the trend for renewables and alternative energy, and that strict environmental standards should be maintained, while continuing the demonstration projects at a secure pace. Ma Mingjie, "Gao youjiage xiadi meizhiyou zhengce quxiang" (Policy tendencies of oil from coal under a high oil price). *Zhongguo fazhan pinglun (Discussion of China's development)*, Vol. 10 (2008), No. 3, pp. 45–51.

[20] Tian Hua, "Development of China's Coal Bed Methane Resources." *Zhongguo nengyuan (Energy of China)* 2007, No. 4, pp. 30–33.

[21] The state strategic reserve should reach 12 million tons (22 days of net imports) in 2010, with in addition some state-subsidized company commercial reserves. After 2010, the government will oblige companies to maintain additional reserves. By 2020, there should be 90 days of public reserves (60 of which institutional, and 30 governmental) and 30 days of commercial reserves. Assuming net imports of 300 million tons in 2020 (60% of consumption), it might mean reserves amounting to 165 million tons. Jin Sanlin of the NDRC provided some background information in *Zhongguo Guoqing Guoli (China National Conditions and Strength)* 2007(9)14–19.

well designed. Moreover, inspection is weak and some are shut down intentionally to lower cost.[22]

One is forced to conclude on the basis of these and other examples that even when central policies are clear, local governments and power companies do not necessarily follow them. Erica Downs captured China's energy policymaking apparatus well in a few words: 'ineffective institutions and powerful firms.'[23] Without an open debate and democratic decision-making, it is hard to tell the weight of various arguments and stakeholder interests within the coordination agency, the National Development and Reform Commission and State Council. The foreign policy dimension of energy security is even less transparent. There we find private daughter companies of state-owned oil companies operating in dozens of foreign countries with or without overt backing by Chinese economic bureaucracies—but apparently uncoordinated in an overall foreign policy framework.

The contrast between the huge impact that China (whatever its policies) will have on the international energy market and our very limited understanding of what drives Chinese decision-makers could not be greater. The need for better insight has motivated us to conduct a survey among educated young people in Beijing, on the assumption that they reflect public educated opinion now and may even become China's policy-makers in the future. In January 2008, 230 advanced (mostly M.A. and Ph.D.) students and staff members at various universities and research institutes were asked to fill in a questionnaire, and their views on energy security are presented and discussed in the last and main part of this article. Only very few professed to have expertise in energy matters. We expected them to be nationalistic, conservationist, and proponents of state regulation rather than market forces, and these expectations were largely confirmed.

---

[22] "Woguo huodianchang tuoliu hangye 2006–nian fazhan baogao." (Report on the development of the desulphurization sector for China's thermo-power stations) *Zhongguo huanjing chanye (China Environmental Protection Industry)*, 2007(10)9–11. At the end of 2006, one-third of all thermo-power generating capacity had been equipped with desulphurization installations, and another 40 percent was under construction. Electricity from power plants using FGD is subsidized, yet according to this report and also the Steinfeld et al. MIT survey many operators do not use the installed facilities.

[23] Erica Downs, *China*, p. 16. However, she also notes that several oil executives, such as Zeng Qinghong, have moved to positions in the national leadership. See Erica Downs, *Business Interest Groups*...

2. How to Assess China's Energy and Oil Security: Do Prices Matter?

China makes an increasing demand on global natural resources and has been a contributing factor to world market price increases for oil, ores, metals, grains, timber, dairy, et cetera. For fossil fuels, there are two main concerns. Resource depletion will affect future availability for the entire world, and high prices affect such availability for low-income countries and poor people already today. China does not seem very concerned about resource depletion (as a developing country it is entitled to 'sustainable' development, and it always points out that *per capita*, China's demand on fossil energy resources and contribution of greenhouse gases is below the world average), nor does the IEA.[24] Apparently, China (as so many other countries) trusts that new technologies will make new sources of energy available in due time.[25]

However, China shows great concern about the effects of rising oil and electricity prices on economic growth: products become more expensive, consumer demand is affected, and farmers have to pay more for diesel oil and fertilizer. Indeed, in 2008 China's costs of (180 m.t. of) crude oil imports went up by 62 percent to 129 billion US$ and of (39 m.t. of) refined oil imports by 83 percent to 30 billion US$.[26] Less growth and price inflation will bring higher unemployment and threaten social stability.

Therefore, electricity producers and oil refineries have been allowed to pass on only part of the coal and crude oil price increases to the consumers and were forced to accept losses.[27] In May 2008, in China

[24] "The IEA doesn't see China as a dangerous burden on the world's energy resources. There are ample fossil fuel resources in the world. We are not going to be restrained in our consumption of oil by resources, but either by economic vulnerability or by carbon dioxide." *Energies* No. 9, March 2006, Interview with William C. Ramsey, Deputy Executive Director of the IEA.

[25] e.g. Liu Shijin (2008), "In front of human wisdom and in face of the rapid development of S&T, the resource and environmental constraints should be understood in a scientific way and from a long-term strategic perspective. There is no reason to be pessimistic. A positive and optimistic attitude does not contradict with the sense of urgency… which we should emphasize." (pp. 3–4).

[26] http://www.customs.gov.cn/publish/portal0/tab1/info156576.htm.

[27] The December 2004 proposal by the NDRC to link the electricity price to the electricity coal price (with 30 percent of coal cost increases to be absorbed by electricity companies) has been accepted in principle, but implementation has been postponed over and again, and now will have to wait until the end of 2008 because of the effect it would have on inflation, according to the China Electricity Regulatory Com-

diesel-oil and gasoline sold for 5,520 and 5,980 Yuan per ton, respectively, much below the world prices of 7–8,000 and 8–10,000 Yuan.[28] Natural gas prices were underpriced even more.[29] But when world crude oil prices dropped from US$140 to below US$40 per barrel by December, and diesel and gasoline prices could be lowered to below the May level, the Chinese government may have felt it did the right thing in protecting its domestic market from violent fluctuations.

Clearly, China supports the general view that affordable prices are part of the definition of energy security. High oil prices are not in China's interest, because they drive up prices, increase foreign dependency, reduce exports, and affect productivity in some industries.[30] China wants to establish a 'stable, economic, clean, and safe' energy system. However, one might argue the opposite: based on its revealed comparative advantage in world trade and positive trade balance, China is more capable of absorbing higher energy costs than most other countries are. Also, higher energy prices would multiply the economically proven energy reserves, particularly in China because cleaner coal technologies and liquefaction would become more economical. Moreover, they would contribute to energy savings.[31] Thus, high energy prices could also be seen as a *positive* contribution to greater energy security in China. However, in interviews we conducted

mission on May 13, 2008. The Electricity Bureau has been pushing for price increases of electricity supply to the nets, which averaged 0.36 Yuan per kWh in 2008. This holds particularly for hydropower, because it receives only 0.22 Yuan per kWh, as against 0.55 Yuan for wind and solar, although its costs have risen. (Electricity Bureau Director Wang Guobao in *Zhongguo nengyuanwang (ChinaEnergyNet)*, May 8, 2008). China's VAT tariff on hydro-electricity is higher than that on thermo-electricity (17% as against 10%), the rationale of which is unclear.

[28] *China Daily*, May 13, 2008.

[29] "A discussion of natural gas prices." *Zhongguo shiyou jingji (China's oil economy)* 2007(4) 46–47. In 2007, Russia told China it should pay at least the same price for imported gas as the EU did (then over 0.18US$ per cu.m.). Beijing residents paid only slightly more, viz. 1.9 Yuan per cu.m. This left no room for distribution cost.

[30] Li Pumin et al., "Zhongguo yingdui gaoyoujiade zhanlüe xuanze." (China's strategic options in dealing with high oil prices) *Hongguan jingji yanjiu (Macroeconomic research)* 2005(12) 8–14.

[31] Yu Shengxiang et al., "Contrary thoughts about the effect of oil prices on the economy and society." *Zhongguo Nengyuan* 2007(4)34–38 point out the negative effects of low oil prices for China: a too heavy reliance on oil, slow development of alternative energy, and weak incentives for energy savings and efficiency. Higher prices would also contribute to more social equality, as car owners are comparatively rich. Falling prices of oil made Shell withdraw from a planned shale-oil joint venture in Jilin in October 2008.

in December 2007, none of the interviewed Chinese experts were ready to accept this point of view.

Even if long-term availability and prices are two main concerns, in China energy security is defined and discussed in much wider terms, as part of an overall national development strategy. A recent authoritative book on China's energy security stressed that it is not only the key to economic growth, but also the basis of social development and national security, and relates to social and political stability. Security goals are not limited to supply, but also include production, transport, and ecology. In a comprehensive global strategy, energy security should be sought on both domestic and international fronts. The main threat to energy is insufficient clean energy supply, and for oil, insufficient domestic oil resources and supply. According to its authors, oil security is threatened most of all by high dependence on foreign oil, competition over oil with other countries such as Japan and the US, insufficient access to foreign oil exploration and exploitation, and difficult cooperation with Russia. Oil imports are mounting rapidly, come from a too limited number of countries, and most are transported in foreign vessels along dangerous shipping lanes. China has almost no strategic reserves, and there is no mechanism to avoid oil price fluctuations.[32]

Remedial policies are supported by a broad political consensus. China should develop and improve its use of domestic coal as a substitute for oil imports, increase its domestic oil and gas production, invest in the development of overseas oil resources, speed up nuclear power development, develop renewable energy, economize on oil, and build up oil reserves. Moreover, its energy planning should be based on integrated development of coal and electricity.[33] The exception is the scope for expansion of hydropower, which is a contested issue because it creates intractable environmental and social problems in minority areas (e.g. in the Nujiang basin in Yunnan), and needs sizeable central government investment in long-distance power lines.[34]

[32] Ni Jianmin et al. (Eds.), *Guojia nengyuan anquan baogao (Report on National Energy Security)*. Beijing: Renmin chubanshe 2005. China's tankers transport only 10–20 percent of imports, because of lack of capacity, quality and certification. China has not yet entered the MARPOL convention. *Zhongguo shiyou jingji* 2007 (9)17.

[33] *Zhongguo nengyuan fazhan baogao 2007 (China Energy Development Report 2007)*. Beijing: Zhongguo shuili shuidian chubanshe, 2007, pp. 241–244.

[34] See e.g. former Minister of Water Conservation Yang Zhenhuai, "Zai baohu huanjing yu shengtaidi jichushang dali fazhan shuidian." (Greatly develop hydro-power on

China is also rapidly extending its gas and oil pipelines, as far as to Turkmenistan, and making up for past underinvestment in a national power grid. Chinese investments are facilitated by short planning procedures, rapid increases in the state budgetary income, easy lending by state commercial banks, and low dividend pay-out requirements of the oil and power companies.[35]

While the above views reflect central policy makers' concerns, some technical studies have tried to assess China's energy security on the basis of more objective indicators. In one such study officials of the Ministry of Land and Resources compiled an integrated index of six indicators of oil security (out of an original 24), with reference values (and *weights*) based on assessments by fifteen Chinese energy specialists: domestic reserve/production ratio (30; *0.2735*), domestic reserve/replacement ratio (1; *0.1116*), dependence on import (30%; *0.2582*), concentration rate of import (50%; *0.1028*), international crude oil price (37US$/barrel; *0.0744*), and domestic reserve storage level (90 days; *0.1794*). They concluded that China's situation was slightly more secure than the USA's: China scored higher in the first and third factors, about equal in the second and fourth, but much lower in the last factor. Both China and the US scored lowest in the most heavily weighted domestic reserve/production ratio (only 15 and 7 years, respectively), and both had falling levels of oil security after 1999.[36] China's largest oil company CNPC compares favorably with Exxon Mobile, BP, and Shell in its reserve/exploitation ratio (20) and reserves/replacement

the basis of protection of the environment and the ecology) in: 'Shuidian kaifa yu lüse weilai' Xinhua luntan wenji (Collection of articles of the Xinhua Hydro-power Investment Company conference 'Development of hydro-power and a green future'). Beijing: China Waterpub., 2007, pp. 5–7, and Andrew C. Mertha, *China's Water Warriors: Citizen Action and Policy Change*. Ithaca: Cornell University Press, 2008.

[35] Calculated by the author on the basis of company financial statements, in 2007 PetroChina and Sinopec's combined capital expenses were 15 percent higher than their net profits, while ExxonMobile, BP and Shell's were 30 percent lower than their net profits. Of these, the Chinese companies spent 78 percent on exploration and development, but the three IOCs only 67 percent. In the absence of share holder pressures, China's oil companies can invest more and accept higher risks. This point is made in more detail by Trevor Houser, "The Roots of Chinese Oil Investment Abroad." *Asia Policy* No. 5 (January 2008), pp. 141–166.

[36] He Xianjie, Wu Chuguo, Liu Zengjie et al., "Shiyou anquan zhibiao tixi yu zonghe pingjia." (Indicator system and comprehensive appraisal of petroleum security) *Internal paper* provided to the author in December 2007.

ratio (1.69).[37] One may note the high weights given to import dependency and concentration rate, the low reference value and weight for the crude oil price, and the absence of domestic distribution networks and system resilience from their index of oil security.[38] An improved and updated index is needed.

3. China's Quest for Overseas Oil

In 2001, the chairman of China's State Planning Commission called for a strategy of 'going out' to acquire overseas oil resources.[39] Other Chinese experts pointed out that the main supply of oil would have to come from the international market rather than from equity oil from overseas wells, and thus they stressed the need for international cooperation. China had enough money to buy oil from abroad.[40] China faces a tough world, dominated by other large powers, and should seek cooperation.[41] China's quest for overseas oil and gas has drawn increasing criticism, because of its apparent disregard for UN and US sanctions, and western political sensitivities about Sudan, Iran, Burma, and other regimes.

---

[37] Luo Zuoxian and Zhang Limao, "Zhongwai shiyou gongsi shangyou chanye jingzhengli pingjia." (An evaluation of the competitiveness of the upstream sector of Chinese and foreign oil companies) *Guoji shiyou jingji (International Oil Economy)* 2007(9) 62–66. CNOOC (mainly off-shore) and Sinopec (mainly downstream) do not do as well, and have been squeezed between rising costs and state-dictated low refined oil prices.

[38] For instance, it does not capture the gasoline shortages and long waiting lines in front of gas stations last summer, when refineries halted production because of their losses due to government price freezes, or the safety of LPG and LNG transport and installations.

[39] Chen Jinhua, "Shiyou zhanlüe yu Zhongguo nengyuan zhengce." (Oil Strategy and China's Energy Policy) *Zhongguo Shiyou*, 2001, No. 2.

[40] Chen Huai, "Guanyu jianli shiyou fengxian caigou pingzhangdi jiben shexiang." (Fundamental thoughts on the establishment of an oil risk state purchase reserve), and "Zhongguo xinde shiyou zhanlüe ying zhongdian yanjiu guoji shichang." (China's new oil strategy should concentrate on international market research) in Ma Hong and Wang Mengkui (eds), *Zhongguo fazhan yanjiu: Guowuyuan fazhan yanjiu zhongxin yanjiu baogaoxuan 2001* and *2004 (China Development Research: Selection of research reports from the State Development and Research Centre 2001* and *2004)*, pp. 139–147 and 223–227, respectively.

[41] Zha Daojiong, *Zhongguo shiyou anquan guoji zhengzhi jingji fenxi (China's Oil Security: International Political and Economic Analysis)*. Beijing: Contemporary World Press, 2005.

In particular, China's Africa policy has come under scrutiny. Foreign trade with Africa and direct Chinese investments in oil and mining have increased greatly, with Angola overtaking Saudi Arabia as China's largest supplier of crude oil and Sudan supplying over 10 million tons in 2007. It bought a 45 percent stake in an offshore oil field and a drilling license in Nigeria for 2.27 billion, and 4 billion US dollars, respectively in 2005–6. According to some African scholars, in its exploitation of African resources "China is merely reproducing the dangerous equation between Africa and its former colonial powers."[42] China established an Africa Development Fund, and concluded several treaties with African countries that in addition to investment and trade agreements contained sizeable foreign assistance elements, particularly in the development of infrastructure and hydropower stations. The resulting presence of so many Chinese companies and construction workers in African countries characterized by corruption, civil war, and failed states has made some foreign governments and existing stakeholders in Africa worry about China's intentions. China is a formidable competitor, with national oil corporations and Sovereign Wealth Funds with very deep pockets, seemingly willing to go to any lengths[43] to secure overseas supplies of oil and mining resources and capture the African market with its textiles, machinery, and consumer products. China attributes its success in Africa partly to its non-interference in domestic policies. According to a senior Chinese ambassador to Africa: "Unlike some Western countries we do not attach any conditions to our aid. We do not give criticisms about who is 'good government' and who is 'bad government.' "[44]

Chinese analysts see nothing unusual in China's oil activities in Africa. Most other regions (Middle East, Southeast Asia) were already dominated by western presence, so China's oil companies had easier

---

[42] Firoze Manji and Stephen Marks (Eds.), *African Perspectives on China in Africa*. Oxford and Cape Town: Fahamu, 2007; Olukoya Ogen, "China-Nigerian Economic Relations: Chinese Imperialism or South-South Partnership?" *China Aktuell*, No. 3, 2008, pp. 78–102.

[43] "In their quest for overseas equity oil, Chinese oil companies regularly pay 20 to 50 percent more than other companies do," *Zhongguo Nengyuan Fazhan Baogao* 2007, p. 256. In 2008, Sinopec used 70 US$ per barrel as a yardstick for its overseas investment in oil exploration and development (private communication to author, December 10, 2008).

[44] Cui Yongqian, quoted in "Zhuanjia fangtan: ZhongFei nengyuan hezuo jinru xinjieduan." (Specialists' interview: energy cooperation between China and Africa enters a new stage) *Guoji shiyou jingji (International Oil Economy)* 2007(3)1–6.

access to Africa where upstream markets are open to foreign investors and the light oils suit China's demand. One analyst claimed that China had accommodated the US and was willing to cooperate with the US in Africa. He expected rivalry with the US, but not a clash.[45] Equity oil is a minor part (some 15 percent, half of which from Sudan where CNPC has a 40% stake in GNPOC) of China's oil imports. Anyway, two-thirds of Africa's oil exports go to Europe and the USA, and only 9 percent to China. China's equity oil interests are all small (and most often, purchased from other IOCs). Analysts have wondered in how far China's foreign policy makers were able to set conditions for and influence the behavior of China's national oil companies CNPC, CNOOC and Sinopec in various African countries.[46] Different and sometimes competing bureaucracies pursue their own economic interests, and China's political image in the west is of no concern to them.

Discussion of China's 'oil diplomacy' has become more common in recent years, particularly since the 2005 energy cooperation agreement between Russia, China, and India, the agreements with Angola and Kazakhstan earlier that year, and Vladimir Putin's visit to Beijing the next year. Russia and China concluded a gas and electricity supply agreement and the Siberian oil pipeline should go not only to Japan, but also to Daqing—but will it? Chinese views are usually couched in rather general terms, even if criticism of other countries (particularly the other major powers the US and Russia) can be strident and China's national interests presented in a very blunt manner.[47] A lack of international experience of its NOCs, high price risks, lack of coordination between departments, and conflicts with the US about Iran, Sudan, Venezuela and even ('the China scare') about investments in Canada and Latin America are perceived to be main problems for China.[48] So China's NDRC demands that the Chinese oil and gas enterprises

---

[45] Hong Zhao, "China's oil venture in Africa." *East Asia* (2007) 24: 399–415.

[46] Erica Downs, "China's Quest for Overseas Oil." *Far Eastern Economic Review* Sept. 2007, pp. 52–56.

[47] Zhang Jianrong, "Zhong'E nengyuan hezuozhongdi wentiji dui Zhongguodi yingxiang." (Problems in Sino-Russian energy cooperation and their effect on China) *Shehui kexue (Social Science)* 2006(1) 59–69. See e.g. an editorial commentary on the reasons why Russia proposed a non-governmental Energy Club at the Shanghai Cooperation meeting in 2006, and why this club serves China's interests, too. *Guoji shiyou jingji (International Oil Economy)* 2007(6)1–5.

[48] *Zhongguo nengyuan fazhan baogao 2007 (China Energy Development Report 2007).* Beijing: Zhongguo shuili shuidian chubanshe, 2007, pp. 245–257. China's activities and US concerns have been presented in LCDR Cindy Hurst, *China's Oil*

become enterprises with real international competitiveness that seek to internationalize their entire industrial chain. "The industry lacks a concrete strategic plan, a sound system, a flexible mechanism, a strong policy and sufficiently available services for internationalization. A special plan should be formulated as soon as possible." It calls for stronger coordination of overseas investment, greater flexibility, and greater state financial support including preferential US$ loans and risk guarantees.[49] However, at the same time, interdependence and international cooperation have become key messages of China now, too.[50] This is also because of the foreign technical support China hopes to receive, and its continued dependence on shipping through sea lanes patrolled by the US Navy. Apparently, the Chinese government has to live with conflicting goals.

4. A Survey of Chinese Perceptions of Energy Security

In January 2008, we conducted a survey on perceptions of energy security among some 200 Chinese M.A. and Ph.D. students and 30 professors and researchers in Beijing.[51] Our thesis is that the opinions of such a group of educated young people in China's capital are important as an indicator of current intellectual thinking and thereby, future Chinese policies. We agree with Fewsmith and Rosen that public opinion does matter in China's foreign policy-making.[52] Most respondents were selected because they happened to attend classes (mostly in economics and business) at leading research institutes and

---

Rush in Africa, and *China's Global Quest for Energy*, Washington, D.C.: Institute for the Analysis of Global Security, July 2006 and January 2007.

[49] Hou Yongzhi and Sun Zhiyan, "Change of Oil and Gas Industry's Mode of Development is Badly Needed for Coping with Energy Challenges." *China Development Review* Vol. 10 (2008) No. 3, pp. 63–70.

[50] e.g. Zha Daojiong and Hu Weixing, "Promoting Energy Partnership in Beijing and Washington." *The Washington Quarterly* Vol. 30(4)2007:105–115, and Shu Xianlin, "Shiyou huoban shengyu jingzheng duishou." (Better oil partners than competing opponents) *Zhongguo shiyou qiye (China Petroleum Enterprise)* 2007(1)120–124.

[51] Eduard Vermeer drafted the questionnaire, which was then improved through discussions with Shi Dan (CASS Institute of Industrial Economy) and Yang Guang (CASS Institute of Asia and Africa). Shi Dan organized the survey and data input.

[52] Joseph Fewsmith and Stanley Rosen, "The Domestic Context of Chinese Foreign Policy; Does 'Public Opinion' Matter?" In: David M. Lampton (ed.), *The Making of Chinese Foreign and Security Policy in the Era of Reform, 1978–2000*. Stanford: Stanford UP, 2001, pp. 151–190.

universities. Only ten percent were directly involved through education or work in energy questions. The average age was 28.2 years old (male 28.6, female 27.4). The 106 students with a M.A. generally were 23 to 25 years of age and the 88 Ph.D.'s 25 to 28 years. One quarter of our respondents (hereafter called 'the older') were above 29. One-third were women. Ten percent had studied abroad. One result of having such an articulate group of highly-educated students and researchers was that very few (percentage wise, mostly women) gave the answer 'I don't know.' And another, probably because of the teacher-student relationship, was that all questionnaires were filled in completely, with few if any invalid answers or blanks.

As far as we know, there is only one other study of Chinese student views on China's energy problems, based on a considerably larger 2006 survey of mainly undergraduate students at seven universities in China and Hong Kong. Where appropriate, we will compare our findings for our youngest age group with those of this Zweig and Ye study.[53]

We started with a test of knowledge of the most important international dimension of China's energy security: China's dependence on oil imports (which had risen to 47% of consumption in 2006). Over one half of all interviewees were not well informed: 29 percent said they didn't know about the percentage of China's oil imports, another 8 percent thought it was less than 30%, and 18 percent (but 36 percent of the older, above 29 years of age) thought it was over 60%. The rapid rise of such imports in recent years, the current projections of 60% and more imports for the medium-term future, and possibly some confusion about gross and net oil imports may have been responsible for most erroneous estimates. Almost half of the interviewees had reasonably close estimates of between 30% and 60%, demonstrating that oil import dependency is a common topic of interest among young intellectuals. This latter group (47% of men, 40% of women; 47% of the younger and 39% of the older group) we labelled the 'correct estimates' C, as against 'incorrect' I.[54]

---

[53] David Zweig and Shulan Ye, "A Crisis is Looming: China's energy challenge in the eyes of university students." *Journal of Contemporary China* (2008), 17 (55), pp. 273–296.

[54] The Zweig and Ye study found a 60% correct score for a question asking for China's largest supplier of foreign oil in 2006, but only a 3.4% correct score for China's external energy dependence. Apparently, their norm for the latter was much stricter than ours.

One third of respondents held that China should be basically self-sufficient in oil in the future. This is quite a large percentage, in view of the fact that neither politicians nor experts in China believe this to be even remotely possible. Among the women and the young, the percentage was slightly higher than among men and the older. 58 percent thought China did not have to be basically self-sufficient. Only 9 percent said they had no ideas about it.

However, when people were asked to give a percentage of imports that should be acceptable for China in the years of 2010 and 2020, most came up with much higher percentages than implied by 'basically self-sufficient.' Results are shown in Table 8.1. We have subdivided respondents on the basis of gender (MF, n = 152 and 78), age (YO, n = 174 and 56) and (more or less) correct answer to our previous question about China's current dependency on oil imports (CI, n = 103 and 127).

Over one half of those who gave a percentage felt that for 2010 and 2020, a range of 40% to 70% would be acceptable. The average acceptable percentage for 2010 was 48% (among the older 54%) and for 2020 55% (among the older, 61%). For 2020, a ratio of over 70% dependency on oil imports was acceptable for a significantly larger group (almost 30 percent) than it was for 2010 (16 percent), which seems to reflect a belief in the inevitability of increasing oil import dependency

Table 8.1 What percentage of China's oil demand could be imported in 2010 without China's energy security being threatened? And in 2020*?

(% of respondents)

	2010 all	M	F	Y	O	C	I	2020 all	M	F	Y	O	C	I
Don't know	30	24	41	31	25	18	39	30	25	41	32	27	18	41
20–30%	17	10	19	20	9	11	22	9	10	9	11	2	7	11
40–45%	11	12	11	12	9	18	6	6	7	4	6	4	5	6
50%	18	21	10	16	23	23	13	16	16	17	16	18	21	13
55–65%	11	11	12	9	16	19	6	16	17	15	14	22	23	9
70–80%	11	14	5	10	14	9	13	18	21	10	17	21	23	13
other	3	3	2	1	4	2	2	4	4	6	4	4	1	5
Average given %	48	49	45	46	54	49	47	55	57	52	53	61	57	53

*Assuming China's GDP will double between 2010 and 2020.
M = male F = female Y = age below 30 O = age 30 and over C = oil import estimate 30% to 60% of consumption I = other estimates & don't know
Source: Author's findings

after 2010. All groups accepted an increase between 2010 and 2020 by about 7 percentage points. Men, the older, and the better-informed were more ready to accept high dependency ratios than others. Women were more likely to answer 'I don't know' than men were, and this pattern continued with all subsequent questions. As was to be expected (as half of the I group consists of those who gave a 'Don't know' answer before), the greatest differences were between the C and I groups (18 vs. 41 percent).

Asked about the effects of increased competition between China, the USA, and Japan on the world oil and gas market, only 13 percent of respondents felt political relations would not be affected, and another 23 percent foresaw only minor problems. A clear majority of 65 percent believed their competition for oil would produce conflicts. Correctness about China's oil import percentage made no difference, but gender and age did. The men and the younger were more convinced that conflicts would occur than the women and the older were. 35 percent of the older thought problems would be minor, as against 19 percent of the younger. These findings for the younger group correspond very well with the Zweig-Ye survey, which found that 68 percent of the students believe that "disagreements over energy are (very) likely to lead to military conflicts between countries."

A similar majority of 63 percent (among the older, 59 percent) thought that Russia would use its natural gas exports to Europe and other regions as a political weapon, and 32 percent (among the older, 38 percent) thought that was at least a possibility. Women were much less certain than men, and only half of them answered yes. See Table 8.3.

Table 8.2 What will the effect of competition between China, Japan, and the USA on the world oil and gas markets be for China in the next decade?

(% of respondents, except 4% 'don't know' answers)

	all	M	F	Y	O
It will not affect political relations	12	10	18	14	7
It will produce conflicts	65	68	58	67	57
It will cause only minor problems	23	22	24	19	35

Source: Author's findings

Table 8.3 Will Russia use its export of gas to Europe and other countries as a weapon to influence their policies?

(% of respondents)

	all	M	F	Y	O	C	I
n=	230	152	78	174	56	103	127
Yes	63	69	50	64	59	66	60
No	4	3	5	4	2	5	2
Maybe	32	28	41	31	38	29	35
Don't know	2	1	4	2	2	0	3

Source: Author's findings

An important factor (apart from the USA and Russia's actual behavior) may have been that China has a tradition of state intervention in the economy as well as in foreign economic relations, and its oil companies are state-owned. Apparently, most respondents expect similar interventionist behavior from other governments, defending their country's political or economic position in the world.

More surprisingly, as many as 92 percent of respondents were worried that energy shortages and high oil prices would destroy the stability of the world economy, of whom 38 percent were even *very* worried. The younger and older groups scored almost the same, but men showed themselves slightly less concerned than women. The answer 'somewhat concerned' was given most often by those who made an incorrect estimate or no estimate of China's percentage of imported oil. See Table 8.4.

Three explanations might be offered for this very high percentage. One is that in domestic political propaganda, social stability is a paramount goal that needs constant nurturing by the CCP and government. If

Table 8.4 Are you concerned that energy shortages and high oil prices will destabilize the world economy?

(% of respondents)

	all	M	F	C	I
Very concerned	38	35	44	43	34
Somewhat concerned	53	55	50	46	59
Not concerned	8	10	5	12	5
Don't know	1	1	1	0	2

Source: Author's findings

left to market forces without proper political guidance, many disadvantaged people would suffer. Thus our mentioning of shortages and high prices provoked a strongly negative reaction. A second one is that respondents had begun to worry how long China's good fortune of ever increasing exports and double-digit economic growth would last, and saw high oil prices as a negative factor, possibly resulting in China being cut off from overseas oil or losing its competitive edge. The third one, and least likely, because most Chinese seem more preoccupied with China than with the world, is that they were really concerned about the world economy as such, for instance about nationalist energy politics and the effect of oil price hikes on poor countries. Indeed, the question of whether the subsequent violent rise in energy prices in the first half of 2008 has been an important factor underlying the present financial crisis is beyond the scope of this article.

Only 11 percent of the respondents in our survey (and 14 percent of the 'correct estimate' group) agreed with our suggestion that high oil prices might have a positive effect on China's energy security, because they would reduce the industrial rate of growth, and thereby energy demand. 81 percent disagreed.[55] This supports the second explanation mentioned above. A high rate of industrial growth is of prime importance to employment and income generation, and few respondents were ready to sacrifice these goals for the sake of greater energy security.

In recent years, China has imported increasing amounts of oil from Sudan, Iran and other countries criticized by the UN. Western criticism has focused on oil exploration and imports from Africa. In order to ascertain the awareness of Africa's growing position as an oil supplier to China, we asked respondents to rank ten countries from where China's crude oil imports originated (145 million tons in 2006, excluding 6 million tons of exported crude and 28 million tons of imported refined oil). We left some major countries out, such as Oman, Yemen, and Venezuela, and included in our list some well-known non-oil producing countries such as Egypt and Vietnam.

Respondents did rather well with Saudi Arabia, Russia, and Iran, which had been responsible for 16%, 11%, and 12% of oil imports, respectively, in 2006. About three-quarters of all respondents put them in the top four. Russia was put in first place twice as often as Iran,

[55] The Zweig and Ye study also found rather lukewarm support for raising oil prices to save energy.

Table 8.5 Top rankings of oil exporters to China given to ten listed countries. (% of respondents)

Country	1st	2nd	3rd	4th	5th	Sum 1–4	Actual rank in 2006 (and % of imports)
Saudi Arabia	34	24	15	5	5	78	2 (16%)
Russia	26	20	15	13	7	73	4 (11%)
Iran	14	22	24	13	6	73	3 (12%)
Angola	7	7	3	11	10	29	1 (16%)
Libya	6	7	7	9	18	29	6 (2%)
Kazakhstan	4	10	8	18	10	40	7 (2%)
Sudan	3	3	7	11	12	24	5 (3%)
Nigeria	2	3	11	12	14	18	– (0%)
Egypt	0	1	6	6	13	13	– (0%)
Vietnam	0	0	1	1	3	2	– (0%)

Source: Author's findings

although China imported less crude from the former in 2004–2006. Thus Russia was slightly overestimated.[56] However, the recent rise of the number one exporter to China, Angola, was completely overlooked.

Angola caught up with Saudi Arabia as the largest oil exporter to China in 2006, but only 17 percent of all interviewees put Angola among the first three. Even back in 2004 Angola was almost at par with Saudi Arabia as oil exporter to China (only one percent below Saudi Arabia's 14% of total Chinese imports in that year). Sudan was underestimated, and Kazakhstan and Libya were overestimated. These estimates may have been influenced by positive publicity about the oil pipeline from Kazakhstan, and not reporting about Angola and Sudan in a context of oil. (In Sudan, CNPC holds a 40% share in a consortium with GNOPC and has a concession in Southern Darfur and West Kordofan, but Chinese reporting concentrates on China's contribution to the peace-keeping effort).

[56] Maybe its sizeable exports of refined oil to China in 2006 (5 million tons), second only to Korea (7.4 million tons), but before Venezuela and Singapore (3 million tons each) played a role. If we include refined oil, Russia exported more to China than Iran did. For more detailed data, see Tian Chunrong, "2006–nian Zhongguo shiyou jinchukou zhuangkuang fenxi." (Analysis of China's 2006 oil import and export situation) *Guoji shiyou jingji (International Oil Economy)* 2007(3)14–21.

One of the basic elements of energy security is reliable sources of supply. Which oil-supplying countries can be trusted, and which cannot? Asked to grade six listed countries on a scale from 1 to 5, our respondents expressed the highest average trust in Kazakhstan (3.34) and the lowest trust in Vietnam (2.50). The next least trusted was Russia (2.87). Iran, Saudi Arabia, and Angola were all slightly above average (3.10–3.15): less than ten percent distrusted these three countries completely (score 1), and over one-third had rather high or high trust in them. Overall, the level of trust was rather low. Kazakhstan and Iran received a 'high level of trust' (5) from only 18% and 16%, respectively, of respondents, Vietnam from 8 percent, and the other three from 12 to 13 percent. As Table 8.6 shows, Saudi Arabia and Angola were trusted more, and Russia much less, by the older respondents.

Because so far, China has not suffered from oil supply disruptions from any of these countries, these trust levels should be attributed to past history (such as the 1979 war with Vietnam, or earlier conflicts with the Soviet Union), or perceptions of power relations (Kazakhstan as dependent on China for its oil exports), or stability of their regimes and the like. It might be interesting to know whether Chinese state importing companies take such Chinese perceptions into account in their importing decisions, but we do not have public statements on the matter.

The low level of trust is related to a general conviction that China is at a disadvantage in the world oil market relative to western countries and Japan. Only six percent of respondents believed China's position in the world market to be equal. 80 percent thought it was unequal. There was little difference between all groups, except for a higher per-

Table 8.6 Levels of trust in security of oil supply to China from six countries among the two age groups.

Country	Low trust (1–2) (% of respondents)		High trust (4–5) (% of respondents)		Average (1–5)
	Y	O	Y	O	all
Kazakhstan	22	27	43	46	3.34
Iran	29	32	38	33	3.15
Angola	30	34	34	45	3.10
Saudi Arabia	30	23	32	44	3.10
Russia	34	48	32	30	2.87
Vietnam	52	50	7	19	2.50

Source: Author's findings

centage 'don't know' answers for women (21 percent as against 8 percent for men).

In our questionnaire, we suggested four remedies, if people felt such a situation of unequal access existed, and asked them to indicate the best and second best options for China. However, all people decided to answer this question, irrespective of their answer to the previous question. See Table 8.7.

Cooperation with international oil companies, concluding political agreements and acquisition of foreign oil resources scored about equal as the best choice. Concluding foreign production sharing agreements was the least popular first choice (12 percent), but the most popular second choice (40 percent). Added up, acquisition of foreign oil resources scored lowest (46 percent) and cooperation with international oil companies highest (54 percent), which may be a comforting thought for those who are wary of China's intentions to control overseas resources. However, men (who are more likely to become politically influential), and the older group favor acquisition of foreign oil resources most. Women, in contrast, prefer cooperation and sharing.

Many factors have an effect on oil energy security. Level of reserves, substitution rate, exploitation/reserves ratio, dependency on foreign imports, the concentration of imports, and international crude oil prices are six important indicators. We asked our respondents to assign a value between 1 (least important) and 10 (most important) to each of these six factors. On average, the level of reserves and substitution rate scored highest with 7.5 and 7.1, respectively, and the international oil price and import concentration, lowest, with 5.9–6.0.

Table 8.7 What should China do in order to improve its access to the international oil market? Choose the best and second best from four options: a. Cooperate with international oil companies. b. Conclude political treaties. c. Buy foreign oil resources. d. Take a share in production agreements with foreign countries.

	First choice (%)							Second choice (%)						
	all	M	F	Y	O	C	I	all	M	F	Y	O	C	I
a	30	26	39	31	29	30	31	24	23	27	25	23	18	30
b	28	28	30	29	25	29	28	20	20	21	23	11	21	19
c	30	38	14	28	36	33	27	16	19	10	16	16	19	13
d	12	9	17	12	11	8	15	40	38	42	36	50	42	38

Source: Author's findings

The low valuation of the world market price of oil may have been influenced by Chinese subsidy policies of keeping domestic oil prices low and stable. The level of reserves received a maximum score of 10 from 27 percent of respondents, but only ten percent gave this score to the exploitation/reserves ratio. The latter also scored a significantly lower average (6.8) than the former, which may indicate a short-term view of resource exploitation. An alternative (not necessarily contradictory) explanation is that some respondents may have interpreted 'the level of reserves' more narrowly as the primary strategic oil reserve storage, or more widely as China's huge existing *coal* reserves, both of which have received much publicity.

Between respondents, views on dependency on foreign imports differed most widely, with nine percent giving the lowest score of 1 and 16 percent giving the highest score of 10 (and a standard deviation of 2.86). Views on the reserve/exploitation ratio (with most scores between 6 and 9), international oil price (most between 5 and 8) and substitution rate (most between 5 and 10) varied least (standard deviations 2.20–2.38). An interpretation of these results might be that foreign dependency is a contested issue in China, with some accepting the economic dictate of reliance on oil imports, but others upholding traditional (communist or nationalist) ideals of self-reliance. More technical issues show less divergence in views. In comparison with the weighting of six indicators by officials of the Ministry of Land and Resources quoted above, our respondents gave less weight to import dependency, and higher weight to the level of reserves and oil prices. We suggest this may be explained by the fact that our survey reflected more recent developments of increasing imports and oil storages and high prices.

In addition to the above-mentioned overall average valuations, we show the high scores in Table 8.8. It confirms the high marks given to the level of reserves, and rather low marks for the price of oil on the international market. Gender and correctness about China's oil imports did not make for significant differences in valuation, but age did. The older group gave higher scores for the oil market price and diversification of sources of imports, and a lower one for the ration reserves/exploitation.[57]

[57] The propensity to give scores of 7–8 and 9–10 hardly differed between groups, and does not affect results.

Table 8.8 High valuations (7–10) of six fundamentals of oil security (scale 1–10), by age group.

(% of respondents)

	all		Y		O	
score	7–8	9–10	7–8	9–10	7–8	9–10
level of reserves	31	42	32	41	25	43
substitution	31	32	30	33	34	29
ratio reserve : exploitation	36	24	35	28	39	13
import dependence	24	29	24	29	23	29
import diversification	28	17	30	14	20	27
oil market price	30	11	29	9	34	20
Average	30	26	30	26	29	27

Source: Author's findings

What are considered to be the main factors furthering energy security in general? We mentioned eleven factors, and asked for an evaluation of each. Respondents gave the highest average scores to R&D and technological progress (7.9), and diversification of types of energy and sources of import (7.5). The former may reflect the optimistic belief in science common to university students. Unsurprisingly, the lowest score (5.5) was given to China and India joining the IEA system. Mobilization of the functions of the market and complete information had the next lowest average scores (6.1), maybe because they are rather abstract concepts. The remaining six concrete factors scored between 6.9 and 6.5, in descending order: emphasizing energy savings and climate change, enhancing resilience by technical and managerial methods, better government planning and regulation, completion of production and transport infrastructure, adequate investments in energy production and supply, and cooperation between exporting and importing countries. The rather modest score of adequate investments in production and transport of energy is surprising, and might have been different if the survey had been conducted a few weeks later. Then, serious break-downs and disruptions of the power network in south China after heavy snowfall (just before many millions of migrants intended to board the train for Spring Festival) greatly increased the public awareness of the inadequacies of China's electricity distribution network.[58] For many years, experts have given warnings about

[58] Lan Xinzhen, "Power to the People," *Beijing Review*, March 20, 2008, reported that most power lines were designed only to hold 10–15 mm. of ice. Almost 200

the fragility of China's interregional power grid, construction of which had been lagging behind, while provinces and localities kept adding on new power plants.[59]

Looking only at the high scores, we see more pronounced differences. Technological progress received the most high scores (69 percent of respondents) and diversification came second with 60 percent. The three issues of energy savings/climate change, technical and managerial resilience, and government planning and regulation scored the same, 44 percent. The lowest scores were given to IEA participation and a properly functioning market. MF and CI scores were rather similar, except that the I group valued good information more highly than the C group did (6.4 as against 5.7). In Table 8.9, we show average scores and high scores (8–10) for both age groups. On average, the younger

Table 8.9 Valuation of eleven fundamentals of energy security, all and by age group.

(average scores, and % of 8–10 scores)

	All	% 8–10	Y	% 8–10	O	% 8–10
R&D, technological progress	7.9	69	8.3	76	6.7	48
Diversification of types and origin	7.5	60	7.7	61	6.7	55
Energy savings, climate change	6.9	44	7.0	46	6.7	39
Technical & managerial resilience	6.8	44	7.1	48	6.0	36
State planning and regulation	6.6	44	6.7	44	6.2	43
Investment in production supply	6.5	35	6.7	38	5.9	25
Adequate structure of production and transport	6.5	34	6.8	37	5.8	23
Cooperation between exporting and importing countries	6.5	33	6.7	39	5.8	18
Good information	6.1	37	6.3	40	5.3	29
Mobilization of market function	6.1	29	6.3	29	5.8	29
China and India in IEA	5.5	20	5.7	22	4.8	13
Average score/percentage	6.6	41	6.9	44	6.0	33

Source: Author's findings

transformer substations were knocked out of commission. The State Grid Corporation of China, one of the two grid operators in the area, estimated it would need 39 billion Yuan for repairs. Subsequently, higher design standards will be introduced, for a once-in-100–years calamity.

[59] Between 2003 and 2007, power stations received over twice as much investment as the power grids, while in developed countries, the latter receive more than the former. Government policies to close down less efficient small regional power plants have exacerbated the electricity transmission difficulties, *ibidem*.

group gave a 0.9 higher score. If we adjust valuations for this difference (methodologically, this may be justified), then only one factor shows significantly different valuations: R&D is valued more highly by the young. The great belief in R&D was also found by Zweig and Ye.

On the question what domestic policies the Chinese government should pursue in order to enhance energy security, we offered a dozen options. We asked to choose the best four, and also to indicate the three least favored options. Promoting investment in R&D in alternative energy sources, such as wind and solar, scored highest (75 percent) and optimization of the energy structure and development of nuclear energy both also scored high (67 and 65 percent). Building a strategic oil reserve was next (40 percent), followed by higher taxes on energy use and emissions, and investments in clean coal technologies and liquefaction (34 and 33 percent). Respondents were least supportive of inviting foreign oil companies to explore and produce in China, letting the market decide on investments in infrastructure, and subsidization and oil futures as means to stabilize oil prices (8 to 10 percent). Subsidization of exploration of oil and gas (19 percent), restricting growth of energy-intensive industries (23 percent) and concentration on hydropower (25 percent) enjoyed below-average support.

These results were consistent with those for the three options deemed least effective. Inviting foreign oil companies into China was rejected by 62 percent of respondents, letting the market decide by 53 percent, and price subsidies and oil futures by 41 percent. Restricting growth of energy-intensive industries was rejected by 33 percent. Only 8 percent rejected development of nuclear power. With the exception of support for building a large strategic oil reserve, which was favored significantly more by the older group (49 percent as against 34 percent of the younger), preferences were not age-sensitive or CI-sensitive. Women professed less support for nuclear energy development (54 percent as against 71 percent of men), and were slightly less negative about encouraging foreign oil companies to operate in China.

One important finding is that policy preferences of the respondents differed from the actual policies of the Chinese government in some respects (notably investments in alternative energy sources, support for clean coal technologies, and raising taxes on energy use and emissions). Much of that could be explained by the fact that actual policies do not correspond with official propaganda, and students tend to be influenced by the latter. In most cases, respondents' choices conformed to official policies. To what extent their preferences had been

Table 8.10 The four most effective and three least effective domestic policies for the Chinese government to enhance energy security.

(% of respondents)

Most effective:	All	M	F
1. Invest in R&D in alternative energy	75	73	78
2. Optimize China's energy structure	67	66	69
3. Develop nuclear power	65	71	54
4. Build large (120 days) strategic oil reserve	40	40	39
Least effective:			
1. Invite foreign oil companies to China	62	65	57
2. Let market decide and improve energy transport infrastructure	53	51	56
3. Stabilize oil price by subsidies and futures	41	39	44
Moderately effective, in order of preference:			
5. Higher taxes on energy use and emissions			
6. Invest in clean coal technologies and liquefaction			
7. Speed up development of hydropower			
8. Subsidize exploration and exploitation of superior resources, such as oil and gas			
9. Restrict energy use by energy-intensive industries of steel, chemical industry, etc.			

Source: Author's findings

formed by government propaganda, by their professors, or reached in a more independent way from the media is difficult to say, even for the respondents themselves. In any case, China has excellent websites on energy questions, such as china5e.com. However, the general rejection of foreign oil company operations in China does conform with actual restrictive policies in China for on-land activities. These may be attributed to nationalist attitudes that do not wane, even if China's economy becomes more integrated with the world economy. This distrust of IOCs is not justified by their record, and one might even argue that the Chinese government would be able to exert greater control over them than it would over its own oil companies. The interviewees all lived in the Chinese capital of Beijing, and that may have been a factor in their opinion about the latitude for different local energy policies. Should provincial and local governments be allowed to diverge from central policies, as long as it was in the interest of local development? A clear majority of 59 percent said no, but 34 percent said yes. Women were more ready to allow room for local policies than men were (42 versus 29 percent), and so were the younger (36 versus 25 percent). These results underscore the difficulties of local implementation of central policies. All too often, central policies and targets have been handed down without adequate prior discussion with all interested parties.

As for global warming, a majority of respondents (62 percent, with only minor differences between groups) thought it would have bad effects on China, and one-third thought its effects on China would be partly positive, partly negative. The latter group may be correct insofar as higher winter and spring temperatures (China experienced a rise in average annual air temperature of 1.5 degrees Celsius between 1950 and 2000, and a further increase of about 2 degrees is expected for Northwest and Northeast China)[60] have prolonged the growing season and increased yields in mountainous regions and the north. However, while some agricultural regions benefited from more rainfall, in the dry north, precipitation declined. Northern areas suffered from more droughts and south China from more floods. We didn't ask to strike a balance. Rather, the question was meant as an introduction to the problem of greenhouse gas emissions by China's energy industry. Views differed more widely on what contribution, if any, the Chinese energy industry should make.

Table 8.11 What contribution should the Chinese energy industry make to a reduction of emission of greenhouse gases? And should it be voluntary, imposed by government, or demanded by the masses (of energy users)?

(% of respondents)

	all	M	F	Y	O	C	I
Very large	43	42	45	49	25	46	41
Moderate	30	28	35	28	39	32	29
Very small	20	22	15	17	30	20	20
No need	1	1	0	1	0	0	2
Don't know	6	5	5	6	5	2	9
Voluntary by industry	7	7	8	7	9	9	6
Imposed by government	67	66	69	68	64	65	69
Demanded by public	22	22	22	23	18	23	21
Don't know	4	5	1	2	9	3	4

Source: Author's findings

[60] NDRC, *China's National Climate Change Programme*, June 2007.

43 percent of respondents felt that China's energy industry should make a *very large* contribution to the reduction of greenhouse gas emissions, 30 percent felt an ordinary one was enough, and 20 percent felt its contribution needed to be only *very small.* (It should be explained here, that in Chinese usage the word *very* does not carry much meaning). The younger group, the better informed ones, and women had higher percentages opting for a very large contribution than the others. The older group preferred a moderate contribution to either a large or a small one. Considering our common future, the high level of support for a large Chinese contribution, particularly among youths, is a hopeful sign.

Only 7 percent thought the industry's contribution to the reduction of greenhouse gas emissions should be a voluntary one. Most felt it should be imposed by the government (67 percent) or demanded by the public (22 percent). Steinfeld et al. have noted the inability of China's government *and* power plants to translate stricter environmental regulations and technological investments into sizeable reduction of greenhouse gas emissions.[61] Nevertheless, two-thirds of respondents supported sizeable government-imposed contributions from the power industry. This is an encouraging finding, as eventually electricity consumers and taxpayers will have to foot the bill of reduction measures. All groups had similar percentages.

China's energy savings law sums up a number of government measures. We asked interviewees to choose three out of six possible energy savings that might be regulated by law. Three scored equally high (70–72%): energy-saving building standards, measures related to power stations, urban heat supply systems and factories, and fuel use of cars and trucks. Maximum heat controls in private houses was the least preferred option (11%), and next was energy savings for household appliances (25%). Reduction of greenhouse gas emissions scored average with 50%. These results indicate a preference for regulation of large systems, rather than going down to the household level, and rather high support for tackling global warming.

[61] "Market pressures seem to be driving substantial upgrading on the combustion technology side, and regulatory pressures seem to be bringing about widespread installation of environmental cleanup systems, but neither of these forces appears to be driving sound environmental practice at the plant level." Steinfeld et al., 2008, p. 30.

Table 8.12 Which areas of energy savings should be emphasized by the government for regulation in the energy law? Choose three out of six areas.

(% of respondents)

	all	M	F	Y	O	C	I
Construction industry standards	72	73	72	63	81	74	69
Power stations, factories, urban heating	71	71	71	74	61	77	66
Fuel use by cars and trucks	70	65	81	72	64	67	73
Emission of greenhouse gases	50	54	42	48	57	52	49
Household appliances	25	25	26	25	27	18	30
Maximum allowable heating temperature	11	11	12	12	9	14	9

Source: Author's findings

Women more often mentioned car fuel use, which may be attributed to their greater propensity to economize. The older group more often mentioned construction industry standards, possibly because more of them were homeowners. Household appliances were mentioned more often by the I (incorrect estimate and don't know) group, who may have found this item easier to grasp than some of the other options.

When asked specifically about their own homes and which equipment deserved to be subsidized, most chose insulating materials (89%), highly-efficient boilers (80%), and solar panels (78%), with energy-saving light bulbs coming next (60%). Only a minority felt electricity and heat meters should be subsidized (35%) and even less double-plated glass windows (28%).

As for foreign policy measures that China should strengthen in order to raise its security of oil and gas supply, we asked respondents to choose the best three out of eight options. By far the most preferred were first, support China's oil companies in acquisition and development of foreign oil and gas fields (62%), and second, participation in and development of international cooperation in renewable energy, clean energy, and new technology (60%). The third best was concluding long-term supply agreements with countries such as Angola and Iran (49%). Three other issues, promoting peace in the Middle East and Africa, avoidance of international conflicts and seeking better relations with the US and Russia, and strengthening regional agreements such as the Shanghai Cooperation and APEC scored below average (30–33%). Stimulating cooperation of Chinese oil companies with foreign oil companies such as Exxon-Mobile and Shell, and building a navy to protect shipping from the Middle East and Africa, were the least preferred (both 18%). One might characterize this outcome as economically nationalistic and aggressive, with a belief in cooperation where it suits Chinese interests (technology, supply agreements), and otherwise peaceful.

As Table 8.13 shows, men and women differed considerably in their policy preferences. Women expressed much more support for participation in international cooperation on renewable and clean energy (77%), and much less support for acquisition of foreign equity oil by Chinese companies (53%) and for building a navy (5%). Clearly, women preferred a cooperative approach and disliked aggressive approaches. The older group was more supportive of acquisition of foreign equity oil by Chinese companies than the younger one was, and less supportive of international cooperation in renewable and clean energy.

Table 8.13 Policy measures that China should emphasize in its quest for greater security of oil and gas supply (Choose the best three out of eight listed measures).

(% of respondents)

	all	M	F	Y	O	C	I
support China's oil companies in acquisition and development of foreign oil and gas fields	62	67	53	59	73	62	62
participate in and develop international cooperation in renewable energy, clean energy, and new technology	60	52	77	63	53	57	63
conclude supply agreements with Angola, Iran, etc.	49	46	54	48	50	48	50
promote peace in the Middle East and Africa	33	30	38	33	35	32	36
strengthen regional agreements, such as Shanghai Cooperation and APEC	30	33	24	31	29	30	30
avoid international conflict, improve relations with US and Russia	30	32	26	29	30	28	31
build navy to protect shipping from Middle East, Africa	18	25	5	19	16	23	14
stimulate cooperation with foreign oil companies	18	14	23	19	13	20	17

Source: Author's findings

5. Conclusion

The rapid growth of energy consumption in China and the weak effect of policies aimed at reducing China's energy intensity have made it increasingly dependent on foreign oil imports. Chinese planners assume that coal will remain the dominant source of energy in the next decades. China's central government and part of the public are concerned about the environmental effects of coal and increased dependency on oil imports, but unwilling to slow down industrial growth or sacrifice the use of cars and other energy-consuming amenities of modern life. Thus, surely if reluctantly, the Chinese have become more willing to accept high levels of foreign dependency as inevitable. This conflicts with the general nationalist attitude, evident amongst the populace even more than in government circles, which feels that China should be self-reliant and that foreign countries and markets in general cannot be trusted. Such a situation has led to a considerable gap between official government propaganda and actual policies, and between technical expert advice and public views as reported in the media.

One way out of the dilemma is a high level of confidence in future technological solutions. Another is the belief that government policies and measures can be effective in creating a more secure supply of energy and mitigating the environmental effects of increased energy use. The Chinese government favors an all-out approach, under which as many as possible sources of energy are tapped simultaneously. This means that the state invests heavily in nuclear and wind power, in oil wells owned by Chinese companies in Africa and in cleaner coal in the Chinese interior, in giant pit power stations in North China and medium-size integrated heat-power stations in coastal cities, in large hydropower stations, and in LPG tanker facilities. The apparent lack of selectiveness may have to do with the size of the country, differences in local geographical and economic conditions, and competing bureaucracies and energy companies. In any case, the choices for particular sources of energy and technologies do not reflect much coordination, and the absence of public debate makes it difficult to understand the processes of decision-making. That goes even more for China's foreign energy policy, nowadays dubbed 'oil diplomacy.'

Our opinion survey of advanced students and researchers conducted in Beijing in early 2008 was meant to provide more insight into

Chinese attitudes towards domestic and foreign energy issues, and particularly energy supply security. We found significant attitudinal differences between men and women, and between the younger and older groups. Less than one-half of respondents could give a fair estimate of China's dependency on foreign oil. Saudi Arabia, Russia, and Iran were well-known as major oil suppliers to China, but Angola was not. One-third held that China should be basically self-sufficient in oil, an obvious impossibility. At the same time, the average estimate of China's dependency on foreign oil by 2010 was 48 percent of domestic consumption, increasing to 55 percent by 2020. The older group gave even higher percentages. Two-thirds thought that the competition with Japan and the USA on the world oil market would produce political conflicts, and almost all were concerned or very concerned that energy shortages and high oil prices would destabilize the world economy. Generally, it was felt that China's access to the world oil market was worse than that of western countries. The best remedial actions were held to be purchase of foreign oil resources (by men and the older group) and cooperation with international oil companies (by women). Concluding political treaties was believed to be more effective than taking a share in production agreements with foreign countries. Average trust in the security of oil supply from foreign countries was neither high nor low, with Kazakhstan and Iran trusted most and Russia and Vietnam least.

A hopeful sign for our planet's environmental future is that most respondents felt that China's energy industry should make a substantial contribution to the reduction of emissions of greenhouse gases, and that these should not be voluntary, but imposed by government. However, almost none were willing to agree with our suggestion that high energy prices might have a positive effect on China's energy security by reducing its rate of industrial growth and energy demand in general and stimulating investment in new energy sources. In comparison with a previous weighting of six indicators of oil energy security by interviewed officials, our respondents gave less weight to import dependency, and greater weight to the level of reserves and oil prices. This may have reflected their more recent awareness of increasing imports and oil storages and high oil prices. Energy security in general was believed to be furthered most by R&D, diversification of energy sources by type and origin, energy savings, and increasing managerial and technical resilience of energy supply systems. Good information, proper functioning of markets, and China's participation in the IEA were valued least. The younger group in particular placed high hopes on R&D, also after correction for their average higher scoring behavior.

The best domestic policy for enhancing energy security, according to our respondents, was investment in alternative sources of energy. Optimization of the energy structure and nuclear power came second and third (men were more positive about nuclear power than women, but only very few people were negative). The least favored policies, out of twelve listed options, were: inviting foreign companies to China, letting the market decide while supporting infrastructure, and subsidization of oil prices and futures as means of price stabilization. These preferences were partly at variance with actual government policies, notably in their strong support for investment in alternative energy sources, clean coal technologies, and raising taxes on energy use and emissions. One reason may be that economic risk and budgetary and managerial constraints played only a minor part in our respondents' preferences. Moreover, most were not willing to give room to local policies that differed from the national ones. On both counts, the Chinese students were more authoritarian than the central government was. But the rejection of participation by the IOCs in developing land oil resources in China conforms to official protectionist policies.

Finally, most believed that China should give more support to its oil companies to purchase and develop foreign oil and gas fields. At the same time, women in particular favored China participating in international cooperation for renewable and clean energy technologies. Concluding supply agreements with producing countries was seen as the third best option. Strengthening its navy in order to protect shipping lanes, and cooperation with foreign oil companies, were least preferred. The students' foreign policies might be characterized as economically aggressive and independent, technologically cooperative, and placing confidence in governments rather than in markets. Almost all viewed oil supply as a major problem in China's 'peaceful rise.'

What are the implications of these findings for EU policies towards China? We can suggest a few. First, the EU should try to dispel general fears about the insecurity of market forces and international oil companies. In the long run, a well-functioning international market in oil and oil products does enhance energy security for all countries, even if this message may be hard to sell in these days of violent price fluctuations and financial break-down. Secondly, the Chinese public should be given a better understanding of the need to involve all main stakeholders in government decisions about the power sector. Our respondents were unduly optimistic about the effects of one-sided imposition of rules and regulations, or provision of government incentives. Thirdly, in addressing Chinese energy issues one should take into account the very limited capacity of its national bureaucracies (the National Energy Administration and Ministry of Foreign Affairs) to influence the actual behavior of coal mining and electric power companies and the overseas operations of China's large oil companies. Fourthly, the EU as a living political example of international cooperation should stress the negative aspects of overly nationalistic attitudes. These can limit a country's or company's political and economic options in energy security, or drive it to suboptimal solutions. Finally, the apparent readiness of educated Chinese to invest in alternative energy sources, cleaner coal combustion technologies, and their related environmental concerns could be exploited by China and the EU to a greater extent in joint programs.

CHAPTER NINE

JAPAN'S EVOLVING NUCLEAR ENERGY POLICY AND THE POSSIBILITY OF JAPAN-CHINA NUCLEAR ENERGY COOPERATION

Raquel Shaoul

Abstract: *Given the current domestic and regional instability within major oil and natural gas producing countries, the fierce competition between consumers for access to these resources, the accumulated environmental damage due to the extensive use of coal, and the uncertain remaining quantity of fossil fuels available to meet world's energy demand, development of reliable alternative energy resources, such as nuclear power, is essential. Amid these circumstances, Japan's insecurity of energy supply is even more acute, as it has almost no energy resources of its own. Since the early 1970s, Japan has committed itself to enhance nuclear energy use within its electricity production. Currently, approximately 30% of Japan's electricity supply is provided by nuclear power—with expectations to reach 40% by 2017. This article aims to evaluate Japan's nuclear energy policies and explore the potential cooperation these may have with China's energy market. It concludes that Japan-China energy relations are characterized by reciprocity, especially in the nuclear power realm: While Japan's nuclear energy expertise and capital are of a significant benefit to China's commencing nuclear energy development, Japan's ability to maximize its nuclear capabilities depends, to a large degree, on expanding its nuclear energy cooperation with China.*

Keywords: Japan, China, nuclear energy, energy cooperation

1. Introduction

The current international energy market, based mostly on fossil fuel production to meet global energy demand, is facing severe hardship. According to the International Energy Agency (IEA), world demand for primary energy is expected to grow by 53% by 2030 and world electricity demand, which is increasing much more rapidly than overall energy use, is likely to almost double from 2005 to 2030 (IEA, 2006; Energy Information Administration [EIA], 2008).

The tight global energy demand at present is the result of rapid development and economic growth in developing countries[1]—countries that are characterized largely by energy inefficiency and governmental misuse of energy subsidies as a mean of sustaining political regime stability. Tight demand is leading to energy nationalism, especially for oil and gas resources, and consequently to an acute rise in energy prices. Moreover, although world oil production is currently peaking, it will soon start dropping, even as the demand for energy continues to soar (Deffeyes, 2001). A mismatch between energy demand and supply is therefore expected. Hence, the ability of nations to sustain their economy will depend on reducing their reliance on fossil fuels sooner rather than later (Elhefnawy, 2008).

Amid the mentioned energy circumstances, nuclear energy is expected to contribute to fill the gap between energy demand and supply, thereby playing a significant role for the achievement of security of energy supply. As for 2008, there are 439 nuclear reactors worldwide, accounting for 15% of the world's power electricity generation (26 trillion kWh). Globally, 36 reactors are currently under construction and another 99 are planned[2] (World Nuclear Association [WNA], 2008a). The Medium Term International Atomic Energy Agency (IAEA) Report of 2007 projects global growth in nuclear power of between 447 GW(e) and 691 GW(e) for 2030 (IAEA, 2007). Global electricity generation from nuclear power is expected to increase to 3.8 trillion KWh by 2030 (EIA, 2008b)—an increase of 46% compared to the FY2008 figure.

The first part of this article explores how Japan's nuclear strategy is designed to achieve energy security, exploring the main gains and limitations impacting policy implementation and the extent to which Japan's approach to nuclear energy is dictated exclusively by energy dependency. The second part evaluates China's present nuclear energy situation and highlights potential trends and prospect in this field. The third and final part assesses the circumstances and conditions that will incline Japan and China to engage in, develop and deepen their nuclear

[1] China and India accounted for 18% of the world's total energy consumption in FY2005, and their share is expected to grow to one-quarter in 2030.

[2] 'Under construction' are reactors for which the first concrete has been poured, or major refurbishment is under way. 'Planned reactors' are those for which approvals, funding or major commitments are in place and the plant is generally expected to be in operation within eight years; otherwise, the construction is well advanced but has been suspended indefinitely. World Nuclear Association, 2008. World Nuclear Power Reactors 2007–08 and Uranium Requirements.

energy cooperation. It investigates whether nuclear energy cooperation can significantly improve the two countries' energy security equation, and in what way nuclear energy cooperation influences the political frameworks within which these two states operate.

2. Japan's Nuclear Policy and Capabilities

Japan is dependent on imports for 81.9% of its energy supply, importing 99.8% of its oil consumption, 96.6% of its natural gas, 98.4% of its coal and 100% of its uranium. In FY 2006, Japan's total primary energy supply was 23,770 petajoules, which consisted of 47.1% petroleum, 20.5% coal, 15.1% natural gas, 3.2% hydropower and 11.2% nuclear power[3] (Statistical Handbook of Japan, 2008). Of its total electric power generation, including private power generation, of 1,161 billion kWh in FY 2006, thermal power[4] accounted for 65%, hydropower for 8.4% and nuclear power accounted for 26.1% (Statistical Handbook of Japan, 2008). Nuclear power generation therefore enhances Japan's energy self-sufficiency from 6.1% to 17.3%, and as such it is a central means of ensuring Japan's national energy security.

2.1 *Japan's Evolving Nuclear Policy*

As a result of the oil crises of the 1970s, the Japanese government set nuclear energy as a national strategic priority. The Advisory Committee for Energy Report of 1975, under the auspices of the former Minister of International Trade and Industry (MITI, at that time), suggested that the development of a stable supply of energy should be a top policy priority. On the basis of this Report, five policy principles were established, namely: reducing oil dependency, diversification of non-oil energy supplies, securing a stable supply of oil through petroleum reserves, exploration and development of oil by Japanese companies, promotion of energy conservation and promotion of new energy R&D (MITI, 1976).

Since then, the Japanese government has stressed the need to develop, promote, and enhance nuclear power generation, focusing particularly

[3] Other energy sources include waste, geothermal, and natural energy (solar and wind), accounting for 2.9% of total primary energy supply.

[4] Thermal power consists of coal, petroleum, and natural gas.

on achieving its 'nuclear fuel cycle' goal—reprocessing and recycling used fuel from light water reactors (LWR).[5] Accomplishment of the 'nuclear fuel cycle' was seen as one of the central means to increase Japan's self-reliance regarding nuclear fuel.

Yet some argue that despite the government's 'nuclear energy rhetoric,' in reality, Japan's nuclear fuel cycle policy failed (Smith, 2007). During the 1980s and 1990s, development of fast breeder reactors (FBR) and advanced thermal reactor (ART), necessary to achieve the 'nuclear fuel cycle' goal, was mainly characterized by delays of plants' construction and continual rescheduling of commercialization of the plants (Smith, 2007; Pickett, 2002).[6] However, despite difficulties, nuclear energy policy was to receive additional governmental attention from the early 2000s, as a result of energy price increase, especially oil and liquefied natural gas (LNG).

On 14 June 2002, the government promulgated the Basic Act on Energy Policy, setting the basic principles of energy security and stable supply. This law—the first of its kind—promoted greater efficiency in consumption (a further move away from dependence on fossil fuels) and market liberalization (Ministry of Economy, Trade and Industry [METI], 2002). As a mean to further move away from dependence on fossil fuels in November 2002, METI announced a precedent-setting tax imposition on coal, oil, gas and (Liquefied Petroleum Gas [LPG]), while reducing its power-source development tax, including that applying to nuclear generation, by 15.7% per year. Tax reform was designed to encourage private industry to switch to nuclear power generation.

Japan's first Basic Energy Plan enacted by the Diet on 7 October 2003, in accordance with the Basic Act on Energy Policy of 2002, aimed to reduce further dependence on oil as well as emissions of carbon dioxide (CO2) and confirmed that nuclear energy contributes to a stable supply of electricity and helps to meet environmental standards. The Plan promotes nuclear power generation, including the 'nuclear fuel cycle', as a

[5] A 'nuclear-fuel cycle' means that the spent fuel is reprocessed to remove usable fissile material, which is then fabricated into mixed-oxide fuels (MOX fuel) and placed back in reactor to produce more electricity. It also includes treatment and disposal of radioactive waste.

[6] In 1994, the FBR commercial timeline was pushed back to 2030, and in 2005 commercial FBRs were envisaged by 2050. FBR proved uneconomic in an era of abundant low-cost uranium, so development slowed and the MOX program shifted to thermal LWR reactors. World Nuclear Association, 2008. Nuclear Power in Japan.

key power source, based on the premise that safety must be guaranteed (Federation of Electric Power Companies of Japan [FEPC], 2008).

On 11 October 2005, the Japan Atomic Energy Commission (JAEC)[7] elected the Framework for Nuclear Energy Policy as the basic policy strategy for the government and industries to follow for the upcoming ten years. The political importance and relevance of the 2005 Framework for Nuclear Energy Policy rests on the fact that this document was to be the first long-term plan drafted after government reorganization, whereby the JAEC was incorporated into the Cabinet Office. The process of drafting the 2005 Framework created completely shared goals among the Cabinet Office, the Ministry of Education, Culture, Sports, Science and Technology (MEXT) and METI. Moreover, the Japan Atomic Energy Research Institute's integration with the Japan Nuclear Cycle Development Institute to form the Japan Atomic Energy Agency (JAEA) during that same month of October 2005, confirms the government's steady commitment to develop a nuclear fuel cycle.

The Framework for Nuclear Energy Policy determined that the share of nuclear power in electricity generation after the year 2030 should be at the level of 30 to 40% of total electricity production. It also confirmed that nuclear power production will be focused on LWRs (JAEC, 2005). The 2005 Framework emphasized Japan's commitment to achieve a 'closed nuclear fuel cycle' (JAEC, 2005). For this purpose, the parties involved agreed about the need to replace existent plants with advanced LWR; introduce commercial FBRs and use mixed-oxide (MOX) fuel in LWR;[8] the objective was to acquire 16–18 reactors of this kind by fiscal year 2010.

In May 2006, the ruling Liberal Democratic Party (LDP) urged the government to accelerate development of FBRs, calling this 'a basic national technology'. LDP proposed to increase budget share, improve coordination in moving from R&D to verification and implementation, and to enhance international cooperation.

METI's New National Energy Strategy Report of 2006 reaffirmed policy directions for nuclear power taken by JAEC in 2005. According

---

[7] The Framework for Nuclear Energy Policy was based on 'Development and Utilization of Nuclear Energy', known also as The Long-Term Program of 24 November 2000.

[8] Used fuel will be reprocessed domestically to recover fissile material for use in MOX fuel. FEPC plans first to use plutonium produced overseas (such as in Britain and France) at the pluthermal plants and to start burning domestically produced plutonium in 2012 or later.

to the Report, three main objectives were formulated: "(a) Establishment of energy security measures that our people can trust and rely on, (b) Establishment of the foundation for sustainable development through a comprehensive, joint approach for energy issues and environmental issues, (c) Commitment to assist Asian and world nations in addressing energy problems"[9] (METI, 2006a). Numerical targets to be achieved by 2030 were set in order to accomplish these objectives. As for nuclear power generation, the New National Energy Strategy called for 30% to 40% or more by 2030.

Policy coordination between government and private entities, in order to accomplish the mentioned policies, was seen on April 2007 when the government selected Mitsubishi Heavy Industries (MHI) as the core company to develop a new generation of FBR—a decision backed JAEA and FEPC.

Recently, Japanese nuclear energy policy has been emphasizing not only its importance in terms of national energy security but also in terms of Japan's leading role in the international arena. Agence France Presse (AFP) news reported on 21 March 2008 that a government Advisory body urged the government to take the lead in promoting nuclear energy worldwide as part of efforts to fight global warming (AFP, 2008). At the 41st Japan Atomic Industrial Forum Annual Conference in Tokyo on 15 April 2008, Japan's, Prime Minister at the time, Fukuda Yasuo, addressed the importance of nuclear power in reducing worldwide greenhouse gas emissions. In PM Fukuda's words:

> For Japan—for whom almost all its resources and energy depend on overseas sources—it is very important not only to promote energy conservation and new energy, but also to steadily promote nuclear power generation as the main power source, including technology development in the future.... The 'Nuclear Renaissance' is the proof that Japan's efforts to consistently promote nuclear energy development are not wrong. (METI, Agency for Natural Resources and Energy [ANRE], 2008).

PM Fukuda's speech was the first, for a Japanese prime minister, to refer publicly to the effectiveness and essentiality of nuclear power generation. His speech also indicates how the domestic debate on nuclear power generation is currently tilting in favor of nuclear power as a core energy resource within Japan's energy policy.

---

9 METI, 2006. The New National Energy Strategy, 10.

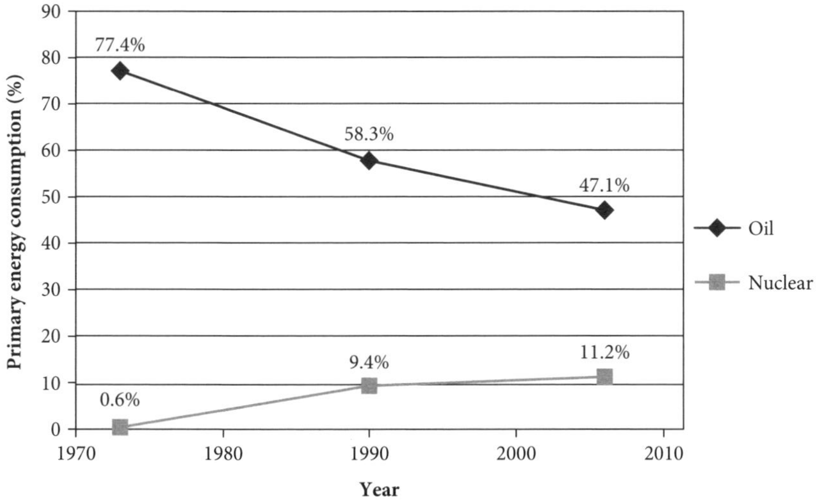

Source: METI ANRE 2008.

Figure 9.1 Graph Primary Energy Consumption (%).

2.2 *Japan's Nuclear Power Generation: Gains and Limitations*

Despite the global energy circumstances that characterized the 1980s and 1990s—that is, the full accessibility and economic affordability to fossil fuels as a result of the fall and near collapse of oil prices in 1986 and in 1997–99—Japan kept a steady commitment to its nuclear policy aims. This policy is evident in the expansion of its nuclear facilities, from five operating reactors in 1973 to 32 in 1990 and 55 in 2006—an impressive 71.9% increase. By 2006, nuclear energy primary consumption grew more than eighteen fold (1,866%) in comparison to 1973 (from 0.6% in 1973 to 11.2% in 2006) (METI, 1974; Statistical Handbook of Japan, 2008). Figure 9.1 below illustrates nuclear share increase compared to oil share decrease in Japanese primary energy consumption.

In 2007, 55 reactors provided 27.5% (47577 MWe) of Japan's total electricity generation.[10] Two reactors under construction and an additional 11 reactors planned will provide 37.5% of total nuclear electricity generation (64807 MWe) by 2016–7. Japan's future nuclear

[10] Nevertheless, the 27.5% electricity generation figure has different significance for some areas in Japan; for example, Fukui prefecture is dependent on nuclear power for about 60% of its electricity generation, hosting 15 nuclear plants within the region.

electricity generation capacity indicates a potential growth of 25% in nuclear power generation over the period of the next eight years, and additional energy efficiency as a result of technological improvement: in 2007, Japanese reactors' power generation output ranged between 439MWe to 1325MWe,[11] while Advanced Boiling Water Reactors (ABWR) under construction are expected to generate 1373MWe and Advanced Pressurized Water Reactor (APWR) under the 'planned reactor' category are expected to generate 1538MWe (WNA, 2008a; WNA, 2008b).[12]

An additional aspect of Japan's nuclear power generation involves a nuclear fuel cycle, comprised of three elements: Fast Breeder Reactors, nuclear-fuel-reprocessing plants and mixed oxide (MOX) use. Though to date, development of FBRs had a negative impact on nuclear fuel cycle development[13] as a result of several accidents[14] that took place while implementing the nuclear fuel cycle program, progress became evident regarding nuclear-fuel-reprocessing plants. Until 2005, reprocessing of spent fuels from Japanese LWRs and FBRs, to obtain MOX fuel, was done largely in Europe by companies such as British Nuclear Fuels (BNFL) and AREVA, at a rate of 4,200 tons and 2,900 tons respectively.[15] Since March 2006, Aomori prefecture, a nuclear-fuel-reprocessing plant run by Japan Nuclear Fuel Ltd. in Rokkasho, started trial operations to extract plutonium for power-generation. The Rokkasho plant—the first large-scale reprocessing facility in Japan—began its commercial operation in the summer of 2008. The

[11] Among the 55 operating reactors, more than 30 reactors provide power capacity under 850MWe.

[12] For definition of 'under construction' and 'planned reactors', see note 2 above. Calculated from World Nuclear Association data, 2008. World Nuclear Power Reactors 2007–08 and Uranium Requirements. The Electric Power Development Corp. has received permission to build a 1383 MWe Advanced Boiling Water Reactor (ABWR) in Aomori Prefecture.

[13] An accidental sodium leakage in the secondary heat transfer system of the Monju Prototype FBR during performance tests in 1995 caused that FBR's shutdown since then. Although the plant is to restart operation in February 2009, its commercialization was delayed until around 2050.

[14] These include the accident at the prototype FBR Monju in 1995, the Tokaimura reprocessing plant fire in 1997, an accident at the Tokaimura nuclear fuel processing plant operated by JCO Company Ltd while producing fuel for the Jojo experimental FBR in 1999 and the secondary system pipe rupture accident at the Mihama Unit 3 nuclear power plant in 2004.

[15] Spent fuel reprocessing in Japan was implemented by the small pilot reprocessing plant at Tokai (1977–2006), which reprocessed 210 tons per year of used fuel.

plant extracts plutonium and uranium from spent fuel for recycling. In 2008, it is expected to reprocess 395 tons, from which it will recover 1.9 tons of fissile plutonium (in reactor-grade material) (WNA, 2008b). The maximum reprocessing capacity of the plant is 800ton U annually, sufficient to reprocess the spent fuel from 40 reactors at 1,000 MWe-class nuclear power plants—an amount nearly equal to 80% of annual spent fuel generation in Japan (Japan Nuclear Fuel Limited, 2008). The Japanese government plans to have one third of its 55 reactors using some MOX by 2010 (WNA, 2007).

2.2.1 *Nuclear Gains*

Taking into consideration Japan's unique energy circumstances, nuclear power has numerous advantages, as analyzed, below:

2.2.1.1 Nuclear power as a mean to expand energy self-sufficiency and enhance security of energy supply

Today, Japan's primary energy self-sufficiency accounts for 17.3% of its energy consumption largely because of nuclear energy. Self-sufficiency is expected to further increase as soon as nuclear fuel cycle will attain commercial operation by 2050—through use of recycled uranium.

Nuclear power generation requires the use of imported uranium. However, suppliers are stable and diverse compared to oil suppliers—since the mid-1980s, the volatile and unstable Middle East provides approximately 90% of Japan's imported oil.[16] Extended oil transportation sea-lanes and their inherent risk, such as piracy in checkpoints, increases energy cost and vulnerability.

Uranium used to generate nuclear power is, in relative terms, widely available in politically stable countries, making it a highly stable energy source. In FY 2004, Japan's uranium requirements (8,350 short tons U3O8) were supplied by the following exporters: Australia[17]—33%, Canada—27%, Namibia—16%, Niger—13%, United States—7%, and others—4% (METI, ANRE, 2006b).

---

[16] Though generally referred to as the Middle East region, most of the oil is supplied by three Persian Gulf countries: Saudi Arabia, United Arab Emirate (UAE) and Iran.

[17] Australia is the world's biggest uranium reserves holder. However, in terms of uranium production from mines, Canada ranks first in world supply, accounting for 23%, followed by Australia with 21% and Kazakhstan with 16%. World Nuclear Association, 2008.

Despite the cartel structure within the current uranium production industry,[18] Japan has bilateral nuclear cooperation agreements with Australia, Canada, China, France, the UK and the US. In April 2007, Japan began negotiations with Russia on a bilateral nuclear agreement, by which, among others, Russia will re-enrich uranium that was extracted from Japanese spent fuel and reprocessed in the UK (Citizens' Nuclear Information Center, 2007). Likewise, on 28 August 2006 a Memorandum of Understanding (MOU) was signed by the governments of Kazakhstan and Japan to promote cooperation in the peaceful use of atomic energy. Japan is actively cooperating in the development of Kazakhstan's uranium reserves[19]—the worlds second-largest—and it is expected that the two countries will soon sign a bilateral nuclear cooperation agreement (Ministry of Foreign Affairs [MOFA], 2008a).

2.2.1.2 Environmental acceptability and international political recognition

Nuclear power, characterized by its low greenhouse gas emissions, plays an important role in reducing Japan's carbon dioxide emissions. Moreover, safe use of nuclear power prevents a wide range of environmental problems caused by coal-fired and oil-fired generators. As such, Japan has adopted and identified itself with Kyoto Protocol, the international regime for the reduction of global greenhouse emissions, from its initiation and through the present.

Nevertheless, despite its commitment to the Kyoto regime, Japan is currently facing difficulties to meet the greenhouse gas emissions target set by the Protocol. The Protocol, which is to expire in 2012, mandates Japan to reduce overall greenhouse-gas emissions to 6% below 1990 levels. However, Japan's current emissions are nearly 7% higher than those of 1990.[20] Though Kyoto's 'cap and trade' system, which created a market for the buying and selling of carbon credits, might

[18] In 2007, 85% of the world's uranium mine production was in the hands of 7 companies, as follows: Cameco, Rio Tinto, Areva, KazAtomProm, ARMZ, BHP Billiton and Navoi. World Nuclear Association, 2008.

[19] Major Japanese trading and energy firms are seeking investment opportunities in uranium mine projects. In 2006, Itochu agreed to purchase 3,000 tU from Kazatomprom over ten years and, In 2007, Japanese interests led by Marubeni and Tepco bought 40% of the Kharasan mine project in Kazakhstan and will receive 2,000 tU per year of its production. See, Masaki, H., 2006. Japan Joins the Race for Uranium Amid Global Expansion of Nuclear Power. Japan Focus.

[20] Japan is planning to purchase an extra 320 million tons of emissions credits from other countries, at a cost of US$9.85 billion, and yet it will be difficult to meet the mentioned goal.

help Japan to achieve its emission reduction goals, in practice 'cap and trade' is difficult to monitor and easy to manipulate (TIME, 2008). Given this situation, the Atomic Energy Commission (AEC) established in June 2007 the 'Vision for Nuclear Energy Policy for Global Environmental Protection,' whereby it proclaimed the role of nuclear energy in contributing to halving Japan's greenhouse-gas emissions by 2050 (AEC, 2008).

The leading Japanese role within Kyoto Protocol regime and other climate change conferences serves not only Japan's environmental goals but also its international political aspirations—to fulfill a leading political role in the international arena, which befits its economic superpower status. As such, Japan was the first nation to endorse US President George W. Bush's Global Nuclear Energy Partnership (GNEP) of 13 April 2007. GNEP focuses, among others, on technical cooperation in civilian nuclear energy, collaboration on policies and programs that support the construction of new nuclear power plants and regulatory and nonproliferation-related exchanges (MOFA, 2008b). Japan's participation in a political framework of this kind is valuable for two reasons: it helps to cultivate further cooperation with Japan's most important ally and also may weaken arguments against nuclear power, due to proliferation concerns.

2.2.1.3 Economic efficiency

Japan's nuclear energy policy is largely based on energy security rather than purely economic criteria. Nevertheless, nuclear energy's economic feasibility is a major element taken into consideration when developing nuclear power.

An OECD 2005 comparative study showed that nuclear power had increased its competitiveness over the previous seven years. Though nuclear *electricity generating cost projections for the year 2010, on a 5% discount rate*, was highest by far in Japan in contrast to other OECD countries[21] (OECD/IEA NEA, 2005; WNA, 2008a), from the Japanese domestic perspective, nuclear electricity generation at a cost of ¥4.8–6.2 per kwh, is an economically competitive energy source, cheaper and efficient compared to all other power sources, as seen from table 9.1 below.

[21] In Japan, the cost accounted for $4.80 US 2003 $/kWh, in comparison to the lowest such as Czech Republic $2.30, Korea $2.34, and France $2.54 (Discount rate of 5%, over a 40 year lifetime and 85% load factor.) OECD/IEA NEA 2005.

Table 9.1 FY 2007—Estimated Unit Cost of Power Generation by Power Source.

Power Source	Power Generation Unit Cost (yen per kWh)	Capacity Generation
Hydroelectric power	8.2–13.3 yen	45%
Oil-fired power	10–17.3 yen	30–80%
LNG-fired power	5.8–7.1 yen	60–80%
Coal-fired power	5–6.5 yen	70–80%
Nuclear Power	**4.8–6.2 yen**	**70–85%**
Photovoltaic power	46 yen	12%
Wind power	10–14 yen	20%

Source: METI, ANRE, FY2007 Annual Energy Report, May 2008.

There are various economic pros and cons vis-à-vis nuclear energy's economic effectiveness. For example, the high frequency of routine safety inspections of nuclear plants is said to reduce cost effectiveness—especially in a country like Japan that has a high frequency and magnitude of earthquakes. Another argument involves nuclear power plants' nominal design lifetime: while these plants' original nominal design lifetime was of 40 years, due to engineering assessments over the last decade, Japan's plant lifetimes currently comprise up to 70 years (WNA, 2008a). This longer span contributes to a massive reduction of primary investment cost. Conclusively, Japan's economic decision to develop nuclear power depended on wider economic considerations—that is, Japan had to choose between importing large quantities of fuel or spending huge capital at home, due to the fact that nuclear power is very capital-intensive. Japanese governments over the years have therefore contended that nuclear power development is economical, affordable and indexes economic competitiveness.

2.2.2 *Limitations and Impediments on Japan's Nuclear Energy Development*

Despite the varied gains offered by nuclear power discussed above, there are some potential and actual obstacles to further development and implementation of nuclear power generation, as analyzed below.

2.2.2.1 Growing popular opposition to nuclear power stations due to safety considerations and proliferation concern

Several accidents, such as the Three Mile Island nuclear power plant accident (Pennsylvania, 1979) and the Chernobyl nuclear power plant accident (Ukraine, 1986) had a negative impact on Japanese attitudes towards nuclear power development. However, it was not until the mid-1990s that the Japanese public in general turned against the nuclear power enterprise—though it is imperative to differentiate between national anti-nuclear opposition and opposition within local communities (i.e., prefectures hosting nuclear power plants).

Local communities' acceptance of nuclear power plants through the early 1990s was the result of the negotiating strategy employed by the utility (persuasion and compensation), strength of fishermen's local cooperatives, and economic funding to prefectures hosting nuclear facilities, among others, in the form of the Three Laws of Power Source Development[22] (Pickett, 2002). Even today, in prefectures such as Fukui, Niigata and Fukushima, where a high concentration of nuclear plants are placed, nuclear power is still perceived as major economic revenue. Power plants provide not only employment opportunities for local residents, but also secure governmental subsidies: of all subsidies allocated by METI and MEXT in FY 2007—a total of ¥124.4 billion for localities hosting nuclear and non-nuclear power plants—more than ¥50 billion was earmarked specifically for those localities hosting nuclear plants (Johnson, 2007). In contrast, at the national level, anti-nuclear weapons groups, labor unions and consumer groups consolidated against governmental nuclear policy (Pickett, 2002).

Since the mid 1990s, unification of national and local critics against nuclear power was the result of several events: the critical accidents within FBR Monju reactor (Fukui prefecture) in 1995 and at the Tokai-mura Plant in 1999[23] drew attention to the potential environmental problems that can result from nuclear power generation; the magnitude 7.2 earthquake in 1995 raised concerns about safety of nuclear plants[24] and public trust was lost in 2002, when Japan's largest power utility-TEPCO's (Tokyo Electric Power Co.) engaged in a safety-data cover-up scandal and was forced to acknowledge falsifying nuclear

[22] Pickett, S., 2002. Japan's nuclear energy policy. *Energy Policy* 30, 1349.

[23] These accidents, like the significant majority of the nuclear accidents, were outcomes of human error rather than natural disasters.

[24] Due to Japan's seismic situation, the Atomic Energy Commission (AEC) formulated the 'Seismic guidelines of September 1978'—revised by the Japanese Nuclear Safety Commission (NSC) in July 1981, again in March 2001 and more recently in September 2006. In contrast to the 1995 earthquake, the recent June 2007 earthquake in Niigata prefecture, with a magnitude-6.8, caused damage to the Kashiwazaki-Kariwa reactor number seven—the biggest nuclear-power-producing site in the world.

safety records in three of its 17 reactors. In May 2003, TEPCO shut down all its reactors for inspections, and by the end of 2003 only seven had been restarted.

Under the given circumstances of the 1990s and early 2000s and the renewed focus on FBR development, regarded to be Japan's nuclear strategy, various anti-nuclear weapons lobby groups merged with the already existing opponents of nuclear power.

Risk of nuclear proliferation was a concern especially from FBRs, as they produce more fissile materials than they consume by converting non-fissile uranium isotopes into fissile plutonium (Elhefnawy, 2008). Moreover, extraction of plutonium from spent nuclear fuel for recycling in the Rokkasho reprocessing plant is contributing to the increase of the already existent Plutonium reserve. In 2004 there were 41 tons of separated reactor-grade plutonium stored and awaiting use in MOX fuel. At the end of 2007, there was 26.4 tons of fissile plutonium (Puf) held by Japanese utilities: 13.9 tons in France, 11.3 tons in UK and 1.2 tons held domestically (WNA, 2008b).

Although in 1976 Japan became a party to the Nuclear Non-Proliferation Treaty (NPT) and in 1999 it ratified the Additional Protocol with IAEA accepting intrusive inspections, its plutonium reserves are adverse to the IAEA principle of not holding surplus plutonium, i.e. plutonium for no specific purpose.

The fact that Japan's current nuclear power capabilities enable nuclear weapons development if so desired becomes a matter of concern in the eyes of anti-nuclear groups. To a certain extent, their concern is validated by the public expressions of senior politicians in favor of Japan's acquisition of nuclear weapons in the name of the country's right to self-defense.[25]

3. China's Evolving Nuclear Policy and Capabilities

Whether or not China's increase of its nuclear power capacity in the near future will cover expected increases in energy demand, the Chi-

[25] See for example Chief Cabinet secretary Fukuda Yasuo's remark to the press on 31 May 2002: "Japan's peace constitution does not prohibit acquisition of nuclear weapons…depending upon the world situation, circumstances and public opinion could require Japan to possess nuclear weapons." French, H., "Nuclear Arms Taboo Is Challenged in Japan." New York Times, 9 June 2002.

nese government currently concedes that China has no alternative but to develop nuclear power. Development of nuclear power, among other measures, will allow China to realize sustainable energy development to meet its energy needs.

According to the Three Stage-Strategy in the National Medium- and Long-term Plans for Science and Technology Development[26] announced by the Chinese Government in February 2006, throughout the Second stage (2021–2035), China‘s goal is to increase the nuclear share to the current global average of 16% (Maeda, 2007). Furthermore, in the 11th Five-Year Plan 2006–10 of March 2006, the Chinese government declared its support for nuclear-power-plant development and construction.[27] China's National Plan for Coping with Climate Change,[28] released by the government in June 2007, confirmed once more that the government's plans to address climate change by 2010 include, among others, boosting nuclear power plant construction (State Council of the People's Republic of China, 2008).

Several factors have encouraged China to pursue nuclear energy, as analyzed below.

The first factor is China's deteriorating energy situation, as manifest in electricity shortages. China has been facing electricity shortages since the mid 1980s,[29] as its economy has grown.

China's primary energy use has increased more than fourfold from 413 Mtoe in 1980 to 1,863.4 Mtoe in 2007, with an average annual rate of growth of 12% (British Petroleum [BP], 2008). Moreover, China's energy consumption is estimated to increase 30-fold by 2030 (from

---

[26] The Three Stage-Strategy in the National Medium- and Long-term Plans for Science and Technology Development present a vision for energy development comprised by 3 stages: 2006–2020, 2021–2035 and 2036–2050.

[27] Governmental support for the construction of nuclear reactor was given for the first time in 1985, with the construction of the 300MW pressurized water type reactor Qinshan and from 1986 onward with the import of Dayabay station (2 units, 984MW).

[28] "China aims to achieve the goal of basically curbing the trend of ecological deterioration, reducing total emissions of major pollutants by 10%, and gain visible results in the control of greenhouse gas emissions during its 11th Five-Year Plan period (2006–2010). China.org, 26 December 2007. White paper on energy, <http://www.china.org.cn/english/environment/236955.htm>.

[29] The first serious electricity shortage occurred in 1985. In 2002, a gap between power production and demand of 30% occurred in Guangdong. In 2004, severe power shortages in coastal provinces led local governments to adopt emergency measures. In 2004, electricity shortage was estimated at 35 million kW and in 2005 it accounted for 25 million kW.

1,560 million tons of oil equivalent in 2005 to three billion tons per year in 2030) (IEA, 2007). Amid these circumstances, China's electricity demand in 2007 was the second largest in the world, and is expected to continue to increase significantly for the next twenty years at an annual rate of as much as about 140TWh (Maeda, 2006). Chinese electricity demand has been growing at more than 8% per year and electricity shortage occurred especially in the coastal provinces, such as Guangdong and Zhejiang, where industrialization is booming. Though nuclear energy currently accounts for only 1.9% of the country's total electricity consumption, in coastal provinces it currently accounts for 13% (Xu, 2008).

Second, some inherent problems exist within coal-fired energy production: coal constitutes 68% of China's current primary energy consumption. The Chinese economy's reliance on coal-fired energy has been problematic since the early 1990s: industrialization was centered in the coastal regions, far away from the coal mines, and as a result coal became a capital intensive and time consuming energy source, accompanied by several supply-logistical difficulties. Moreover, the dangerous conditions and mining accidents within the coal industry induced energy security problems and negative social effects.

In addition, China became a net-coal importer in the first quarter of 2007[30] (Xinhua, 2007). This fact is seen as complementary to the recent government decision to abolish import tariffs on coal in 2006, as a mean to achieve a reduction of coal's share in total energy use to 54% by 2020 (Wu et al., 2008). Even though China's status as a net-coal importer was short-lasting, as a whole coal imports and exports were even at the end of FY 2007, though the still heavy reliance on coal has led to serious environmental pollution.

In 2007, "burning coal contributed to 90% of the national total sulfur dioxide (SO^2) emissions, about 70% of the national total dust, nitrogen oxide (NOx) emissions and carbon dioxide (CO^2) emissions"[31] (Zhang, 2007; Xu, 2008).

Coal burning also contributes to global warming. In 2006, China overtook the US as the world's number one producer of greenhouse gases (Bloomberg News, 2007). Hence, the 11th Five-Year Plan 2006–

[30] In the first quarter of 2007, coal imports exceeded exports by 2.89 million tons.

[31] Zhang, Z., 2007. China is moving away from the pattern of 'develop first and then treat pollution. Energy Policy 35 (7), 3547.

10 environmental aim 'preventing the spread of pollution' states, "reduction in major pollutant emissions by 10% and control of GHG emissions" (Maeda, 2007). According to the Three Stage-Strategy in the National Medium- and Long-term Plans for Science and Technology Development, during the Third stage (2036–2050), China's energy goal is to reduce coal share in primary energy consumption to less than 50% and increase the combined renewable and nuclear share to more than 30% (Maeda, 2007).

Third, China's gradual loss of energy self-sufficiency caused concern within the government. China's growing economy and industrialization have had significant implications on its oil demand and consumption. In 2004, China was to become the world's second largest oil consumer after the United States, with a total demand of 6.28 million barrels per day. China's oil consumption in FY2000 more than doubled that in FY 1990, and reached almost threefold in 2005, in comparison to FY 1990[32] (BP, 2008; The Institute of Energy Economic, Japan [IEEJ], 2007). As a result of its' growing demand for oil, China has become a net importer of crude oil since 1993 and of oil products since 1996.

In FY 2007, crude oil imports reached a record 163 million tons of oil equivalent (Mtoe), a rise of 12.4% over the previous year (China Briefing News, 2008). The FY 2007 figure compared to the FY 2000 figure reveals a 132% increase in oil imports over the course of only seven years. China's dependence on oil imports therefore increased to 40% in 2003 and it is projected to rise to 50% by 2010 and to 60% by 2020 (Cheng, 2008). China's dependence on imported oil not only obstructs energy self-sufficiency but also poses an economic burden for a developing economy.

China's nuclear energy program dates back to the early 1970s. However, it was only in December 1991 when the first commercial nuclear power was connected to the grid. As of 2007, 11 operating reactors provided 1.9% (8,587 MWe) of China's total electricity generation. Seven reactors are under construction and an additional 26 are planned over a period of the upcoming eight years; together, these reactors will provide a nuclear capacity of 50GWe to 60GWe by 2020. This is a sixfold

[32] In FY 1990, primary consumption of oil was 110 million tons of oil equivalent (Mtoe), in FY 2000 it accounted for 247.4 Mtoe, in FY 2005 it accounted for 327.8 Mtoe and continued to increase to 368 Mtoe in FY 2007.

increase compared to 2007 figure (WNA, 2008a). According to the China National Nuclear Corp. (CNNC),[33] one of the main governmental institutions responsible for the nuclear-power program, China plans to invest 400 billion Yuan (equivalent to US$48 billion) to build 30 nuclear reactors by 2020 (Stakelbeck, 2006). The 76 'proposed reactors' (62600MWe) which will contribute to an additional three to fourfold energy production increase (120–160GWe) by 2030 represent further expected growth in China's nuclear electricity generation. Consequently, the annual projected nuclear generated electricity consumption of China could rise to 66 billion kWh in 2010 (from 16 billion kWh in 2000) and to 142 billion kWh by 2020 (EIA, 2008).

To power its new generation of nuclear power plants, China needs uranium. However, China's known resources of 70,000 tons of uranium are said to be sufficient to meet only short-term needs. For this reason, the Sino-Australia Uranium Agreement (SAUA) was signed on 29 April 2006.[34] According to SAUA, Australian producers will be able to export 20,000 tones of uranium to China annually for power generation. However, the SAUA is more than an energy deal. It indicates, from China's perspective, the country's political acceptance by the international community, in contrast to the previous 'containing China' policy as held by the West (Wu et al., 2008).

China's nuclear energy achievements are the result of a policy comprised of two major pillars: self-reliance and international cooperation in the form of foreign partnership (Xu, 2008; State Council PRC, 2006). Nine nuclear power plants, out of the 11 reactors currently operating in China, are designed with technology imported from France, Canada and Russia—with local development based on French technology. More recently, technology transfer has come from the US and Japan as well.

[33] CNNC is the main investor in and the biggest owner of all the nuclear power plants across China. It is also the major force of nuclear technology development, the main supplier of nuclear power design and nuclear fuels and the most important provider of technical service for the operation of nuclear power plants in the country.

[34] China's other energy agreements for uranium supply include those with Kazakhstan, Russia and Nambia. Uranium has also been imported from Australia as of late 2008. Talks on this matter have commenced with Canada as well.

4. Is Nuclear Cooperation Feasible? Evaluation of Potential and Actual Nuclear Energy Cooperation

When discussing Japan-China potential for nuclear energy cooperation and limitations on this potential, it should be noted that energy cooperation between the countries had existed over the years regarding different issues and at different levels.

There are precedents of Japan-China bilateral cooperation in the oil realm during the 1970s (Liao, 2007). Even today, amid competition over fossil fuels, the two countries share various major interests vis-à-vis security of oil supply concerning third parties, such as Middle East's oil suppliers.[35]

Since the mid 1980s, Japan has been transferring technology to China via different semi-governmental organizations such as the New Energy and Industrial Technology Development Organization (NEDO), both to achieve energy efficiency[36] (R&D) and to cope with environmental pollution problems arising as a result of growing energy consumption.

The potential future establishment of IEA, the Asian version of the International Energy Agency, as an outcome of the IEF (International Energy Forum) meeting in September 2002 in Osaka, provides evidence of political cooperation between China and Japan in the energy field.[37] At present, the Asian Pacific Economic Cooperation (APEC) and the Association of Southeast Asian Nation (ASEAN) plus China, Japan and ROK (10+3) Energy Cooperation are the leading energy schemes designed to promote cooperation between Asian nations.

The establishment of the Asian Cooperation Dialogue (ACD) served as additional political recognition of the need to promote regional

---

[35] China' and Japan's high dependence on Middle Eastern oil may lead the countries to seek cooperation as a result of their common interests: it is in both countries' interest to promote regional stability in order to maintain an uninterrupted flow of oil. Middle Eastern regional stability would also eventually enable greater access to local markets for the two countries. Collective oil strategy in the form of security of oil shipping routes (sea-lanes) and ground energy exploration and utilization might take place. See, Shaoul, R., 2005. An Evaluation of Japan's Current Energy Policy in the Context of the Azadegan Oil Field Agreement Signed in 2004. Japanese Journal of Political Science 6 (3), p. 22.

[36] In 2005, adjusting for Purchasing Power Parity (PPP) China used 21% more energy than Japan to produce an equivalent unit of GDP (The figures for every dollar of GDP in 2005 were 7,906 British thermal units for China and 6,539 for Japan,) EIA, 2006.

[37] *The Washington Times*, Tokyo, 21 April 2004. Online at: <http://www.washtimes.com/upi-breaking/20040422-033916-4264r.htm>

energy cooperation.[38] The ACD meeting of Ministers of Foreign Affairs and other Heads of Delegations, involving over 20 representatives of different Asian oil consuming countries, set a joint declaration to establish an energy forum to regularly discuss key energy aspects.

Furthermore, Japan and China recently signed a bilateral agreement on energy cooperation at the Japan-China Energy Conservation Forum on 29 May 2006[39] (METI, 2006b). The countries agreed to create a new framework for 'Policy Dialogue' between METI and the National Development and Reform Commission (NDRC)[40]—which is responsible for the formulation of the 'Five-Year Plan. Though apparently this energy agreement calls for the continuation of existing energy cooperation trends, it paved the way for the signature of a Memorandum of Understanding (MOU) between METI and NDRC on the implementation of Japan-China energy conservation and an environmental business promotion model project. The 2nd Japan-China energy conservation and environmental Issues Forum held in China on 12 September 2007, mentioned nuclear energy cooperation for the first time: "The two parties recognize that the expansion of nuclear power generation in Asia and across the world will contribute to alleviating the tight energy supply situation and arresting global warming... [Japan and China] will continue cooperation related to the construction and safe operation of nuclear power plants" (METI, 2007).

The circumstances and conditions that will incline Japan and China to engage in, develop and deepen their nuclear energy cooperation are analyzed below.

4.1 *Japan and China's Nuclear Power Interests: Reciprocity and Mutual Benefit*

Both the Japanese and Chinese nuclear energy markets are multifaceted and composed of several and diverse players. Therefore, identifi-

[38] The idea of the Asian Cooperation Dialogue was first initiated by Thailand's Prime Minister Thaksin Shinawatra and formally established during the 34th ASEAN Foreign Minister's Meeting in Hanoi in July, 2001.

[39] According to this agreement, Japan will assist China to achieve energy efficiency targets set forth in the 11th five-year plan.

[40] The National Development and Reform Commission is a macroeconomic management Agency directly under the State Council. It is responsible for assessment and approval of major nuclear projects.

cation of 'nuclear energy interests' should take into account sectorial approaches, each representing particular interests.

Recent studies on the Chinese nuclear energy market's structure conclude that economic reforms and bureaucratic restructuring in China for the past two decades "have resulted in a shift of power and resources away the from the central government to the state-owned energy companies... [this created a market structure characterized by] strong enterprises and weak government"[41] (Downs, 2006; Xu, 2008). This argument is illustrated by CNNC—currently the major investor and operator in China's nuclear power market, responsible for the operation of seven nuclear power plants out of a total of 11. Other major companies are China Guangdong Nuclear Power Company (CGNPC) and the four power generation companies that play a central role in nuclear energy development. Therefore, at present, NDRC is in a position to set policy direction rather than dictate policy.

The main energy purpose of the government is therefore to upgrade the capability and quality of the structure of energy production, procurement, supply, and demand, while taking into consideration the actual limitations posed by the different domestic players when bargaining over policy formulation.

Two pairs of shared interests currently exist between Japan and China. The first is already leading to nuclear cooperation largely at the company-to-company level and the second bears potential in terms of future nuclear cooperation, mainly at a government-to-government level. These interests are analyzed below.

4.1.1 *Capital investment and Business Opportunity*

Actual nuclear energy cooperation between China and Japan is taking place as a result of the Chinese government's need for capital[42] to develop and expand its nuclear power industry. Furthermore, Japanese companies are looking for external energy business opportunities—especially at a time when the Japanese economy is facing difficulties.

[41] Xu, Y., 2008. Nuclear Energy in China: Contested Regimes, Energy 33 (8), 1198.

[42] The Chinese government does not need to take foreign loans to finance nuclear energy development due to its' large budget surplus. China, however, regards foreign FDI as having a major alleviating impact over her economy. Huge capital required to finance growing energy demand might be disturbing for an economy still in development stage. Japan has become China's third largest source of foreign capital, accounting for US$6.572 billion in 2005.

In order to attract foreign direct investment, China has been making active efforts to improve laws and policies related to the opening up of its market.[43] Specific reference to the energy market is found in the 2004 revision of the 'Catalogue of Advantageous Industries for Foreign Investment in the Central and Western Regions'. This revision was designed to encourage foreign investment in the energy sector, including energy and energy-related exploitation, production, supply, transportation and energy equipment production (NDRC, 2007).

In face of the massive potential of the Chinese energy market, the three big Japanese businesses dealing in power systems—Toshiba Corp., Hitachi Ltd and Mitsubishi Heavy Industries Ltd (MHI)—have in recent years started nuclear energy businesses in China.

On 16 December 2006, NDRC Chairman Ma Kai and U.S. Department of Energy (DOE) Secretary Samuel W. Bodman signed a Memorandum of Understanding (MOU) which paved the way for Toshiba's Subsidiary Westinghouse Electric Company[44] to construct four AP1000 nuclear power reactors in China (CNNC, 2006). In July 2007, the agreement was signed between Toshiba's Subsidiary Westinghouse along with consortium partner Shaw and the several Chinese companies.[45] Two plants are to be constructed at Sanmen and another two at Haiyang, with an estimated cost of about $US3 to 4 billion for each plant. Full construction of these 3rd generation plants is to start in 2009 and the first power plant is expected by 2013. In other words, more than 50% of current Chinese nuclear power plant construction (four out of seven plants) is under the management of a Japanese company.[46]

[43] This includes promulgating of laws such as the Law on Sino-Foreign Equity Joint Ventures, Law on Sino-Foreign Cooperative Joint Ventures and Law on Foreign Capital Enterprises. These laws are designed to create an open environment for foreign investment.

[44] Toshiba purchased Westinghouse for US$5.4 billion in 2006.

[45] The Chinese companies are State Nuclear Power Technology Corporation (SNPTC), Sammen Nuclear Power Company, Shangdong Nuclear Power Company (for Haiyang) and China National Technical Import & Export Corporation.

[46] SNPTC selection of Westinghouse Consortium and the AP1000 technology, over France's Areva Group, to construct power reactors has to do with safety and technical reasons. The AP1000 features passive safety systems used for core cooling, containment isolation and containment cooling, are regarded to be, according to Westinghouse, "the safest, most advanced, yet proven nuclear power plant currently available in the worldwide marketplace." See, Staff Report, Modern Power Systems, September 2007. Online at: <http://www.westinghousenuclear.com/docs/ContractChina.pdf>.

An additional nuclear energy cooperation agreement between MHI and the Harbin Power Equipment Co., a major Chinese heavy machinery concern, was signed on 28 September 2007. This agreement opened for MHI a full-scale entry into the Chinese market for turbines and other equipment used in nuclear power plants. MHI is to supply key power components for the Sanmen and Haiyang nuclear power plants—two orders worth more than 100 billion Yen. MHI also intends to contract with Harbin Power Equipment Co., to build a future nuclear reactor[47] (The Nikkei Weekly, 2007).

However, it is critical to note that the existence of common interests does not mean that the parties agree on the means of materializing them.[48]

4.1.2 *Reliable Energy Supply and Maximization of Nuclear Energy Utility*

The second shared interest between China and Japan that might materialize into nuclear energy cooperation involves China's interest to cope with and eventually eliminate electricity shortages, and Japan's interest to maximize its nuclear power capacity.

Electricity cannot be stored; therefore, supply and demand must be kept in balance in real-time. Supply of nuclear electricity is inflexible to momentary energy demand change due to its lack of substitutability. For this reason, nuclear electricity generation in Japan is powered by mixed energy sources (including oil and natural gas together with nuclear power) which compensates for nuclear energy's inflexibility. Amid this situation, nuclear energy's share of the total power plant's electricity generation can rise to a maximum of about 40%[49] This fact consequently limits overall nuclear energy share in total electricity production in Japan.

Therefore, Japanese nuclear energy cooperation with China can theoretically materialize in the form of transmitting electricity surplus from Japan to China in times of energy demand fluctuation. This will

[47] In 2007, MHI also won a contract to supply ten gas turbines to be installed at four thermal plants in China. Although the contract relates to thermal power generation, it will help MHI to broaden its economic presence in China's energy market.

[48] See Article's Section 4.2.2.Domestic Political Limitations: an Obstacle for Potential Nuclear Energy Cooperation.

[49] Interview with Sudo Shigeru, International Development Center of Japan, Tokyo, 25 July 2007.

make possible for Japan to increase its nuclear electricity generation from its current approximately 30% and provide China a solution for its electricity shortage, even if for the short to mid term.

The concept of electric power grid interconnection in North East Asia is not new. This has been evaluated in the context of the rising electricity demand in the region, especially in China. (Yun and Zhang, 2006).

Analysis of electricity grid interconnections between Japan and Korea, under the assumption that the interconnection is achieved by a High Voltage Direct Current (HVDC) undersea cable technology, reveals positive energy cooperation results (Kanagawa and Nakata, 2006).

The idea behind trans-national electricity supply grids as a means to promote nuclear energy cooperation between Japan and China relies on the recent HVDC technology breakthroughs. One breakthrough relates to the world's longest transmission link from the Xiangjiaba hydro power plant, located in the southwest of China, to Shanghai, China's leading industrial and commercial center, located 2,071 km away[50] (ABB Corporate, 2007). Another technological HVDC improvement regards to undersea electricity transmission. Since September 2008, the NorNed HVDC transmission link, the longest undersea in the world with a 580 km-long and with a 700MW transmission capacity, connects Netherlands and Norway[51] (ABB Corporate, 2008).

Though several problems might arise when trying to apply current HVDC undersea technology to the Japanese-Chinese energy context, due to distances that exceed 580 km, future technology development might make feasible an undersea connection, for example, between Fukuoka and Shanghai, located 878 km away.

Nevertheless, as previously mentioned, existence of common interest does not guarantee the parties come to an understanding on how to materialize them. To achieve the potential benefits from power grid interconnection and trade there are many issues that need to be resolved. There could be technological problems due to differences in standards and quality of power by country, such as different frequencies and voltages. There is also serious concern about the reliability of

[50] The link, owned by the State Grid Corporation of China, will transmit 6,400 MW power and is scheduled to go in operation in 2011.

[51] In 2009 there are three further HVDC links scheduled in Europe, including Britned, a 1,300MW sub-sea cross-border connection between the UK and the Netherlands.

interconnected power grids, because their malfunctioning may lead to costly and hazardous blackouts (Yun and Zhang, 2006). All these issues remain pending as energy cooperation between the countries materializes.

4.2 *Potential Impediments for Nuclear Energy Cooperation*

Enhancement and materialization of current and future nuclear energy cooperation will largely depend on how Japan and China manage to associate despite their diverging energy needs and domestic political limitations.

4.2.1 *Self-Sufficiency Equation and Nuclear Energy Development*

China's energy necessities are very different from Japan's. China's expanding energy needs contrasts with Japan's stagnant energy demand. Since 2000, China's energy demand resulted in its energy consumption growth rate exceeding its GDP growth rate. The main energy problem China currently has is therefore the lack of effective governmental mechanisms designed to promote and materialize energy efficiency and conservation. Despite the mentioned state of affairs, the Chinese energy market is self-sufficient for ~ 90%.

The Japanese energy market in contrast, prompted by oil crises in 1973 and 1979, shifted its high energy consumption and environmentally-unfriendly heavy industries to adopt major energy efficiency measures. However, due to it's almost total lack of domestic energy sources, Japan's primary energy self-sufficiency accounts merely for 17.3% of its energy consumption.

Consequently, the two countries have adopted different 'Energy Mix' approaches to assure energy security. While development of nuclear energy is vital to enhance Japan's energy security, China's energy market, in the short term, can continue without significant nuclear energy development.

As previously analyzed, development and utilization of nuclear energy provides many advantages for China's energy market, especially in the mid and long-term. However, the Chinese government may choose to develop alternative energy solutions, such as clean coal technology and renewable energy technology, instead of or in addition to nuclear energy development.

4.2.2 *Domestic Political Limitations: an Obstacle for Potential Nuclear Energy Cooperation*

Mutual distrust derived from historical rivalry; ongoing geo-political and territorial disputes; hostile domestic public opinion and political nationalist sentiments in both countries have the potential, under certain circumstances, to obstruct energy cooperation.[52]

Structural energy market differences are also considered to potentially undermine bilateral energy cooperation. Some argue that the massive Chinese government support of its national energy companies causes significant commercial disadvantages for Japanese private energy companies, in times when they compete for the same foreign energy resources. This argument is partly valid for the fossil fuels energy sectors; it does not apply for the nuclear energy sector where governmental long-term commitment and performance are in many aspects similar in China and in Japan.[53]

An additional factor potentially obstructing nuclear energy cooperation refers to some of the prominent positions kept by players within China's nuclear energy community over the question of how to develop China's nuclear energy policy.

The two key players within nuclear energy expansion discourse are CNNC and NDRC. CNNC is particularly interested not only in developing new technology but also ensuring the opportunities to adopt the technology in the process of nuclear energy expansion, as illustrated by the principle of "self-reliance, assisted by international cooperation" (Xu, 2008). In other words, CNNC regards indigenous R&D and utilization of local designs and technology as crucial for China's nuclear expansion.

The government, in the appearance of NDRC, has agreed with CNNC's approach until 2002. However in 2003, NDRC's decision to put the Sanmen and Yangjiang projects up for international bidding and the publication of the 11th 5-year plan (2006–2010) in 2005, illus-

[52] This argument however runs the risk of simplifying reality, judging from China-Japan's highly complementary economies from the early 1990s to the present. From 1993 to 2003, Japan was China's largest trade partner for 11 years in a row, and now still remains the third largest trading partner of China.

[53] The government of Japan has acted for the development of NPPs continuously since NPP development program started. Among the most recent Cabinet Resolutions supporting nuclear energy development are: Framework for Nuclear Energy Policy (October 2005), Policy packages under the nuclear National Plan (August 2006) and Economic Fiscal Reform (June 2008).

trates NDRC's abandonment of "self-reliance, assisted by international cooperation" principle.

The prominent position of CNNC, within the nuclear energy market as a whole, led NDRC to demand full transfer of nuclear technology to local partners when concluding nuclear agreements; assuming this will ease CNNC's loud criticism—as previously analyzed in the Toshiba-Westinghouse case, Westinghouse agreement to transfer technology to CNNC could lead to the construction of many more nuclear reactors in China over the next 15–20 years (Xu, 2008).

CNNC's consistent emphasis on "self-reliance, assisted by international cooperation" had led China to demand technology transfer as a pre-condition to conclude nuclear energy deals. As a result, Japanese companies may perhaps remain reluctant to invest in the Chinese market. Companies are concerned about potential technology lost to low-cost Chinese competitors, due to the insufficient protection of intellectual property rights in China.

5. Conclusions

Considering the trans-national character of the current worldwide energy situation, countries like China and especially Japan are already formulating some of their energy decisions while taking into consideration their geopolitical situation and the global scene.

Though nuclear power usage might cause problems that are also of a trans-national character, such as nuclear proliferation, storage-related problems, early warning system for radioactive contamination and long-term management of nuclear waste (Manning, R., 2000), it seems that nuclear power is still perceived by the two countries as a viable alternative to fossil fuels and as a solution for national and especially regional environmental problems.

The current worldwide 'Nuclear Renaissance', is the result of several interrelated external and internal factors to the country, as previously analyzed. Whereas the governments of Japan and China recognize the significant advantages of nuclear power cooperation, current concrete nuclear energy cooperation is materializing mainly at the company to company level. Japanese companies have succeeded in entering China's promising nuclear energy market. However, broad nuclear energy cooperation will only be achieved as a result of governmental support in a capital-intensive field. Hence, company-to-company nuclear energy cooperation at the most provides short to mid-term outputs, on a case-specific basis.

Nuclear energy cooperation can significantly improve the two countries' energy security equation when this initiative will be materialized at the governmental level. In such a case, METI's New National Energy Strategy of 2006, that targeted nuclear power generation to increase to a level of 30–40% of total power production by 2030, has a better chance at being achieved—consequently contributing significantly to Japan's energy self-sufficiency.

As for the Chinese government, extensive nuclear energy cooperation with Japan, as examined earlier, may contribute to its regime stability. China's economic boom, which resulted due to the government's political and economic openness and reduction of its involvement in the market, among other factors, is currently facing intensified urbanization and a sharp increase of electricity demand. Development of nuclear energy, in concert with other energy measures, may permit China to continue to carry out economic development and avoid negative side-effects from accelerating industrialization.

Nuclear energy cooperation occurring at the bilateral and domestic levels can be one of the significantly positive influences on the countries' broad-spectrum political relations, for example issues such as the long-standing territorial and energy dispute over the East China Sea: Diaoyutai/Senkaku islands.

CHAPTER TEN

TRANSITION MANAGEMENT AND INSTITUTIONAL REFORM: THE CASE OF A TRANSITION TO HYDROGEN AS A MOTOR FUEL IN THE NETHERLANDS

Daniel Scholten

Abstract: *This chapter explores whether the Dutch government has so far been able to ensure the institutional reforms necessary for the success of technical changes inherent in an energy transition. Special attention is paid in this respect to transition theory and transition management. The former addresses technical innovation processes within their wider societal context and stresses the importance of institutional change in socio-technical transitions. The latter transforms the theory's insights into practical policy and has been adopted by the Dutch government as the official transition governance framework. Considering a prominent pathway towards an infrastructure for the use of hydrogen as a motor fuel in the Netherlands by 2050, this piece investigates what the achievements of transition management have been so far in regard to institutional reforms in general and for hydrogen in particular. The chapter concludes that transition management has so far ignored institutional reforms and seems to have neglected valuable lessons of transition theory regarding the coevolution of institutions and technologies.*

Keywords: Transition, Hydrogen, Institutions

1. Introduction

Increasing fossil fuel scarcity and deteriorating environmental conditions make transition to a more sustainable energy system urgent. However, behind this simple notion lies a very complex reality: such change does not only involve technical and economic aspects but also institutional reforms. On the one hand, moving to a new energy system will be a daunting task because it will disassemble and reassemble the whole infrastructure from sources and production facilities to transmission, distribution, storage means and end-use applications. Moreover, these technical changes strongly affect vested economic interests. On the other hand, as many scholars have already noted, to be effective with certain new technologies, "a nation requires a set of institutions compatible with and supportive of them. "The ones suitable for an earlier set of fundamental technologies may be quite inappropriate for the new" (Nelson 1994: 58). Indeed, history is full of examples where existing institutional structures posed obstacles to the success of new technologies and complementary industries which "require institutional reform if they are to develop effectively" (Nelson 1994: 58). The question arises how one can achieve such a 'societal transformation process' in which an energy system would change structurally over an extended period of time (Rotmans et al. 2001), especially with regard to aligning institutional changes to the technical ones inherent in an energy transition.

A promising approach to address this complexity is transition management. Based on transition theory, it addresses technological innovation processes within their wider societal context, claiming that changes in one without the other will remain fruitless over the long run. It poses a step forward from the energy policies of the 1990s based on bottom-up, market-oriented policies in that it not only looks at techno-economic aspects, but also tries to create a framework wherein government policy makers, industry stakeholders, non-governmental organizations (NGOs), and scientific institutes actively pursue accompanying institutional changes to ensure the emergence of a new energy system. Realizing its potential, the Dutch government has embraced transition management as the official governance framework for a transition towards a more sustainable energy system in 2000 (Fourth National Environmental Policy Plan (NMP4)). Since then, it has been further developed by the Dutch Ministry of Economic Affairs (EZ), which has taken on the role of 'transition manager' (EZ 2004).

Considering transition management's novelty as a policy perspective, it has yet to prove itself in practice. As such, it needs to be investigated whether transition management has so far been able to ensure that institutions align to the technical changes of the energy transition. This chapter hence aims to provide an in-depth look at Dutch transition policy to the present day with an eye to institutional reform. To do so, it will take a closer look at the transition to the use of hydrogen as an 'alternative motor fuel', one of the possibilities identified by the Dutch government as part of creating a more sustainable energy system (EZ 2004: 22).

The chapter is structured as follows. The first part starts by elaborating transition theory in order to understand the complexity of socio-technical change. It not only introduces institutions into a framework of technical change, but also highlights the importance of institutional reforms complementary to technical changes for a successful infrastructure transition. Afterwards, the attempt of transition management to turn these lessons into a concrete mode of governance with accompanying policy measures will be presented.

The second part introduces a Dutch hydrogen transition pathway (developed by the Energy Centre of the Netherlands (ECN) as part of the wider European Union's (EU) HyWays Project) to illustrate the conceptual framework. It will be preceded by a short introduction to the various technologies that make up a hydrogen infrastructure.

The third part brings both former sections together in an attempt to analyze the role transition management has played until now in governing a transition to hydrogen. This will finally allow us to take a critical look at what the concrete achievements of transition management concerning institutional reform are in relation to our hydrogen case.

2. Infrastructure Transitions

As we saw in the introduction, a transition to a more sustainable energy system implies more than a simple replacement of one energy source by another. To understand the complexity of such a development requires a closer look at socio-technical transitions.

2.1 *Transition Theory*

Governing an energy transition is tricky. Since it has yet to happen, one cannot know for sure what it is likely to look like and how such a transition might occur and be brought about. Fortunately, studies of past transitions provide some guidance as to what can be expected. Transition theory builds upon the discipline of science and technology studies and stresses the interrelation between technical and social change. It analyses "how technology is shaped by social, economic, cultural and political forces as well as how new technologies shape society and the interaction between various actors" (Elzen et al. 2002: 11). Hence it helps to understand a transition process within which both technical and institutional changes are objectives.

Transitions are defined as "processes of socio-technical evolution in which economic, institutional and technological structures develop interactively and change drastically in the long run" (Bruggink 2005:

6). This has led Rotmans, Kemp, Loorbach, van Asselt, Molendijk, Geels, and Verbong to classify such a transition as a societal transformation process of an evolutionary nature lasting at least one generation (twenty-five years) in various publications over the last decade. Special attention is paid to three 'levels' and the interactions among them: energy landscapes, socio-technical regimes, and technological niches. Figure 10.1 captures these three levels and their interaction in one central model.

2.1.1 *Socio-Technical Regimes*

The starting point in any study involving transitions is the socio-technical regime. Socio-technical regimes consist of "a set of technologies embedded in a social, political, and institutional context with its associated regime-specific set of rules, procedures, habits and practices"

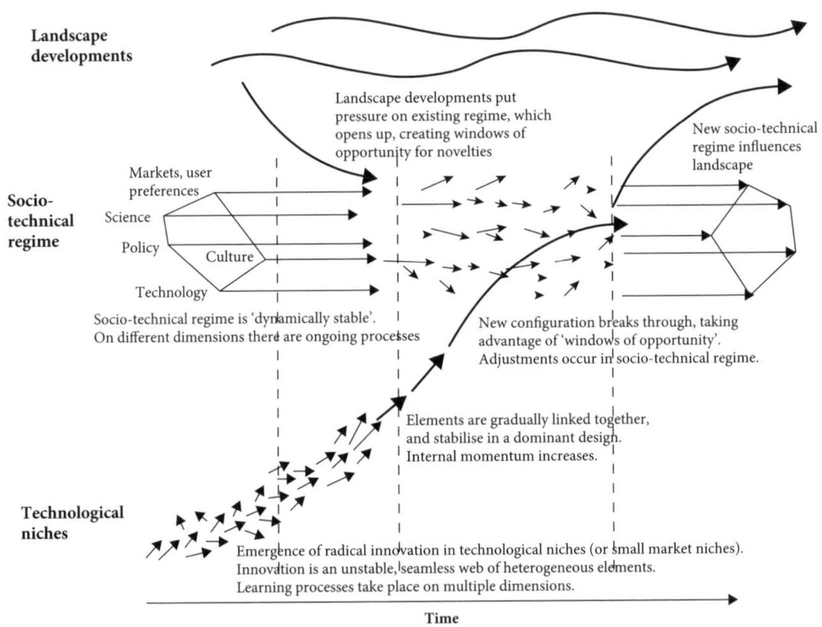

Source: Shackley and Green 2007: 222.

Figure 10.1 The multi-level perspective on technological transitions.

(Shackley and Green 2007: 223).[1] An example of a socio-technical regime for the transport sector is given in Figure 10.2 below. Energy infrastructures are a good example of such socio-technical regimes as they are often constructed around a core set of related technologies in order to function (like a gas pipeline network), while they also necessitate rules of the game to ensure a certain performance (for example, in terms of efficiency and effectiveness, public values, and technical robustness). These rules of the game make up the formal and informal institutions that "structure political, economic and social interactions" and "set limits on human behavior" (North 1990). As we shall see, a possible infrastructure for the use of hydrogen as a motor fuel for personal transportation represents just such a socio-technical regime. What makes the regime work are the actors linking the various technologies, markets, and rules together. These can be technology suppliers (universities, R&D departments, and knowledge institutes), industry incumbents (production, trade, storage, transmission, distribution, and retail), government policy makers (federal, state, regional, and local authorities and agencies), and private actors (consumers and public and private organizations).

A regime is dynamically stable; while the different technological, economic, and institutional aspects are adjusted to each other, bringing stability, they are also constantly changing within the range or context of the regime.[2] However, the regime is quite unable to adapt to changes outside the discourse of the regime, hence the idea of system transition. Thus, it is at this level where "'lock-in' [and path-dependence] may take place, whereby technological regimes emerge alongside

---

[1] This conception differs from the definition of technological regimes Nelson and Winter (1982) use because the rule-set incorporates a wider embeddedness than engineering communities only. Many social groups affect technical trajectories. This led Rip and Kemp to define a technological regime as "the rule-set or grammar embedded in a complex of engineering practices, production process technologies, product characteristics, skills and procedures, ways of handling relevant artefacts and persons, ways of defining problems; all of them embedded in institutions and infrastructures" (Elzen et al. 2002: 12–13).

[2] The interrelations between these dimensions are the key to understanding the performance of the whole regime. Innovative technologies, institutional changes, and changes in consumer demand, for example, can have profound impacts not only on their own dimension, but across the whole regime or infrastructure. Without the technical feasibility to allocate goods and services in physical networks, there can be no economic transaction. Moreover, infrastructures often involve market failures, necessitating institutional governance. However, without proper governance, technological functioning and economic performance will be sub-optimal.

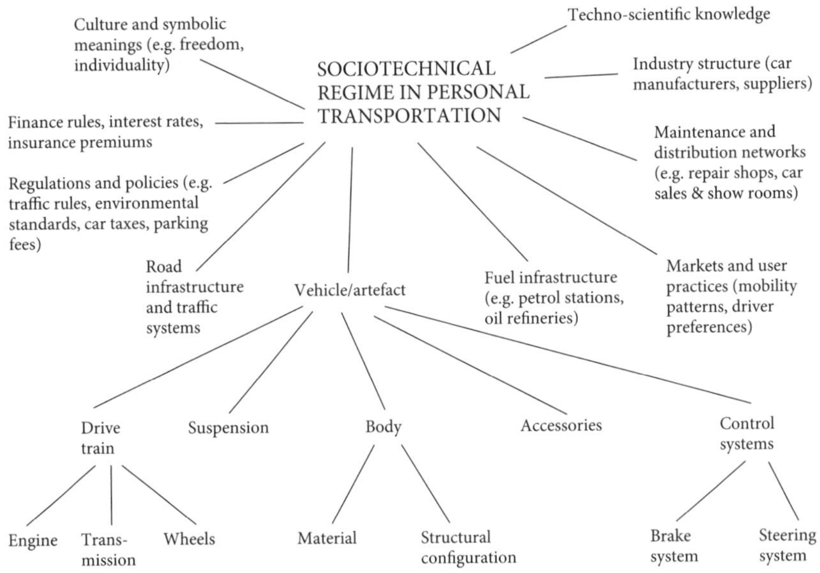

Source: Elzen et al. 2002: 12.

Figure 10.2 Socio-technical regime in personal transportation.

institutional and social changes" (Shackley and Green 2007: 223). This dynamic stability also implies that changes within the regime are of a more incremental nature, whereas changes emerging outside of the regime force the creation of a new regime and hence are of a more radical nature.

2.1.2 *Landscapes and Technological Niches*

The landscape level provides "the dominant assumptions, values and deeply rooted socio-economic trends at a given period of time" (Shackley and Green 2007: 222). This includes the key 'Weltanschauung' or philosophy behind policy-making and reflects "the dominant perception of 'problems' and the way to resolve those problems" (Shackley and Green 2007: 223). These belief systems change very slowly and are hence often considered as the constant external environment or structure in which the socio-technical regime operates but over which its actors have no direct control. To illustrate, "in our own society, the landscape is given by a concept of economic growth which has relied since the industrial revolution on fossil fuels, albeit with major shifts from coal, to oil and natural gas" (Shackley and Green 2007: 223). However, recently two trends have also begun to shape the landscape level: institutional liberalization and increasing environmental awareness. Other important aspects include global fossil fuel scarcity and its geographical concentration, increasing energy demand, consumer preferences, demographic composition, urbanization levels, available technology, cultural characteristics, national specificities, migration, and wars. Changes at the landscape level can have profound impacts on the socio-technical regime as they can alter the rules of the game in which regime actors operate by changing the relevant goals and performance criteria, and even by altering the perception of what is the appropriate or 'normal' way of doing things.

The technological niche is the level at which "new technologies emerge and some develop, protected from the full effects of competition with the dominant technologies in the socio-technical regime" (Shackley and Green 2007: 224). Some examples are the army, industrial vehicles, and shipping. Niches are important to test on a small scale the expenses, safety, and reliability of innovations. They serve as incubators for radical novelties that have a "low technical performance, are often cumbersome and expensive" and operate under different performance indicators or selection criteria (Elzen et al. 2002: 13). Niches also create opportunities to construct social networks among (regime) actors that support specific innovations such as alternative and renewable energy technologies like hydrogen.

2.1.3 *The Multi-Level Perspective and the Transition Process*

The relations among the three levels are characterized as a nested hierarchy. As such, niches are embedded in regimes and regimes are embedded in landscapes. Here, the socio-technical regime "accounts for stability of existing technological development and the occurrence of trajectories. The macro-level of landscape consists of slow changing external factors, providing gradients for the trajectories. The micro-level of niches accounts for the generation of radical innovations" (Elzen et al. 2002: 14). Within this context, innovation mostly occurs within the existing regime, and involves incremental changes towards regime optimization. This gradual pattern results from the interconnectivity of the various regime dimensions that allows only for small changes within the margins of the regime. However, "although gradual change is the common pattern, there are plenty historical examples of more radical transitions" (Elzen et al. 2002: 14), like the transitions from coal to oil and on to gas-based energy systems. Here, novelties emerge and are developed in niches under the old framework—existing regime and landscape—often encountering a mismatch with the established socio-political dimensions (Freeman and Perez 1988). New opportunities are typically countered by certain regime actors threatened by them. In both cases, "it is the alignment of developments (successful processes within the niche reinforced by changes at regime level and at the level of the socio-technical landscape) which determine if a regime shift will occur" (Kemp et al. 2001: 277). This shows the importance of the complementarity of institutional and technical change for a successful transition.

As will be shown later, a transition to hydrogen represents a radical change (that necessitates the emergence of a new socio-technical regime around hydrogen-based technologies for its success). In this respect Geels distinguishes between two general transition patterns, one of substitution and one of transformation. These patterns are closely related to the four phases delineated in Figure 1. These are referred to as the pre-development phase, the take-off phase, the acceleration phase, and the stabilization phase (Loorbach et al. 2008: 296).[3]

In the technological substitution pattern, "the existing socio-technical system is relatively stable, until the new technology breaks through in mainstream markets" (Elzen et al. 2002: 15). One has to consider that while radical novelties are developed in niches outside the existing socio-technical regime, the regime may still be in a phase of incremental change. Because of the technology push character of such a breakthrough, the novelty can remain hidden from the regime for a quite a while. However, when a new technology is able to gain "sufficient internal momentum, and landscape developments put pressure on the existing regime, a breakthrough can occur" (Elzen et al. 2002: 15). As regime actors are caught by surprise, a new socio-technical regime is formed in a phase of radical change. This process is depicted in Figure 1,

[3] Other research defines them similarly as the R&D or invention and innovation phase, the niche and early-market phase, the pervasive diffusion or mass-market phase, and the saturation (and senescence) phase. In the pre-development phase, the regime is stable, although there is increasing bottom-up innovation in niches and the landscape is slowly changing. In the take-off phase, change gets underway and forces a rethinking of the state of the system and the regime begins to shift. In the acceleration phase, structural changes take place as the result of the accumulation of social, technical, economic, and institutional changes. In the stabilization phase, the speed decreases as a new regime is established.

where the third phase represents the shift from the current to the new socio-technical regime and the whole regime is in a state of flux.

In a "technological transformation route, the regime becomes unstable sooner. [...] It is heating up, opening up, because of persistent problems or landscape changes" (Elzen et al. 2002: 16). An important driving factor in this route is the influence of landscape trends that affect the socio-technical regime in the second phase of Figure 1. "The loosening up of the existing regime may create multiple windows of opportunity for novelties and stimulate actors to experiment with many technical options" (Elzen et al. 2002: 16). Niches thus develop and are legitimated by current problems and dynamics in the regime or at the landscape level. This shows itself as a prolonged period of experimentation involving many novelties. Hopefully this leads to a period of 'cooling down,' wherein the number of technical options is narrowed and hopefully one technology emerges as the dominant one, subsequently making the formation of a new socio-technical regime around it possible.

2.2 *Transition Management*

Concerning the governance of the transition, the Ministry of Economic Affairs has embraced the concept of transition management. Transition management is an attempt by Dutch scholars to turn transition theory into policy prescriptions. In the light of transition theory, transition management prescribes the way in which "society-wide and complex system innovations can be guided deliberately towards [certain goals]" (Bruggink 2005: 10). New in transition management is:

> the pretension that long-term, coevolutionary processes in society are not an inescapable fact of life, to be understood in retrospect rather than be controlled proactively. Policy makers are no longer viewed as detached and clinical observers of socio-technical change, they are considered as active participants able to promote technological innovation in the right direction. They cannot only offer a promising perspective towards sustainability; they can actively make it happen (Bruggink 2005: 10).

However, because of liberalization, privatization, and deregulation, the role of the state as the central authority that controls and shapes society has transformed. The traditional nation-state's power is weakening and "instead of enforcing its will on society, the state may help private actors to jointly realize the public interest" (Hisschemöller et al. 2006:

1228). Innovation is associated with collaboration, not competition. Transition management hence represents a shift from government to governance, or put differently, a move from a monocentric interventionist perspective to a polycentric model with many decentralized actors that have only limited coercive capacity (Hisschemöller et al. 2006: 1228). The governance in a market setting is complex—multi-layered, multi-dimensional, and multi-actor—and focuses therefore on "context control" through incentives like regulation, taxes, and subsidies. (Kemp and Loorbach 2006: 14).

The method proposed by transition management for achieving regime change is to inject "goal-directing processes into socio-technical transformations" (Kemp and Loorbach 2006: 22). Two major elements in this respect are the "stimulation of technological innovations in market niches through participative involvement of companies, research institutes and civil society and the creation of challenging visions for a sustainable future as a roadmap towards system innovation" (Bruggink 2005: 10). Key in the proposed method are transition platforms where various public and private stakeholders in the energy transition continuously readdress visions, transition paths, and experiments in an iterative and reflexive manner consisting of four phases: 1) organizing a multi-actor network, 2) developing sustainability visions and transition agendas, 3) mobilizing actors and executing projects and experiments, and 4) evaluating, monitoring, and learning (Kemp and Loorbach 2006: 17). Transition management therefore is "not an instrument as such, but a new perspective for decision-making and governance" that offers practical suggestions on how to manage evolutionary processes and guide them towards predefined policy outcomes (Kemp and Loorbach 2006: 22). This implies that under transition management institutional design is considered as "an innovation process, where market practices and market institutions coevolve" (Midttun 2005: 46).

The first point of action includes structuring problems and envisioning goals. Here multi-stakeholder forums or transition platforms are convened, facilitated by the government, and sector experts come together and exchange their perspectives, expectations, and agendas. It is important to have an adequate mix of incumbent interests and societal actors in innovative activities (frontrunners) in order to bundle expertise and finances with enough innovative potential to overcome lock-in into existing systems. "Emphasis is placed on mutual learning, consensus building and developing a shared problem perception in relation to the goals" (Smith and Stirling 2008: 8). The aim is to come to shared options for long-term goals, like CO_2 reductions, that the transition should achieve. Visions and scenario-building play an important role in this respect. These visions are not intended to be an optimal blueprint with fixed goals to reach, but simply target the creation of images that are desirable to reach. This leaves room for adjustments along the way.

Afterwards, transition pathways are established in the direction of the visions and experiments are conducted in niche markets to explore their possibilities. Sector specific pathways are constructed through a mix of backcasting from visions (to prioritize goals and pathways) and forecasting from current possibilities (to remain realistic). Consequently, the paths involve both short-term and long-term goals. "This identification of transition pathways provides a framework for the subsequent development and support of alternative socio-technical practices in strategic niche experiments" (Smith and Stirling 2008: 8). The success or failure of the experiments then determines which innovations and niches continue and which can be rejected. At the same time, new innovations are continuously added as time progresses, rejuvenating the search for and the development of new novelties.

The next step focuses on processes of adaptation that link short-term actions in niche experiments to long-term goals and pathways. Lessons are drawn not just for instrumental improvement of the niche practices themselves, but also at higher levels concerning required revisions in the framing of associated policies, marketing, user relations, across the entire socio-technical configuration (Smith and Stirling 2008: 9).

This is essentially where scrutiny takes place and policies are evaluated. Here niches also inform about institutional requirements, constraints, and opportunities with regard to the expected outcomes (visions). This makes the process reflexive and allows transition images and paths to adapt over time. To facilitate this learning, transition management favors a gradual step-by-step approach. Supporters argue that "evolutionary change, founded on trial and error, while wasteful in the short-term, is often the most intelligent approach in the long run" (Kemp and Loorbach 2006: 16).

Finally, the matter of institutionalization comes to the fore. Although the least considered in transition management literature, it is also "acknowledged to be the most important, since it overarches all the other activities" (Smith and Stirling 2008: 9). The institutionalization of a new regime out of niches is politically and economically very difficult since it requires "demanding policy reform, infrastructure investment, market restructuring, citizen mobilization, and changed consumer behavior" (Smith and Stirling 2008: 9). It is here where vested interests in the incumbent regime clash with those of newcomers to innovative technology, and where the big test for transition management lies. Hence transition management advocates, next to supporting niches, "control policies to put pressure on the existing regime [...] to bring about transitions" (Kern and Smith 2008: 2). Examples include creating a level playing field and promoting participation by the public and NGOs. This poses a step forward from the energy policies of the 1990s based on bottom-up, market-oriented approaches, because it not only looks at market incentives and technology push policies, but also acknowledges regulatory and cultural barriers to innovation and learning processes.

To summarize, it is important to note that the iterative process is ever-present in all of the four phases depicted in Figure 1. As a consequence, under transition management transitions are always starting and ending in an ongoing evolutionary process, while it is still possible to determine the beginning and ending of the individual socio-technical regime changes based on their particular niche innovation. In the case of the latter, policies shift from keeping options open, holding participatory discussions, and strategic niche management in the early development phase to mobilizing actors and creating visions in the take-off phase. Policy focus then shifts to steering societal and market forces, guarding developments, and choosing promising visions in the acceleration phase to end with policies that consolidate and stimulate the new regime in the stabilization phase.

3. Towards a Hydrogen Infrastructure in the Netherlands

Let us now turn to a practical example of a transition to a more sustainable energy system: a technical pathway towards the use of hydrogen as a motor fuel for personal transportation in the Netherlands. This will add some meat to the conceptual skeleton based on transition theory set up in the previous section. I follow the pathway developed by the Energy Center of the Netherlands (ECN) for the European Union's (EU) HyWays project. In doing so, the focus is on hydrogen as a motor fuel only, neglecting stationary applications, the reason being that hydrogen is mostly considered an alternative to oil in the transport sector and other applications are only contributory.[4] SenterNovem (Hoogma 2005) also comes to similar conclusions in this respect. The case serves to illustrate what a possible transition to a hydrogen infrastructure would look like and basically represents the s-curve of Figure 1 in that it maps the road from a niche technology to the establishment of a socio-technical regime. This illustration will help to visualize the efforts of transition management so far in pursuing institutional reform to support the functioning of an emerging hydrogen infrastructure in part three. Before doing so, some background knowledge on hydrogen is required. To this end several matters will be addressed: why make a transition to hydrogen, what are its technological characteristics, and how are the various parts of the supply chain connected?

3.1 *Hydrogen Technologies and Networks*

In the Energy Report 2004 of the Dutch Ministry of Economic Affairs, the Dutch cabinet restated its ambition to achieve a sustainable energy system by 2050 by means of an energy transition. The use of hydrogen as an "alternative motor fuel" in combination with fuel cell engines is considered a prominent possibility in achieving this goal. Not only can hydrogen be produced and distributed in a CO_2 neutral manner, thereby abating climate change, its use in fuel cell vehicles leaves solely water vapor as an exhaust gas, limiting air pollution in urban areas. Moreover, hydrogen could contribute to supply security as it can be produced in a variety of ways and from a variety of sources, and is hoped to lead to innovation in a number of hydrogen-related technologies, creating cutting-edge export products. As such, hydrogen harbors great promise for dealing with current environmental and geopolitical concerns surrounding the affordability, availability, and safety of energy (McDowall 2006: 1242–1243). However, the sense of urgency in social and environmental spheres for an energy transition is not matched by support in the form of economic and technical opportunities. Currently, road transport is very much based on

[4] According to the EU HyWays (2007: 27) project, in the Netherlands hydrogen only stands a chance in use in transport applications, as competition from electricity and gas for stationary applications is too stiff.

fossil fuels, and in comparison hydrogen and fuel cell technologies are rather costly and immature. In addition, the absence of a hydrogen-refuelling infrastructure on the one hand, and the lack of an infrastructure for fuel cell vehicles on the other also creates a 'chicken and egg' problem because building one without the other serves no purpose (McDowall 2006: 1242–1247). Consequently, policy makers working on the energy transition are trying to develop hydrogen technologies and deploy hydrogen applications by creating collaboration among the relevant actors in an effort to coordinate the introduction of all parts of a hydrogen infrastructure.[5]

Without being distracted by the precise technicalities, a brief presentation of the hydrogen supply chain is in order. Three core issues shape hydrogen as a motor fuel: 1) hydrogen is an energy carrier, not an energy source; 2) hydrogen is very versatile, making its use very attractive because it can be produced from a variety of sources and in a variety of ways, can be transported, distributed, and stored using different means, and can be used for a variety of (motive) end-use applications; 3) efficient hydrogen usage requires fuel cell technologies instead of internal combustion engines for vehicle use. As such a hydrogen infrastructure supply chain contains various parts with various possibilities. See also Figure 10.3 below.

Hydrogen sources are fossil fuels like oil, coal, and gas, alternative sources like biomass and nuclear energy, and renewable sources such as wind, solar, hydro, and geothermal energy. Consequently, the production of hydrogen also involves a wide variety of processes: "using heat and catalysts to reform hydrocarbons or carbohydrates, using [renewable or nuclear] electricity to split water, through the gasification of coal [and biomass] and more experimental processes involving sunlight, plasma discharge and micro-organisms" (Lovins 2003: 5). Currently, the steam reforming of fossil fuels (SMR) and in the medium run, the gasification of coal and biomass, are the most competitive options but also the most polluting in terms of emissions. Of course, carbon capture and storage technologies (CCS) can remedy this problem, but they would increase the costs while some concerns around the environmental impact of sequestrations also remain. On

[5] As we have seen, transition management aims to facilitate cooperation among stakeholders in various platforms.

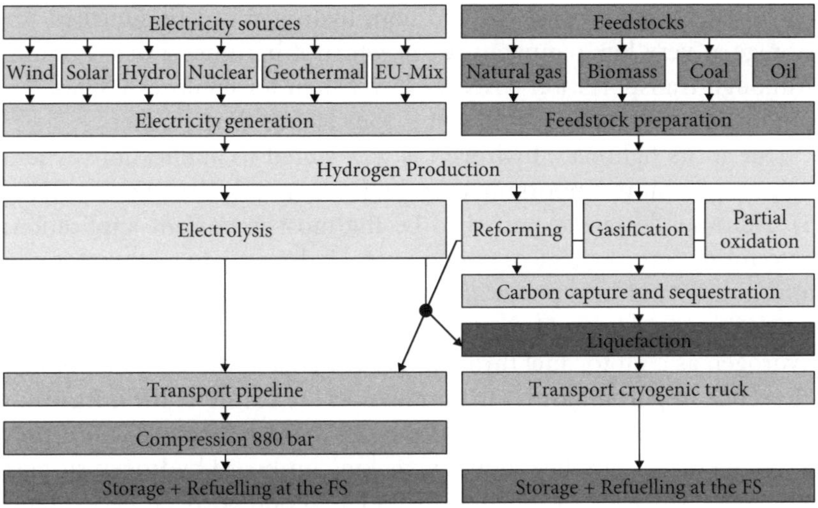

Source: Wietschel et al. 2006: 1286.

Figure 10.3 Hydrogen supply chain possibilities.

the contrary, using renewable sources is clean, but it is only expected to produce the necessary amounts at affordable prices after large investments and in the very long run. Nuclear electricity takes somewhat of a middle position in terms of competitiveness and cleanness, but has to overcome acceptability issues (Dunn 2002: 244–246 and EC 2003).

Currently, the most common way to transport hydrogen currently are tanker trucks carrying liquid hydrogen, using double-walled insulated tanks to limit the amount of boil-off. This means of transport is generally used over large distances. In the future, it will also be possible to transport compressed gas using pipelines, cylinders, and tube trailers. For the latter two options high pressure compression is required, making general use limited to small quantities over short distances. The most efficient way to transport gas is through a network of (underground) pipelines; however, the initial expense of installation needs to be overcome, and hydrogen pipelines must be able to handle the lower density and higher diffusion rate of hydrogen relative to natural gas. Pipelines are ideally suited to handle large quantities and long-distance energy transmission (Dunn 2002: 247–248). Concerning storage, hydrogen is typically compressed and stored in gas cylinders or spherical containers both above and underground, or on vehicles. Compressed hydrogen gas is the simplest and cheapest method for on-board vehicle storage. In addition, hydrogen can be liquefied for storage at very low temperatures, at which it becomes a dense liquid. Although transportation costs are lower than for compressed gas, the production of liquefied hydrogen is four to five times higher.[6]

Due to its lightness, hydrogen is well suited to applications where lightness is more important than bulk, as in vehicles (Lovins 2003: 5). Automobiles are expected to be the most important application, accounting for over 85% of traction use. Hydrogen can be transformed into motive and heat power through fuel cell engines.[7] The hydrogen infrastructure discussed further below will focus on the end-use of hydrogen as a motor fuel through fuel cell engine-driven automobiles. Here, one important aspect that influences the nature of the infrastructure needs to be taken into account: on-board reforming is not preferred by the automobile industry, making on-board hydrogen storage necessary. The use of fuel cell vehicles based on hydrogen necessitates the building of filling stations that can cover the national road network while being preferably close to local distribution centers and networks of hydrogen sources. Additionally, there need to be enough fuel cell vehicles on the road to economically warrant the building of filling stations (the same goes for pipelines). This is the core dilemma facing the economic deployment of hydrogen as a motor fuel.

The versatility that makes hydrogen use so attractive also poses a core problem to formulating a pathway, as it blurs what the exact technical characteristics of a hydrogen economy are. The HySociety project of the European Union, for example, identifies 42 different hydrogen supply chains of which Wietschel et al. identify ten as the economically most likely ones for the EU 25 (2006: 1288–1290). Fortunately, the EU's HyWays project, carried out by ECN, has identified six possible hydrogen supply chains for mobile applications more attuned to the Dutch energy sector specificities (de Groot et al. 2005): see Table 10.1 below. Moreover, this list of possible hydrogen infrastructures is

[6] More futuristic possibilities are the storage of hydrogen in metal and chemical hydrides and carbon structures (Dunn 2002: 246–247). Other means might be storage in abandoned oil and gas fields, aquifers, and salt caverns.

[7] Technically, internal combustion engines and hybrid electrics are also options. In terms of cost per mile driven, the efficiency of fuel cells is paramount to the overall competitiveness of hydrogen vis-à-vis fossil fuels and internal combustion engines. However, as "fuel cells are not subject to the same thermodynamic limits as fuel-driven engines, because they are electrochemical devices," one cannot simply compare the two (Lovins 2003: 5).

Table 10.1 Dutch HyWays hydrogen supply chains.

Chain	Source	Production	1st Conversion	Transport & distribution	2nd Conversion	End-use
1	natural gas	onsite SMR	–	–	CGH2	mobile & stat.
2	natural gas	central SMR	–	pipeline	CGH2	mobile & stat.
3	natural gas	central SMR	liquefaction	LH2 truck	LH2	mobile only
4	(hard) coal	gasification + CCS	–	pipeline	CGH2	mobile & stat.
5	biomass	gasification	–	pipeline	CGH2	mobile & stat.
6	offshore wind	central electrolysis	–	pipeline	CGH2	mobile & stat.

Source: Mulder 2005 and EU HyWays Member State report 2007: 54.
Legend: LH2 stands for liquid hydrogen; CGH2 stands for compressed hydrogen gas; stat. stands for stationary.

narrowed down further when put in a temporal perspective, as we will see in the next section.

3.2 *A Pathway towards Hydrogen in the Netherlands*

The transition to a sustainable energy system differs essentially from past energy transitions according to Kok (2004: 4). Several reasons account for this. First of all, the duration is estimated to be around 50 years instead of 10 to 20 years due to the long-term interests regarding the environment and supply security that drive the current or upcoming transition, and the more concrete short-term focus regarding costs and benefits that drove past transitions. Secondly, whereas the current transition takes place in a liberalized market setting involving many actors, past transitions took place in a regulated setting with few actors, among which the government took the dominant position. Added to that is the increasingly dominant role Europe plays in the energy sector. Finally, the current transition has a very diverse set of technologies and complex solutions in mind, whereas past transitions had comparatively simple technical goals. Consequently the general public was more positive towards past transitions in their time than they are now towards the current one, which is often considered rather vague in terms of goals, policies, and outcomes. Considering the fact that, as we have seen, building an infrastructure for the use of hydrogen as a motor fuel requires whole new production facilities, storage, and distribution means and end-use applications, one starts to realize the large scale and radical nature of the implied changes. This has led Shackley and Green (2007: 225, 232) to define the transition to hydrogen as one of reconfiguration or the "replacement of a set of interlocking technologies by an alternative array of inter-related technologies which fulfill the same, or similar, functions."

The transition to a hydrogen economy, entailing the production, storage and distribution, and use of hydrogen as an energy carrier for personal transportation in the Netherlands is estimated to begin now and last until at least 2050. Based on certain assumptions and Dutch energy sector characteristics,[8] the EU HyWays project, the paper presented by de Groot et al. (2005), and the 2006 Smit et al. article (both ECN) all present "a bottom-up approach for developing a pathway towards hydrogen energy systems" (HyWays 2007: 7) that takes into account the predicted growth in hydrogen demand. The pathway paints the following picture. Three core urban areas are planned as early user centers to start up demand: Rotterdam, Amsterdam, and the Arnhem-Nijmegen area. The introduction of hydrogen in the Netherlands is planned to start with niche applications such as fleets of buses and industrial vehicles involved in demonstration projects.[9] Hydrogen will

[8] Strong determining factors for the hydrogen introduction path are (HyWays 2007 and de Groot et al. 2005): 1) industrial hydrogen in Rijnmond area; 2) extensive natural gas grid and the large share of gas in the national energy mix; 3) strong logistics and transport capabilities (for import of feedstock like coal and biomass); 4) promotion policies for off-shore wind energy; 5) availability of huge carbon storage locations; 6) population and transport density.

As such, it is predicted that: 1) the introduction of hydrogen will begin with on-site reforming based on natural gas; 2) natural gas will play a dominant role until at least 2030; 3) SMR capacity in Rotterdam area will produce additional hydrogen; 4) in time, large-scale production with CCS and renewable hydrogen will be needed; 5) hydrogen production will be cost-competitive early on, but fuel cells only later; 6) hydrogen infrastructure will probably become economically viable in 2022 for the Netherlands; 7) pipelines will grow from the existing industrial pipeline infrastructure.

Some important underlying assumptions are: 1) the cost-margin between large-scale and small-scale hydrogen production is the driver of infrastructure (pipeline or truck) development; 2) the model is highly regionally- and time-dependent; 3) a decrease in fuel cell costs is a sensitive parameter for the development of hydrogen demand.

[9] Niches identified by SenterNovem are: shipping, navy, consumer electronics, batteries and storage, local hydrogen networks (Rijnmond), emergency power generators, internal transports, industrial vehicles, urban transport, and the public sector. Larger

be initially produced on-site at refuelling stations in the urban centers through small-scale steam methane reforming of natural gas, because of the lack of an existing large-scale hydrogen network and presence of a well-developed gas infrastructure. The next filling stations will be erected along the main highways between the user centers and along the gas pipeline network, and will be supplied by tanker trucks or through on-site reforming. The Rotterdam area is being considered as the starting point for a national hydrogen pipeline infrastructure due to the existing hydrogen production facilities and industrial hydrogen pipelines. In essence, the first and third supply chains of Table 10.1 start the hydrogen transition.

By 2030 the transition towards a hydrogen infrastructure will be well under way. After the initial stage local hydrogen communities and refuelling stations will be connected to form larger hydrogen distribution networks and local production will be replaced by large-scale facilities (and CCS) when demand for hydrogen has become high enough. In urban areas in the west of the Netherlands (Randstad) the beginnings of a centralized infrastructure for the use of hydrogen as a motor fuel have emerged out of the initially industrial hydrogen network existing in the Rotterdam area. The core production takes place through central steam methane reforming of natural gas in Rotterdam with CCS, while plants are being built to feed regional networks evolving from the Rotterdam area. The pipeline network will also grow by integrating regional hydrogen networks, production plants, and fuel stations (see Figure 10.4 below). From the possible six supply chains, the first and third are increasingly phased out or incorporated into the expanding network. Natural gas remains the main source while coal gasification, biomass, and wind energy will start to become interesting options due to rising gas prices. If CCS technologies are still expensive then more biomass will be used than coal. It is also assumed that there is a market for the use of hydrogen as a motor fuel; that is, fuel cell cars are slowly but surely gaining a market share and replacing internal combustion engine vehicles. A big issue for governance will be the simultaneous development and deployment of production facilities, pipeline networks, and fuel cell vehicles. Hence, capacity is sometimes too large

applications are: shipping construction, transport, energy producers and distribution, consumer electronics, industrial gas producers, and filling station build-up.

and sometimes too low for the market. In such cases, the natural gas grid functions as an overflow area of hydrogen. This would fit with the Dutch government's greening of gas project.

Summed up, a transition to a hydrogen-based motor fuel infrastructure involves mainly a movement from on-site to central production facilities, from small-scale to large-scale production near user centers, from liquid hydrogen trucks to hydrogen gas pipelines, from fossil to renewable sources, and from SMR to electrolysis. Overall, the transition is from more disconnected local networks to one interconnected national hydrogen infrastructure, or in other words from decentralization to centralization. In this phase the second hydrogen supply chain of Table 10.1 seems to establish itself as the dominant one while the fourth, fifth, and sixth chains play a marginal role.

By 2050, hydrogen will be available throughout the country (based on penetration rates of Smit et al. 2006: 1395). It is expected that a connected hydrogen pipeline system will be in place in 20 to 25 of the 40 regions that cover most urban areas in the Netherlands. The penetration of cars running on hydrogen is expected to be 40% by 2050. Fuel cell vehicles are believed to be the only vehicles in production and are thus slowly replacing internal combustion engine vehicles. Heavy trucks may still drive on biofuels, however. The widespread demand justifies a national pipeline infrastructure.[10] Hence, distribution would take place through a pipeline system much in the same way natural gas is distributed today, although tanker trucks will also be used to deliver liquefied hydrogen in remote regions where they are economically sounder than pipelines. Hence some coexistence of pipelines and trucks will remain. On-site hydrogen production will also exist marginally in local areas further away from hydrogen pipelines. It is also hypothesized that the centralized pipeline spreading out from Rotterdam might be connected to Antwerp, while the German Ruhr area will extend itself to the Arnhem-Nijmegen area. Hydrogen is produced centrally and on a large-scale through the steam methane reforming of

---

[10] The model shows that local small-scale production may generate enough demand to warrant pipelines and large-scale production (Smit et al. 2006). The large-scale centralized system will evolve when it has lower costs than small-scale on-site production. The ECN model also estimates that the cumulative investments will be 11 billion euros, but that a fully developed pipeline system would cost between 12 and 20 billion euros.

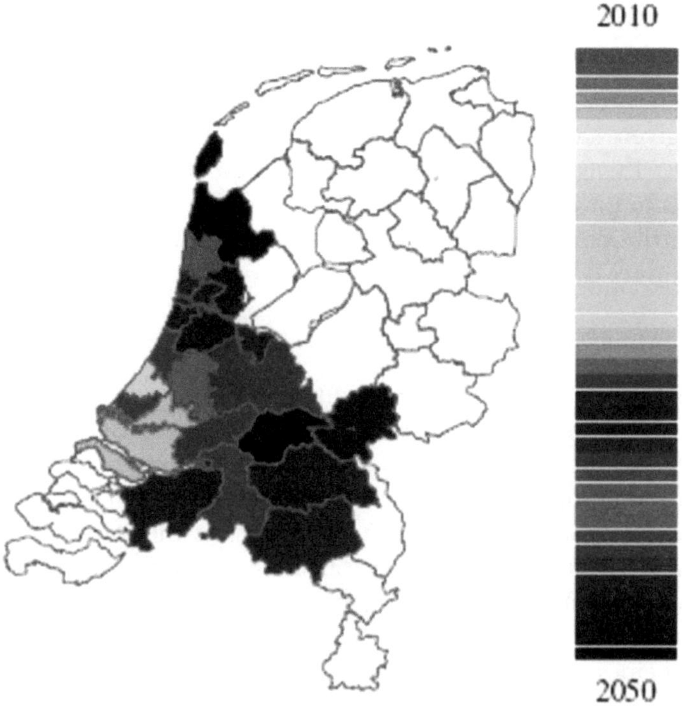

Source: Smit et al. 2006: 1394.

Figure 10.4. Growth of a hydrogen pipeline infrastructure.

natural gas and gasification of coal with the use of carbon capture and storage technologies. In this situation it is hypothesized that oil has peaked and coal is increasingly replacing natural gas as the main source, while the need for renewables steadily increases the share of biomass and wind in the energy mix. Production takes place near the location of the sources, although the existing natural gas pipelines are likely to favor refining natural gas into hydrogen in Rotterdam. Wind-based hydrogen is likely to function as a balancer when too much electricity is produced from wind for the electricity grid. Hydrogen and electricity will comprise the energy carriers powering vehicles and homes, respectively, as there are no breakthroughs expected in electricity storage. In the end, the second hydrogen supply chain of Table 10.1 will remain dominant while the fourth, fifth, and sixth chains start to play an increasingly larger role. Nevertheless, chains one and three will still remain in remote, less populated areas.

4. Transition Management and Institutional Reform in the Hydrogen Case

We now turn our attention to how the energy transition has so far been given shape and the role transition management has played therein. This section will not only show how the Dutch government intends to govern a possible transition to hydrogen, but also analyzes its contribution to institutional reforms complementary to the technical pathway described above.

4.1 *Managing a Transition towards Hydrogen in the Netherlands*

The Fourth Dutch National Environmental Policy Plan (2000) is the starting point for both a transition towards a more sustainable energy system and the introduction of transition management as the governance framework. The Ministry of Economic Affairs (EZ), in charge of energy and innovation policy, has taken on the role of transition manager. The heart of the energy transition project[11] is currently based on seven transition platforms where public and private actors meet to develop shared visions, pathways, and experiments. The platforms and their themes are mainly derived from consultations among existing energy sector incumbents and scenario studies conducted under the long-term energy supply strategy (LTVE) project drawn up in 2000 (Kern and Smith 2008: 3; EZ 2004: 11). The project aimed to create themes of a future energy supply that is clean, affordable, and secure. It basically came up with four of the current seven themes (two more were added in 2006 and one in 2007). They are: new gas, chain efficiency, green resources, sustainable mobility, sustainable electricity, the built environment, and greenhouse as energy source (SenterNovem 2009).

In 2002 EZ started the Project Implementation Transition management (PIT). It aimed to find out whether the various themes would have enough "support, enthusiasm and commitment" from the relevant stakeholders (Kemp and Loorbach 2006: 19). The result was a conditional yes, meaning that stakeholders were willing to invest if the transition management approach were made more concrete, clear

[11] The energy transition project is here used as a term "to summarise all activities initiated by the ministry of economic affairs to implement 'transition management' in energy policy following the National Environmental Policy Plan" (Kern and Smith 2008: 2). These activities are also captured within the term, 'energy transition policy.'

visions were developed and government support, both financially and process-wise, were guaranteed. The beginning energy transition also led to the policy renewal project wherein the government looked for new ways of steering and contemplating about energy policy instruments and changing its relationship with society in general and with business in particular (Kern and Smith 2008: 4). It has led to a more participatory and interactive form of governance.

After these initial developments, the Ministry of Economic Affairs started to develop the seven themes into strategic visions in 2003–2004. To this end, public-private transition platforms were established for each theme. The transition platforms were tasked to work out "possible transition pathways along which an energy transition can be achieved" (Kern and Smith 2008: 3). The platforms existed of stakeholders recruited from existing policy networks. Consequently, businesses became the dominant actor group in these platforms, while government actors, scientific institutes, and NGOs are few in number. In June 2006, for example, the sustainable mobility platform contained three representatives from government, ten from business, three from NGOs, and none from scientific institutes and other intermediaries such as municipalities, provinces, SenterNovem, and advisory boards. (Kern and Smith 2008: 5). The new gas platform similarly included one representative from government, six from business, one from NGOs, three from scientific institutes, and one from intermediaries. Since 2005 the first transition pathways are being explored in technological niche experiments carried out by stakeholder coalitions. The goal is, in line with transition management, to test the possibilities of the platforms.

2005 also saw two major organizational changes to the energy transition project. First was the introduction of an energy transition taskforce consisting of high-level members from Dutch industry and the public sector to complement the existing platforms. Second was the creation of an interdepartmental energy transition directorate encompassing civil servants from various relevant ministries.[12] While the taskforce is essentially an advisory group that oversees the transition process, identifies strategic directions, and aims to "strengthen the role of the

[12] Economic Affairs (EZ); Housing, Spatial Planning and the Environment (VROM); Transport, Public Works and Water Management (V&W); Agriculture, Nature and Food Quality (LNV), Finance (Fin); Development Cooperation (OS) (part of Foreign Affairs).

platforms" (Kern and Smith 2008: 3), the directorate is hoped to integrate transition policy and ongoing policies. Since then the taskforce has become the dominant actor in the process and published a national transition action plan in 2006. As the facilitator, the Ministry of Economic Affairs also created a trendsetters desk in order to remove institutional barriers for initiators of experiments. The desk provided support for policy and legislation as well as financial support. By and large, the activities of EZ have made the transition process mature and ensures the commitment of hundreds of professionals (Loorbach et al. 2008: 308).

On 25 February 2008 the 'regieorgaan energietransitie' or energy transition directing organ succeeded the taskforce as the leading body of the energy transition for the coming five years (SenterNovem 2009). The organ's main task is to advise the government on how to best facilitate the market, create support for the transition among public and private actors, guard the coherence among the various platforms and experiments, and prioritize promising pathways. Whereas the taskforce aimed at sketching ambitions, planning visions, and creating a high profile for the transition, the organ focuses more on its execution and has therefore a more directing character. It is made up from the seven platform chairmen, three independent members, and the chairman of the organ itself. The organ is supposed to function as the independent and most important body that helps the ministers of Economic Affairs and of Housing, Spatial Planning and the Environment, formulate policy. The organ is also supported by the six ministries that are part of the directorate.

The energy transition is funded through public subsidies and investments by companies. The overarching framework is the National Energy Research Strategy (EOS). In it, the energy transition project has led to a new subsidy scheme called the Unique Chances Subsidies Scheme (UKR), which provides specific funding to transition experiments. It succeeds a program for feasibility studies. The UKR tries to bridge the gap between transition experiment requirements and existing energy policy instruments, and is also a reaction to the criticism that the "Dutch innovation system lacks support for high-risk innovation" (Kern and Smith 2008: 5).

So where does hydrogen fit into this picture? In essence there is no specific hydrogen (and fuel cells) program or unified Dutch policy on the matter (Hoogma 2005). On the one hand, hydrogen is currently a potential future discussed in the energy transition project. More precisely, hydrogen is part of two of the seven themes developed by the Ministry of Economic Affairs in light of the energy transition project: 'new gas' and 'sustainable mobility.' While hydrogen plays a role in the former as a cleaner of gas to be used in stationary applications, in the latter it takes a more prominent position as the replacement of gasoline and diesel (although it is definitely not considered the only option in this regard). Actually, in both platforms hydrogen is only one of four working groups (the area of green gas in the new gas platform) and one of four pathways (driving on hydrogen in the sustainable mobility platform). In addition, other ministries (Housing, Spatial Planning and the Environment (VROM) and Transport, Public Works and Water Management (V&W)) are looking into hydrogen for their interests in CO2 emission reduction and alternative fuels, respectively. On the other hand, hydrogen research is conducted outside of the transition project framework. There it continues as part of energy R&D policy at knowledge institutes and universities. The NWO/ACTS Sustainable hydrogen program and H2 dialog are good examples of this. Internationally, the Dutch participate in the CUTE project (demonstration project in Amsterdam with hydrogen buses) and a number of EU projects such as HyWays (pathways), Naturalhy (mixing with gas), Storhy (storage), EIHP (mobile and infrastructure), and HySafe (safety) (SenterNovem 2005).

In general, the focus of research thus far has been on hydrogen technologies. Hydrogen production (mostly reforming), distribution (mixing with gas), storage (diverse), and stationary applications (PEM and SO fuel cells) all receive a fair amount of attention (SenterNovem 2005). Interestingly, the little existing research in applications focuses mostly on stationary fuel cells, not the transport applications one might expect from the hydrogen pathway presented in Part Two (which explicitly leaves out stationary applications (HyWays 2007: 27)). This also contrasts with SenterNovem's stance on hydrogen in the Netherlands: that transport applications are the most viable ones (Hoogma 2005). Non-technical research has thus far largely focused on safety and public acceptance studies on the one hand, and developing pathways and their accompanying policy tools to support the deployment of hydrogen in the market place, on the other. Some demonstration projects have also gotten underway. Most notable is the CUTE project in

Amsterdam. The expectation is that in the near future more local and regional activities will be carried out.

More disturbing is the share of hydrogen experiments in the UKR under the energy transition umbrella. Under the new gas platform only one out of 22 projects relates to hydrogen, according to SenterNovem (2009). In the sustainable mobility platform, none of the seven projects carried out involves hydrogen directly, although their outcomes may affect hydrogen's prospects. One thing is clear, however: all of the platforms, and especially their experiments, are of a strong technological character. A lot of goodwill is necessary to see non-techno-economic use in the results of the experiments conducted under the UKR. Fortunately, more possibilities for non-technical research exist under the larger EOS program, as it allows for a wider range of research and a more long-term focus. These research subsidies, however, fall under 'normal' energy policy, not the energy transition project as such. In any case, hydrogen development in the Netherlands happens predominantly in the form of industrial research.

4.2 *Transition Management and Institutional Reform*

With sufficient background knowledge on the possibilities for a hydrogen pathway for the Netherlands and of the practical track record of transition management so far (the transition project or policy), we can now turn our attention to transition management's contribution to institutional change. To state it outright: institutional reform has thus far largely been neglected in the energy transition project. For instance, in their more general evaluation of energy transition policy, Kern and Smith (2008: 7) argue that there seems to be a strong focus on a techno-based reasoning that should bring about change. Indeed, there is no platform on social or behavioral change or even an institutional pathway for that matter. In a similar assessment of transition management Loorbach et al. (2008: 310) are less critical, but nonetheless find that "although experiments also involve societal and institutional aspects, they are still insufficient to amount to a fundamental debate, let alone change, at the level of societal culture and structures." As such, they also conclude that "the behavioural, institutional, structural and cultural changes needed are more or less ignored" (Loorbach et al. 2008: 309). Let us now take a closer look behind these strong statements. A number of interrelated issues emerge.

The core cause of the absence of institutional reform is mostly ascribed to the overrepresentation in the transition platforms of stakeholders in the existing energy sector in comparison to more innovative outsiders (Kern and Smith 2008: 7; Loorbach et al. 2008: 311–312). As we have seen, in the beginning, stakeholders were mostly selected from among existing energy sector incumbents. Indeed, the share of societal actors, NGOs, or renewable energy companies with no stake in the current energy infrastructure, like industrial hydrogen producers, engineering companies, and local authorities, is low. Similarly, the taskforce also favored big players in the current Dutch energy system. Indicative of this is Shell's prominent position within Dutch energy transition policy as both the chairman of the energy transition taskforce and as a representative in the various transition platforms. This preference for energy sector incumbents is contrary to transition management teachings, which clearly emphasize the important role innovators and other societal actors play in transition platforms.

The reason why the dominance of business actors is considered negative is because it has supposedly led to a focus on technological innovation rather than social or institutional change (Kern and Smith 2008: 7). A likely reason for this is that the Netherlands has an ample number of experienced companies in the petro-chemical industries (Shell, GasUnie, and Akzo Nobel, to name a few), and the port of Rotterdam functions as one of the main refineries and distribution centers for overseas oil products. As such, Dutch industry interests are vested in the further development of energy production, distribution, and storage technologies when it comes to enhancing sustainability. In essence, there is a conflict of interest wherein the taskforce, dominated by representatives from the oil and gas sector, urges for investments that primarily benefit themselves, like CO2 storage technologies. Similarly, the absence of a Dutch automobile industry and little military industry is likely the cause for the relative lack of attention paid to hydrogen transport applications. Hence these interests shift the focus on hydrogen from an 'alternative motor fuel' to a 'cleaner of gas,' drastically reducing its profile. Added to that is the short-termism of business actors. For them, profits need to be made in the foreseeable future, not 2050. This moves the policy focus away from more fundamental social and institutional changes of which the outcome is uncertain, towards options targeting realistic or solvable goals, often more technical and economic in nature.

As the government also needs to achieve results, Loorbach et al. (2008: 311) argue that it might have opted to pursue transition paths more attuned to existing policy goals and relied on already existing stakeholder networks to create the necessary cooperation. The lack of a clear hydrogen vision on the government's part might be a good example of this. Consequently, most of the seven platform themes represent innovations that are complementary to or an extension of existing energy technologies like the greening of gas, and not radical new ones like a switch from oil to hydrogen as a motor fuel. The focus of transition policy hence seems to be on incremental optimization instead of developing radically new technologies. This may not be a problem, as the accumulation of small steps can result in big change, but it is more likely that this gradual process will end with incremental change, that is to say, change that conforms to the existing regime. If so, new technologies are not likely to necessitate institutional reform as they are likely to function under the existing socio-technical regime with its specific set of technologies and institutions. Hence transition policy creates a setting in which institutional reform is, and perhaps can continue to be, neglected. Considering, however, that a transition to the use of hydrogen as a motor fuel represents radical change, transition policy has so far largely failed to address institutional reform and the complex goal of aligning institutional change to the technical changes of a transition to hydrogen.

The focus on technical aspects has another repercussion: transition management seems to add little that is new to transition policy, apart from the fact that it deliberately tries to organize stakeholder consultations to get the technological (and economic) policies going. A comparison with the EU's research programs on hydrogen illustrates this. For example, the Roads2hycom project focuses on finding opportunities for research efforts and early adoption means. It is attuned to R&D support, following a learning-by-searching approach, and is more technology-specific. The Hylights project investigates the preparation of large-scale demonstration projects for transport and hydrogen deployment, following a learning-by-doing approach and maintaining a focus on the economic side. The HyWays project, as we have seen, focuses on creating European and member state pathways towards hydrogen based on country preferences and creating specificities and visions for hydrogen energy systems based on expert opinion. Finally, the HySociety project identifies the broader challenges (opportunities and barriers), actors, and factors, and then aims to develop a concrete action plan. Together, these research projects cover most of the relevant aspects of a transition to hydrogen just as well as Dutch transition policy, if institutional aspects are neglected, as they are in our case. As such, the added value of transition theory and management is lost on the transition project, as it only uses the established model around the technical pathway.[13]

To summarize, it is interesting to note how much of the above is confirmed by Hisschemöller's experiences in the H2 dialog (2008).[14] He points to the three main complaints by various stakeholders in the hydrogen business on transition policy (based on transition management): 1) large power differentials between regime actors and other market parties, 2) the dominance of a monopolistic knowledge network (that neglects radical innovations outright), and 3) the lack of a clear and consistent strategic vision by the government (regarding hydrogen). Indeed, the combination of these factors result in a situation where transition management attempts to manage radical transitions, while the practical focus is on evolutionary change, that is to say, regime optimization paths. As a consequence, institutional reforms are not addressed. It is no wonder, then, that the focus on technology has led to the fact that on the landscape level the process of liberalization seems to outweigh sustainability concerns, that on the regime level the energy transition has limited impact on the current energy regime, and that on the niche level the market with its focus on efficiency and effectiveness remains the dominant selection mechanism, while more long-term social and environmental selection criteria are neglected (Kern and Smith 2008: 6–8). Hence, Loorbach et al. state that "although a major step forward from the fragmented and purely bottom-up approaches in the 1990s, the [energy transition] so far thus

[13] This can also be seen in the fact that the Hylights project has a similar division into four phases shown in figure 1, and distinguishes between policies of a more protective, competitive, regulatory, and obligatory nature in the four phases, respectively. Related ECN studies also distinguish between a focus on R&D and experiments in the first phase, subsidies in the second, a trading system in the third, and a market take-over in the final phase.

[14] In December 2007 the H2 dialog came to an end. It involved some fifty people from public services, business, and scientific institutes who came together to discuss the possibilities of hydrogen for increasing the sustainability of the Dutch energy supply. This dialogue was not carried out under the transition management paradigm, but was funded by the NWO/ACTS program 'Sustainable Hydrogen.'

does not yet fully benefit from the systemic approach [read transition theory] underlying transition management" (2008: 309).

4.3 *Prospects for Institutional Reform*

If we follow the above reasoning, we can put transition management's neglect of institutional reforms squarely into the hands of policy makers. They ignored transition management's focus on newcomers and instead focused on existing policy networks with energy sector incumbents when organizing transition platforms. And it was this overrepresentation of vested interests that offers an explanation for why the energy transition was biased towards incremental improvements over radically new technologies from the beginning. If so, the obvious remedy lies in the expansion of niche market actors (such as renewable energy companies with no stake in the current energy infrastructure), NGOs, and other societal groups. A solid mix of incumbents' expertise and finances and newcomers' innovative potential could overcome lock-in into existing systems.

Moreover, it needs to be realized that it may still be too early to assess transition management's track record on institutional reform. The early phases of a transition focus is on exploring and developing technical alternatives (keeping options open and not picking a winner). This renders institutional change premature, as one does not know which technology is going to be the next big thing and hence which institutional requirements need to be fulfilled. In defense of transition management and policy makers, one needs to recognize that we simply might not have arrived at the institutionalization part of the cycle yet. In other words, we have to wait and see a bit longer before passing judgment. Although this gives some hope that institutional reforms will become the focus of future transition policy, there might be more to it than that.

Hisschemöller et al. (2006: 1234) state that transition management "reveals an institutional bias in that it articulates opportunities for collaboration and competition in a particular way, thereby creating a context for policies, regulations, and instruments, which may at first glimpse look identical but are given a specific meaning by their institutional context." Consider in this respect that the government plays the role of a facilitator and not an authority. They point to the fact that the focus on finding consensus among stakeholders might inhibit quick and responsive decisions, consequently leading to a more incremental change process favoring businesses as usual options. In addition, the lack of competition might yield conservatism rather than innovation. For example, "most observers do not see a substantial impact of the energy transition on 'regular' energy policy. [...] Core energy policy issues like security of supply, liberalization and affordable prices are not being reframed by the energy transition" (Kern and Smith 2008: 6). As such, stakeholder dialogue on innovation might inhibit what it aims for, innovation. Hisschemöller et al. (2006: 1234) hypothesize further that "the more complex infrastructure requirements, the greater the likelihood that major government interventions will be needed." Hence they argue that low-profile governance like transition management will likely "fail to make the transition to a hydrogen based economy, either because of the collective good character of heavy infrastructures or because advanced options are being excluded from competition." Yet specific measures to support breakthrough options are vulnerable to the critique that they undermine the level playing field, even as—or because—they are meant to provide an equal level playing field for newcomers" (Hisschemöller et al. 2006: 1234).

The big threat of Hisschemöller's view seems to be that transition management's dealing with institutional reform might come too late in the transition process. By the time we reach the third (or fourth) phase of Figure 1, the governance choices of the first two phases will have determined the consequent process and might even predetermine the outcome. Most striking in this respect is the focus on keeping options open in the early phases. In essence, the longer this is key government policy, the longer it will be market forces that by and large shape the transition process. Only when one 'best' option appears, can the government take a clear stand and support a certain technical option. However, market forces might by then have predetermined the technical option. Radical innovations that require extensive infrastructure investments, for example, are likely to be excluded already despite possibly promising final results. Kok (2004) also warns that the complexity of any upcoming transition could lead to the slow degradation or abandonment of transition efforts because of the sheer amount of options and the focus of transition management on keeping options open in the initial transition stages. He urges the government to start making clear choices and to stick to them (picking a winner) at an earlier stage. This, however, implies a bigger role for the government than is perhaps currently possible under transition management.

A similar argument is put forth by Kern and Smith (2008). Pointing to the need to take a closer look at the dynamics behind how new regimes emerge from niches, Kern and Smith (2008: 8) state that the politics of transitions are neglected. They call attention to four aspects: 1) long term goals and commitment versus short term success; 2) level playing field versus certainty for investors; 3) focus on frontrunners versus regime incumbents; 4) control policies versus nurturing niches. Interesting in this respect is how all of these tradeoffs cut across stakeholder interests. Whereas the incumbent industry actors are likely to favor all of the latter, new players are likely to favor all the former options, highly politicizing the transition. Transition management addresses these struggles during 'institutionalization' but does not address how these struggles influence and condition policy formulation and implementation beforehand. As such, the transition may well be determined by politics before it begins.

4.4 *Towards a Different Approach: Theories on Coevolution*

The key problem in assessing the lack of institutional reforms seems to be that it is in line with the expectations of its supporters, while causing anxiety over its possible long-term implications among critics. Although this could serve as sufficient closure on transition management's efforts concerning institutional change, there is one thing that I feel is sincerely neglected in the debate thus far: what efforts have transition theory and management made to model the coevolution between institutions and technologies? Although this shifts the focus from practice back to the conceptual, I feel it is necessary to address it as its neglect is reflected in the practices we have so far observed.

Let me return to Nelson. At the end of his 1994 article (page 61) he notes that formally modeling the coevolution of supporting institutions to technical changes is not an easy task. He rightly gives two main reasons for this, which are the key to any answer as well: it is not yet clear how various institutions can be represented, and the (social) evolutionary processes involved in changing institutions are very different from those "built into extant formal evolutionary models." The former reason concerns matters such as what is meant by institutions and how one can compare them to technologies in order to make statements on coevolution or the lack thereof. The latter reason urges asking how a coevolution process between institutions and technologies can be operationalized. My point is simple: there has not yet been a serious attempt by transition theory and management to address either difficulty in aligning institutions to technologies. While transition theory gives a solid description of the overall transition process and defines the various dimensions, actors, and factors and their interplay, it fails to clearly define how the dynamics behind the evolution of technologies (as represented by the s-curve in Figure 1) influences the dynamics behind institutional changes at the regime (and landscape) level. How do, for example, changes in technologies affect institutions, when do technical changes spur institutional reforms, under which conditions do they do so, and what role can the government play in facilitating these processes? Moreover, how can one talk about institutional reforms while it remains unclear which institutions are required? Transition theory fails to answer these questions. As such, it is unable to clearly causally link technological and institutional evolution because one does not have the analytical tools to compare both and hypothesize about their possible coevolution. If transition management is to make any policy progress in the light of institutional reforms complementing technical change, both issues need to be addressed.

Fortunately, some initial starting points for such an exercise exist. Concerning the comparison of institutions and technologies, Künneke et al. (2008) have laid down a method to match institutions to technologies. In their article they make a step-wise progression from technological characteristics to the critical technical functions of infrastructures, then identifying the control mechanisms and critical transactions that should facilitate the former in order to derive the mode of organization for the institutional framework. Künneke et al. are interested "in modes of organization that guarantee the coordination of critical transactions related to critical technical functions of liberalized infrastructures" (2008: 235). The mode of organization is operationalized along a divide between hierarchical integration (top-down, public orientation) and autonomous market processes (bottom-up, private orientation) with various intermediary modes in between. The institutional framework is operationalized as ownership (private or public), the regulatory framework (competition policy or sector specific), and the market structure (monopoly, oligopoly, competition). Four technical functions can be considered critical for safeguarding the technical complementarity and performance of infrastructures: interoperability, interconnection, capacity management, and system management.[15] The key to linking these modes of organization to technical specificities lies in the fact that higher levels of criticality in the four functions require a more top-down governance approach (often matching large-scale centralized infrastructures), while lower levels of criticality in the four functions can be organized through a more bottom-up approach guided by market forces (often matching small-scale decentralized infrastructures). A mismatch hence occurs when the technical and institutional dimensions are not both governed by the same approach.[16]

Concerning the formal modeling of coevolutionary dynamics between institutions and technologies, I feel that we have to go all the way back to basics, that is, theories on (social) coevolution, technological evolution, and institutional evolution. Only then will we be able to understand the key dynamics behind that coevolutionary process. As it stands, transition theory does a fine job in describing past transitions, but falls short if it is applied to future transitions. Yet without a thorough understanding of the dynamics behind (co)evolutionary processes, it will prove difficult to make any statements on when institutional reforms are required or can be expected. For policy makers hoping to derive clear-cut recommendations on how necessary reforms

---

[15] Interoperability focuses on the "mutual interactions between network elements" and as such "defines technical and institutional conditions under which infrastructure networks can be utilized." Interconnection deals with the "physical linkages of different networks [...] As such, interconnection is closely related to the technical system boundaries." Capacity management concerns the allocation of "scarce network capacity to certain users or appliances." Questions as to who should get access, the facilitation of actual access, and the daily operational management are addressed. Finally, system management "pertains to the question of how the overall system (e.g., the flow between the various nodes and links) is being managed and how the quality of service is safeguarded" (all Finger et al. 2006: 11–12).

[16] According to recent studies by Finger, Künneke and Groenewegen the economic, social and technical performance of infrastructures is dependent on the degree of "coherence between the technical and the institutional coordination" or mode of organization (Finger et al. 2006: 13). They showed, for example, by analyzing infrastructures before and after institutional liberalization, privatization and deregulation that the performance differed because institutional changes were not matched by technological ones. Thus where the market structure, regulatory framework and the ownership structure of the energy sector has become decentralized, market oriented and is guided by private values, the technological side has remained to a large extent centralized, top-down organized and guided by public values. As such, non-matching institutions and technologies affected performance.

could be brought about this might be disappointing. It is here where further research is required and upon which we should focus.

5. Conclusion

This chapter started by pointing out that stimulating only the technological (and economic) aspects of a transition to a sustainable energy system is insufficient to achieve socio-technical regime change, because complementary institutional reforms are also necessary. It then asked whether Dutch policy makers, under the transition management governance framework, have thus far been able to facilitate institutional reforms in the case of a transition to the use of hydrogen as a motor fuel.

In attempting to answer this question, a brief overview of transition theory was given, an approach that combines insights on technical change with the social sciences, thereby addressing technical transitions within their wider institutional context. This highlighted the importance of institutional reforms complementing technical changes for a successful transition and showed the complexity of realizing it. Transition management was introduced as both an attempt to transform transition theories' lessons into practical policy and as the governance framework chosen by Dutch policy makers to govern or manage the transition to hydrogen. Afterwards, this conceptual framework was fleshed out by presenting the example of a Dutch pathway towards the use of hydrogen as a motor fuel.

Attention then turned to the track record of transition management in governing the energy transition so far. Using the hydrogen pathway as an illustration, the role of transition management in its realization, with an eye on facilitating complementary institutional reforms in particular, was investigated. Yet instead of providing valuable information on institutional change, the hydrogen case unfortunately highlighted transition policy's neglect of transition theory and the management's concern about the necessity to incorporate institutional reform. Three interrelated reasons accounted for this. The first reason is the overrepresentation of vested interests of existing energy production and distribution companies in the various transition platforms. Secondly, there exists a bias towards a techno-based research and short-term policy focus in the transition process. Thirdly, the options considered in the energy transition as a whole and for hydrogen in particular represent incremental innovations that fit in the existing institutional structures.

In the end, I argue that the failure of transition policy to address institutional reforms supporting the functioning of a hydrogen infrastructure could be attributed to policy makers who might have misread transition management's teachings, and the fact that it might still be too early to assess transition management's track record in general and regarding institutional change in particular, as reforms are envisioned in the later phases of a transition. In contrast, there is also the possibility that transition management is inherently biased as a specific transition governance paradigm towards certain institutional processes and technical outcomes. Transcending the argument, I note that there has not been a serious attempt in the transition literature and practice so far to address the two difficulties in aligning institutions to technologies as developed by Nelson (1994: 61). Firstly, it is not yet clear how various institutions can be represented so as to compare them with technologies in order to align them. Secondly, it is unclear how the (social) (co)evolutionary processes involved in changing institutions, which are very different than those "built into extant formal evolutionary models," can be operationalized. And without a way to formally conceptualize the coevolution between institutions and technologies, it should not be surprising that transition management has not yet been successful in ensuring institutional reforms that complement the technical changes of a transition to hydrogen.

BIBLIOGRAPHY

Books, Book Chapters, Articles, Official Documents

Abrahamian, E., *Iran Between Two Revolutions* (Princeton: Princeton University Press, 1982).

Afary, J., *The Iranian Constitutional Revolution 1906–1911: Grassroots Democracy, Social Democracy and the Origins of Feminism* (New York: Columbia University Press, 1996).

Agnew, J., *Geopolitics: Re-visioning World Politics* (London: Routledge, 1998).

Agnew, J. and S. Corbridge, *Mastering Space-Hegemony, Territory, and International Political Economy* (London & New York: Routledge, 1995).

Akbardzadeh, S., "India and Pakistan's Geostrategic Rivalry in Central Asia." *Contemporary South Asia*, vol. 12, no. 2, (2003), 219–228.

Alam, A., *Yad'dashthaj-e Alam* (Alam's Manuscripts), vol. 1, Tehran: Ketab-e Sara, 1993.

Algar, H., *Religion and State in Iran* (Berkeley and Los Angeles: University of California Press, 1969).

Ali, T., *The Nehrus and the Gandhis: An Indian Dynasty* (Basingtoke and Oxford: Picador, 2004).

Allaire, J., "The Impact of Urban Development in China on Global Warming", *China Perspectives*, vol. 1, (2007), 51–61.

Allison, R., "Regionalism, Regional Structures and Security Management in Central Asia", *International Affairs*, vol. 80, no. 3, (2004), 463–483.

Amineh, M. P., *Die Globale Kapitalistische Expansion und Iran- Eine Studie der Iranischen Politischen Ökonomie 1500–1980* (Münster, Hamburg, London: Lit Verlag, 1999).

——, *Globalization, Geopolitics and Energy Security in the Caspian Region* (Den Haag: CIEP, 2003).

——, "Power and Energy Supply Security," *IIAS Newsletter*, vol. 37, no. 6, (July 2005).

Amineh, M. P. and S. N. Eisenstadt, "The Iranian Revolution: The Multiple Contexts of the Iranian Revolution" in: M. P. Amineh (ed.) *The Greater Middle East in Global Politics: Social Science Perspectives on the Changing Geography of the World Politics* (Leiden, Boston and London: Brill Academic Publishers. 2007), 117–145.

Amineh, M. P. and H. Houweling, "Caspian Energy: Oil and Gas Resources and the Global Market" in: *Central Eurasia in Global Politics: Conflict, Security and Development*, eds. M. P. Amineh and H. Houweling (Leiden, London & Boston: Brill, 2005a).

——, "Introduction: The Crisis in IR Theory: Towards a Critical Geopolitics Approach" in: *Central Eurasia in Global Politics: Conflict, Security and Development*, eds. M. P. Amineh and H. Houweling (Leiden, London & Boston: Brill, 2005b).

Amirahmadi, H., *Revolution and Economic Transition: The Iranian Experience* (New York: State University of New York Press, 1990).

Amuzegar, J., "Iran's Oil Stabilization Fund: A Misnomer", *Middle East Economic Survey*, vol. 48, no. 47 (2005).

An, Weihua, Qian, Xuemei (eds), *Commenting on A New Theory of Gulf Petroleum* (Social Science Documentation Publishing House, 2000).

Arjomand, S. A., *The Turban of the Crown: Islamic Revolution in Iran* (Oxford and New York: Oxford University Press, 1988).

Asgar S., *The Constitution of Iran: Politics and the State in the Islamic Republic* (London: Tauris Publishers, 1997).

Atomic Energy Agency, *IAEA Annual Report*, 2007, http://www.iaea.org/Publications/Reports/Anrep2007/index.html.

Atomic Energy Commission, "Concerning the Actions Japan Should Take for the Expansion of Nuclear Energy Use in the World as a Measure against Global Warming", *Atomic Energy Commission*, 13 March 2008.

Babali, T., *Caspian Energy Diplomacy Since the End of the Cold War* (Ankara: Turkish Foreign Policy Institute, 2006).

Bahgat, G., "Europe's Energy Security: Challenges and Opportunities", *International Affairs*, vol. 82, no. 5, (2006), 961–975.

Bajpaee, C., "The Emerging Cold War on Asia's High Seas", *Power and Interest News Report*, 13 February 2006, http://www.pinr.com/report.php?ac=view_report&report_id=439&language_id=1.

Baker & MacKenzie, *Relaw Assist: Renewable Energy Law in China*, Issues Paper, June 2007.

Bakhash, S., *Iran: Monarchy, Bureaucracy & Reform under the Qajar (1858–1896)*, (London: Ithaca Press, 1978).

——, "Iran's Foreign Policy under the Islamic Republic 1979–2000" in: L. C. Brown (ed.) *Diplomacy in the Middle East: The International Relations of Regional and Outside Power*, (London: I.B. Tauris, 2001), 247–258.

Bakhtiari, B., *Parliamentary Politics in Revolutionary Iran: The Institutionalisation of Factional Politics*, (Gainesville: University Press of Florida, 1996).

Balkir, C. and A. M. Williams, eds., *Turkey and Europe*, (London: Pinter Publishers, 1993).

Banani, A., *The Modernization of Iran: (1921–1941)*, (Stanford: Stanford University Press, 1961).

Bar-Yam, Y., *Dynamics of Complex Systems* (Reading: Addison-Wesley, 1997).

Behdad, S., "The Post-Revolutionary Economic Crisis" in: S. Rahnema and S. Behdad (eds.), *Iran after the Revolution: Crisis of an Islamic State*, (London: I.B. Tauris, 1995), 97–128.

Behera, L. K., "India-Iran Economic Cooperation", *World Focus*, vol. 26, no. 8, (2005), 11–13.

Berlin, D. L., "The Great Base Race in the Indian Ocean Littoral: Conflict Prevention or Stimulation?", *Contemporary South Asia*, vol. 13, no. 3, (2004), 239–255.

Berrah, N. et al., *Sustainable Energy in China: The Closing Window of Opportunity*, (Washington: The World Bank, 2007).

Bill, J. A., *The Eagle and the Lion: The Tragedy of American-Iranian Relation*, (New Haven, Conn.: Yale University Press, 1988).

Bowen, W. Q. and J. Kidd, "The Iranian Nuclear Challenge", *International Affairs*, vol. 80, no. 2, (2004), 257–276.

Bozorgmehr, M. "Iran-France Relations in the first Decade of the Islamic Revolution", *Political and Economic Ettela'at*, vols. 111–112, (Tehran: Ettela'at Institute, 1996/97).

Brent, W., "Cleantech Boom....or Bust?", *The China Business Review*, July–August 2007, 33–35.

British Petroleum, *BP Statistical Review of World Energy*, 2006, 2007, 2008, http://www.bp.com

——, *BP Statistical Review of World Energy*, June 2008, http://www.bp.com

Bruggink, J. J. C., "The Next 50 years; Four European Energy Futures", *Energy Research Centre of the Netherlands*, 2005.

Brummer, M., "The Shanghai Cooperation Organization and Iran: A Power-Full Union", *Journal of International Affairs*, vol. 60, no. 2, (Spring/Summer 2007), 185–197.

Buchta, W., *Who Rules Iran: The Structure of Power in the Islamic Republic*, (Washington: Washington Institute of Near East Policy, 2000).

Buzan, B., "The Asia-Pacific: What Sort of Region in What Sort of World?" in: *Asia-Pacific in the New World Order*, ed. A. MacGrew, (London: Routledge, 1998), 68–87.

Calabrese, J., "Indo-Iranian Relations in Transition", *Journal of South Asian and Middle Eastern Studies*, vol. 25, no. 5, (2002), 60–82.

——, "China and Iran: Mismatched Partners", *The Jamestown Foundation*, Occasional Papers, (August 2006).

Central Intelligence Agency, CIA World Factbook, *Japan*, 2007, http://www.cia.gov/library/publications/the-world-factbook/geos/ja.html

——, *China and India pages*, 2008, https://www.cia.gov/library/publications/the-world-factbook/

Chang, P., "China's Policy Toward Iran: Arms for Oil?", *China Brief* (ed. by the Jamestown Foundation), vol. 8, no. 21, (7 November 2008).

Chen, H., "Guanyu jianli shiyou fengxian caigou pingzhangdi jiben shexiang" (Some fundamental ideas about the establishment of oil purchase risk egalization), and "Zhongguo xinde shiyou zhanlüe ying zhongdian yanjiu guoji shichang" (China's new oil strategy must stress international market research), in: Ma H., Wang M. K. (Eds.), *Zhongguo fazhan yanjiu: Guowuyuan fazhan yanjiu zhongxin yanjiu baogaoxuan 2001* and—*2004. (China Development Research: Selection from research reports of the State Council Development Research Centre 2001* and—*2004)*, Beijing: Zhongguo fazhan chubanshe, (2001, 2004), 139–147 and 223–227.

Chen, J. H., "Shiyou zhanlüe yu Zhongguo nengyuan zhengce" (Oil strategy and China's energy policy), *Zhongguo shiyou (China Petroleum)*, vol. 2, (2001), 31.

Chen, M., *China Depends on Oil of the Middle East*, Transnational Energy Security Forum, Beijing, 19–21 June 2004.

Cheng, J., "A Chinese view of China's energy security", *Journal of Contemporary China*, vol. 17, no. 55, (2008), 297–317.

China Industry Research Information Net, Annual Research Report on China's Wind Power Industry, 2007–2008, http://www.36021.cn.

China National Nuclear Corp., *China awards contracts for new four units to Westinghouse*, (26 December 2006), http://www.cnnc.com.cn/2006-12-26/000227517.html.

China National Petroleum Corporation (CNPC), Research Institute of Economics and Technology of China National Petroleum Corporation, *2008 Report on the Development of Domestic and International Oil and Gas Industry*, (January 2009).

China New Energy Chamber of Commerce: 2007 Annual Report on New Energy Industry.

Chinese State Council, *White Paper on Energy*, Beijing, (26 December 2007).

Chubin, S., *Iran's Nuclear Ambitions*, (Washington: Carnegie Endowment for International Peace, 2006).

Cirincione, J., J. B. Wolfstahl and M. Raikumar, *Deadly Arsenals: Tracking Weapons of Mass Destruction*, (Washington D.C.: Carnegie Endowment for International Peace, 2002).

Citizens' Nuclear Information Center, 2007 Bilateral Nuclear Cooperation Agreements, http://cnic.jp/english/newsletter/nit118/nit118articles/nit118bilateral.html.

Closson, S., "Energy Security of the European Union", *Center for Security Studies*, vol. 3, no. 36, (June 2008).

Contaldi, M., F. Gracceva, A. Mattucci, "Hydrogen perspectives in Italy: Analysis of possible deployment scenarios", *International Journal of Hydrogen Energy*, doi: 10.1016/j.ijhydene.2007.12.035, (2008).

Cottam, R., *Iran and the United States: A Cold War Case Study*, (Pittsburgh: Pittsburgh University Press, 1988).

Cottrell, A. J., "Iran's Armed Forces under the Pahlavi" in: G. Lenczowski (ed.), *Iran under the Pahlavis*, (Stanford, CA: Hoover Institution Press, 1978), 389–432.

Cox, R. W., *Production, Power, and World Order: Social Forces in the Making of History*, (New York: Columbia University Press, 1987).

Cui, Minxuan (ed.), "The Energy Development Report of China", *Social Science Documentation Publishing House*, (2008), 272.

Cui, Y. Q.,"Zhuanjia fangtan: ZhongFei nengyuan hezuo jinru xinjieduan" (Specialists' interview: energy cooperation between China and Africa enters a new stage), *Guoji shiyou jingji (International Oil Economy)*, vol. 3, no. 2, (2007).

Curzon, G., *Persia and the Persian Question*, (London: Longman, Green, 1892).

Cutler, R. M., "Cooperative Energy Security in the Caspian Region", *Global Governance*, vol. 5, no. 2, (1999), 251–271.

——, "Turkey and the Geopolitics of Turkmenistan's Natural Gas", *Review of International Affairs*, vol. 1, no. 2, (2001), 20–33.

——, "US-Russian Strategic Relations and the Structuration of Central Asia", *Perspectives on Global Development and Technology*, vol. 6, no. 1–3, (2007), 109–125.

——, "Russia Deepens Gas Hegemony", *ISN Security Watch*, (30 January 2009), http://www.isn.ethz.ch/isn/Current-Affairs/Security-Watch/Detail/?lng=en&id=95931.

Dan, S., "Regional Differences in China's Energy Efficiency and Conservation Potentials", *China & World Economy*, vol. 1, (2007), 96–115.

——, "*Analysis of China's Renewable Energy Development Goals and Policy Impacts*", Paper presented at the international conference "Global Warming and Climate Change, Implications for Human Security, Visions from Asia and Europe", Madrid, (15–17 October 2008a).

——, "China's Energy Policy and its Development" in: *Energy Security: Visions from Asia and Europe*, ed. Antonio Marquina, (Houndsmills-Basingstoke-Hampshire-New York, 2008; Palgrave MacMillan, 2008b), 135–146.

Daojiong, Z., "China's Energy Security: Domestic and International Issues", *Survival*, (Spring 2006), 179–190.

Deffeyes, K., *Hubbert's Peak: The Impending World Oil Shortage*, (Princeton NJ: Princeton University Press, 2001).

Dijking, G.-J., *National Identity and Geopolitical Visions: Maps of Pride and Pain*, (New York: Routledge, 1996).

——, "Geopolitical Codes and Popular Representations", *GeoJournal*, vol. 46, no. 4, (1998), 293–299.

Dittick, P., "National Oil Companies Invest Beyond Borders", *Oil and Gas Journal*, (17 July 2006).

Doran, C.F., "Economics, Philosophy of History, and the 'Single Dynamic' of Power Cycle Theory: Expectations, Competition, and Statecraft", *International Political Science Review*, vol. 24, no. 1, (2004), 13–49.

Dorraj, M. and C. L. Curier, "Lubricated with Oil: Iran-China Relations in a Changing World", *Middle East Policy*, vol. 15, no. 2, (Summer 2008), 66–80.

Dosi, G., "Technological Paradigms and Technological Trajectories; a Suggested Interpretation of the Determinants and Directions of Technical Change", *Research Policy*, vol. 11, (1982), 147–162.

—— and R. Nelson, "An introduction to evolutionary theories in economics", *Journal of Evolutionary Economics*, vol. 4, (1994), 153–172.

Downs, E., "*The Brookings Foreign Policy Studies Energy Policy Series: China*", (Washington: Brookings Institution, 2006).

——, *China: Grapping with rapid energy demand growth*, The Brookings Foreign Policy Studies Energy Security Series, (December 2006), http://www3.brookings.edu/fp/research/energy/2006china.pdf.

Downs, E. S., "The Fact and Fiction of Sino-African Energy Relations", *China Security*, vol. 3, (Summer 2007), 42–68.

——, "Business Interest Groups in Chinese Politics: The Case of Oil Companies," in: Cheng, L. (Ed.), *China's Changing Political Landscape: Prospects for Democracy*, (Washington, D.C.: Brookings Institution Press, 2008), 121–141.

Dunn, S., Hydrogen futures: toward a sustainable energy system, *International Journal of Hydrogen Energy*, vol. 27, (2002), 235–264.

Economy, E., "The Great Leap Backward?", *Foreign Affairs*, (September–October 2007), 38–59.

——, "China's Environmental Challenge", *Current History*, (September 2005), 278–283.

——, *The River Runs Black: The Environmental Challenge to China's Future*, (Ithaca, NY: Cornell University Press, 2004).

Editorial, *Guoji shiyou jingji (International Oil Economy)*, vol. 6, (2007), 1–2.

Ehteshami, A., *After Khomeini: the Iranian Second Republic*, (London: Routledge, 1995).

——, "The Future of Iran's Defence and Nuclear Policy" in: W. Posch (ed.), *Iranian Challenges*, Chaillot Paper, vol. 78, (2006), 79–81.

Elhefnawy, N., *"The Impending Oil Shock"*, *Survival*, vol. 50, no. 2, (2008), 37–66.

Elzen, B., F. Geels, and P. Hofman, *Sociotechnical Scenarios (STSc); A new Methodology to Explore Technological Transitions*, PRET project final report, University of Twente, Enschede, (2002).

Energy Information Administration, *Iran-Country Analysis Brief*, (February 2000), http://www.eia.doe.gov.

——, *China-Country Analysis Brief*, (August 2006).

——, *European Union-Country Analysis Brief*, (January 2006).

——, *Iran-Country Analysis Brief*, (August 2006).

——, *Iran-Country Analysis Brief*, (October 2007).

——, *India-Country Analysis Brief*, (January 2007).

——, *Country Analysis Briefs: Persian Gulf Region*, (June 2007).

——, *International Energy Outlook*, (January 2008).

Energy Transition Task Force, *Interim Report December 2006*, Energy Transition Task Force, (2006).

Eurasia Group, "*China's Overseas Investments in Oil and Gas Production*", Eurasia Group (prepared for the US-China Economic and Security Review Commission), (October 2006).

European Coalition on Oil in Sudan, *Oil Development in Northern Upper Nile*, Sudan, (May 2006), www.ecosonline.org.

European Commission, Green paper—*Towards a European Strategy for the Security of Energy Supply*, (November 2000), http://ec.europa.eu/comm/energy_transport/doc/2005_green_paper_report_en.pdf.

——, *Communication from the Commission to the European Parliament and the Council: EU Relations with the Islamic Republic of Iran*, (7 February 2001), http://ec.europa.eu/external_relations/iran/doc/com_2001_71en.pdf.

——, *Green Paper: Towards a European Strategy for the Security of Energy Supply*, Luxembourg: Office for Official Publications of the European Commission, (November 2001), http://ec.europa.eu/comm/energy_transport/doc-principal/pubfinal_en.pdf.

——, *Green Paper on Energy Efficiency: Doing More with Less*, Brussels, COM (2005) 265 final, (22 June 2005), http://eur-lex.europa.eu/LexUriServ/LexUriServ.do?uri=COM:2005:0265:FIN:EN:PDF.

——, *Green Paper: A European Strategy for Sustainable, Competitive and Secure Energy*, COM (2006) 105 final, Brussels, (8 March 2006), http://eur-lex.europa.eu/smartapi/cgi/sga_doc?smartapi!celexplus!prod!DocNumber&lg=en&type_doc=COMfinal&an_doc=2006&nu_doc=105.

——, *An Energy Policy for Europe*, Communication from the Commission to the European Council and the European Parliament, Brussels, (10 January 2007), http://ec.europa.eu/energy/energy_policy/doc/01_energy_policy_for_europe_en.pdf.

——, *Energy Corridors*, Community Research, Luxembourg (2007), http://www.energy.eu/publications/KINA22581ENC_002.pdf.

——, *Energy for a Changing World: An Energy Policy for Europe-the Need for Action*, (2007), http://ec.europa.eu/energy/energy_policy/doc/2007_03_02_energy_leaflet_en.pdf.

——, "*An Overview of the Sectoral Dialogues between China and the EU*", Brussels, (2008a), http://ec.europa.eu/external_rel tions/china/sectoraldialogue_en. htm#Energy (accessed on October 12, 2008).

——, *An EU Energy Security and Solidarity Action Plan. Second Strategic Energy Review 2008*, Commission Working Staff Document: Europe's Current and Future Energy Position, Demand-Resources-Investments, Communication from the Commission to the European Parliament, the Council, the European Economic and Social Committee of the Regions, Brussels, (November 2008b), http://europa.eu/rapid/pressReleasesAction.do?reference=MEMO/08/703&format=HTML& aged=0 &language=en&guiLanguage=en.

European Commission Special Report, *Hydrogen Energy and Fuel Cells: A Vision of Our Future*, EUR 20719 EN, (2003), http://ec.europa.eu/research/energy/pdf/hlg_vision_report_en.pdf.

European Council, *Presidency Conclusions*, Brussels, (8 March 2007), http://www.consilium.europa.eu/ueDocs/cms_Data/docs/pressData/en/ec/93135.pdf.

Facts on File 1987, New York: Facts on File Inc., (8 February 1987).

Fateh, M., *Fifty Years of Oil*, (Tehran: Payam, 1358/1979).

Federation of Electric Power Companies of Japan (FEPC), *Japan's Nuclear Power Program*, (2008), http://www.japannuclear.com/nuclearpower/program/dereg.html.

Fesharaki, F., *Development of the Iranian Oil Industry: International and Domestic Aspects*, (New York: Praeger, 1976).

Fewsmith, J., Rosen, S., "The Domestic Context of Chinese Foreign Policy; Does 'Public Opinion' Matter?" in: Lampton, D.M. (Ed.), *The Making of Chinese Foreign and Security Policy in the Era of Reform, 1978–2000*, (Stanford: Stanford University Press, 2001), 151–190.

Finger, M., J. Groenewegen and R. Künneke, *The Quest for Coherence Between Institutions and Technologies in Infrastructures*, discussion paper, (2006).

Finger, M., and R. Künneke, "The Coevolution Between Institutions and Technology in Liberalized Infrastructures: the Case of Network Unbundling in Electricity and Railways", Paper presented at the *11th Annual Conference of The International Society for New Institutional Economics (ISNIE)*, in Reykjavik, Iceland, (21–23 June 2007).

Fuller, G. E., and I. Lesser, eds., *Turkey's New Geopolitics*, (Boulder: Westview, 1993).

Ganguly, S., "India's Foreign Policy Grows Up", *World Policy Journal*, (Winter 2003–04), 41–43.

Garver, J. W., *Protracted Contest: Sino-Indian Rivalry in the Twentieth Century*, (Seattle and London: University of Washington Press, 2001).

——, *China and Iran: Ancient Partners in a Post-Imperial World*, (Seattle and London: University of Washington Press, 2006a).

Gasiorowski, M. and M. Byrne, *Mohammad Mosaddeq and the 1953 Coup in Iran: Modern Intellectual and Political History of the Middle East*, (Syracuse: Syracuse University Press, 2004).

Geels, F., "Technological Transitions as Evolutionary Reconfiguration Processes: a Multi-level Perspective and Case Study", *Research Policy*, vol. 31, no. 8–9, (2002), 1257–1274.

—— and J. Schot, "Typology of sociotechnical transition pathways", *Research Policy*, vol. 36, no. 3, (2007), 399–417.

Gerschenkron, A., *Economic Backwardness in Historical Perspective*, (Cambridge: Harvard University Press, 1962).

Global Security, "*Indian Navy" and "People's Liberation Army Navy*", (2007), http://www.globalsecurity.org/military/world/india/navy.htm and http://www.globalsecurity.org/military/world/china/navy.htm.

Goltz, T., *Azerbaijan Diary: A Rogue Reporter's Adventures in an Oil-Rich, War-Torn, Post-Soviet Republic*, Armonk, (NY & London: M.E. Sharpe, 1998).

Gong, G. J., "Wu kan 'meizhiyou'" (Five views on 'oil from coal'), *Zhongguo shiyou jingji (China Oil Economy)*, vol. 7, (2007), 20–21.

Götz, R., *Russlands Erdgas und Europas Energiesicherheit* (SWP-Studie S 21), (Berlin: August, 2007).

Government of China, "*Facts and Figures: China's Main Targets for 2006–2010*", (2006), http://english.gov.cn/2006-03/06/content_219504.htm.

Government of Pakistan, *Brief on China* and *Brief on USA*, Government of Pakistan Board of Investment, undated, http://www.pakboi.gov.pk/Country_Brief/countries_brief.html.

Groot, A. de, R. Smit, and M. Weeda, Modeling Large-scale Hydrogen Infrastructure Development, paper presented at the *Hydrogen Energy Congress & Exhibition*, Istanbul, Turkey, ECN-RX—05–148, (13–15 July 2005).

Guiyang, Z., "How Will China Move Towards Becoming a Low Carbon Economy?", *China & the World Economy*, vol. 16, no. 3, (2008), 93–105.

Guliyev, F. and N. Akhrarkhodjaeva, Transportation *of Kazakhstani Oil via the Caspian Sea (TKOC): Arrangements, Actors and Interests*, (Lysaker: Fridtjof Nansen Institute, 2008).

Guo, X., "The Energy Security in Central Eurasia: The Geopolitical Implications to China's Energy Strategy", *China and Eurasia Forum Quarterly*, vol. 4, no. 4, (2006), 117–137.

Guowuyuan (State Council), "Zhongguodi nengyuan qingkuang yu zhengce" (Situation and policies of China's energy), *Shuili xinxiwang (Water News Net)*, (28 December 2007).

Hale, D., "China's Growing Appetites", *National Interest* (Summer 2004), 137–147.

Hasan-Beygi, A., *Oil, Buy-back and National Interests*, (Tehran: Away-e No Publications, 2002), 79–106.

Hassan, A., "*Pakistan's Gwadar Port—Prospects of Economic Revival*", MA Thesis, US Naval Postgraduate School, Monterey, (California, June 2002).

He Wenping, "*The Darfur issue and China's role*," Symposium on Chinese-Sudanese Relations, Beijing, (26 July 2007).

He, X. J. et al., "Shiyou anquan zhibiao *tixi yu zonghe pingjia" (Indicator system and comprehensive appraisal of petroleum security), Internal paper*, Beijing: Ministry of Land and Resources, (2007).

Hen-Tov, E., "Understanding Iran's New Authoritarianism", *The Washington Quarterly*, vol. 30, no. 1, (Winter 2006/07), 163–179.

Hibbs, M., "Bonn will Decline Tehran Bid to Resuscitate Bushehr Project", *Nucleonics Week*, (2 May 1991), 17–18.

High Representative/European Commission, *Climate Change and International Security*, S113/08, Brussels, (14 March 2008).

Hirschl, R./Schlaak, T./Waterlander, O., *International Gas Markets. Security of Demand and Supply: Different Sides of the Same Coin?*, BOOZ-Allen-Hamilton, USA, (May 2005).

Hisschemöller, M., *De Lamentabele toestand van het energietransitiebeleid; Observaties en kennisvragen vanuit een interdisciplinair beleidswetenschappelijk perspectief*, Vrije Universiteit Amsterdam, W-08/10, (2008).

——, R. Bode, and N. van de Kerkhof, "What governs the transition to a sustainable hydrogen economy? Articulating the relationship between technologies and political institutions", *Energy Policy*, vol. 34, (2006), 1227–1235.

——, R. Bode, M. van de Kerkhof, and T. Stam, *Trajecten voor verduurzaming van de Nederlandse Energievoorziening: met een accent op de bijdrage van waterstof*, Rapportage uit de H2 Dialoog, NWO/ACTS, R-07/06, (2007).

Hodgson, G. M., and T. Knudsen, "The Nature and Units of Social Selection", *Journal of Evolutionary Economics*, vol. 16, (2006), 477–489.

Hoogma, R., "Transition to hydrogen & fuel cells in the Netherlands", *SenterNovem presentation*, Maastricht, (3 November 2005).

Hou, Y. Z., Sun, Z. Y., "Change of Oil and Gas Industry's Mode of Development is Badly Needed for Coping with Energy Challenges", *China Development Review*, vol. 3, (2008), 63–70.

Houser, T., "The Roots of Chinese Oil Investment Abroad", *Asia Policy*, No. 5 (January 2008).

Howard, R., *Iran Oil: The New Middle East Challenge to America*, (London and New York: IB Tauris, 2007).

Hsiao, R. "Energy Security the Centerpiece of China's Foreign Policy", *China Brief*, vol. 8, no. 16, (1 August 2000).

Huang Minxuan, *China Energy Development Report 2008*, (Beijing: Social Sciences Academic Press, 2008).

Hurst, C., *China's Oil Rush in Africa*, (Washington, D.C.: Institute for the Analysis of Global Security, 2006).

——, *China's Global Quest for Energy*, (Washington, D.C.: Institute for the Analysis of Global Security, 2007).

Huwaidin, M. bin., *China's Relations with Arabia and the Gulf*, (London: Routledge-Curzon, 2002).

HyLights—ECN, *Policy support for large scale demonstration projects for hydrogen use in transport*, ECN-E—06-064, (2006).

HyWays, *Member States' Report: Vision on the Introduction of Hydrogen in the European Energy System*, European Commission, Number: SES6 502596, (2007).

International Crisis Group, *India/Pakistan Relations and Kashmir: Steps Towards Peace*, International Crisis Group Asia Report No. 79, (2004), http://www.crisis group.org/.

——, *China's Thirst for Oil*, International Crisis Group Asia Report No. 153, Brussels, (9 June 2008), http://www.crisisgroup.org/.

International Energy Agency, *World Energy Outlook 2004*, IEA: Paris, (2004), www .iea.org.

——, *World Energy Outlook 2005*, IEA: Paris, (2005), www.iea.org.

——, *IEA/OECD Nuclear Energy Agency Study on Security of Energy Supply for Electricity Generation*, (2005), www.iea.org.

——, "Factsheet: Iran", In *World Energy Outlook 2005*, IEA: Paris, (2005), www.iea .org.

——, "Factsheet: Saudi Arabia", In *World Energy Outlook 2005*, IEA: Paris, (2005), www.iea.org.

——, *World Energy Outlook 2006*, Paris: IEA/OECD, (2006), www.iea.org.

——, Statistics 'Share of total primary energy supply in 2006', (2006), http://www.iea .org/textbase/stats/pdf_graphs/29TPESPI.pdf.

——, *World Energy Outlook*, IEA: Paris, (2007), www.iea.org.

——, *World Energy Outlook 2007*, Chinese version, Paris (2007).

——, *World Energy Outlook 2007: Fact Sheet China*, IEA: Paris, (2007), http://www .iea.org.

——, *World Energy Outlook 2007: China and India Insights*, Paris: IEA/OECD, (2007), www.iea.org.

——, *World Energy Outlook 2008*, Paris: IEA/OECD, (2008), www.iea.org.

International Monetary Fund, *World Economic Outlook*, Washington: International Monetary Fund, (2008), http://www.imf.org.

Japan Atomic Energy Commission, *The Framework for Nuclear Energy Policy*, (2005), http://www.aec.go.jp/jicst/NC/tyoki/tyoki_e.htm.

Japan Nuclear Fuel Limited, *Reprocessing*, 2008, http://www.jnfl.co.jp/english/index .html.

Japanese Ministry of Economy, Trade and Industry, *Basic Act on Energy Policy*, (2002), http://www.cas.go.jp/jp/seisaku/hourei/data/ep_2.pdf.

Japanese Ministry of Economy, Trade and Industry, Agency for Natural Resources and Energy, *the Challenges and Directions for Nuclear Policy in Japan: Japan's Nuclear Energy National Plan*, (1 December 2006a).

——, *Energy in Japan*, (2006b).

——, *FY2007 Annual Energy Report*, (May 2008), http://www.enecho.meti.go.jp/english/report/19fy_outline.pdf.

Japanese Ministry of Economy, Trade and Industry, *New National Energy Strategy*, (May 2006).

——, *Agreement on Cooperation Reached by Japan and China at Japan-China Energy Conservation Forum*, Press Release, (2 June 2006).

——, *A Joint Statement by the Japanese Ministry of Economy, Trade and Industry and the People's Republic of China's National Development and Reform Commission on Enhancement of Cooperation Between Japan and the People's Republic of China in the Energy Field*, (11 April 2007).

Japanese Ministry of Internal Affairs and Communications, *Statistics Bureau*, Statistical Handbook of Japan 2008, (2008).

Japanese Ministry of Foreign Affairs, *Second Round of Negotiations to Conclude a Japan-Kazakhstan Agreement for Cooperation in the Peaceful Uses of Nuclear Energy*, (2008a), http://www.mofa.go.jp/announce/event/2007/9/1175496_856.html.

——, *United States-Japan Joint Nuclear Energy Action Plan*, (25 April 2007). http://www.mofa.go.jp/policy/un/disarmament/arms/action0704.pdf.

Japanese Ministry of International Trade and Industry, *Report of the Advisory for Energy Committee*, (1974).

——, *Report of the Advisory for Energy Subcommittee*, (1976).

Jiang, W., "Beijing's New Thinking on Energy Security", *China Brief*, vol. 6, no. 8, (12 April 2006).

Jin, S. L., "Jianli fuhe guoqingdi guojia sheyou zhubei tixi." (Establish a state oil reserve system that fits with our national conditions), *Zhongguo quoqing guoli (China's national conditions and strength*), vol. 9, (2007), 14–17.

Jung, N., *Technical Assistance for the Turkmenistan-Afghanistan-Pakistan Natural Gas Pipeline Project Phase II*, (Manila: Asian Development Bank, 2003).

Kallis, G., "A co-evolutionary institutional approach to sustainable development: theory and potential applications" in: *Proceedings of Frontiers in Ecological Economics*, (Cambridge: Cambridge University Press, 2001).

——, "When is it coevolution?", *Elsevier*, doi: 10.1016j.ecolecon.2006.12.016., (2007).

——, "Socio-environmental coevolution: towards an analytical approach", *International Journal of Sustainable Development and Global Ecology*, vol. 14, no. 1, (2007), 9–19.

Kanagawa, M., Nakata, T., "Analysis of the impact of electricity grid interconnection between Korea and Japan: feasibility study for energy network in North Asia", *Energy Policy*, vol. 34, (2006), 1015–25.

Karbassian, A., "Islamic Revolution and the Management of the Iranian Economy", *Social Research*, vol. 67, no. 2, (Summer 2000), 621–640.

Karshenas, M., *Oil, State, and Industrialization in Iran*, (Cambridge: Cambridge University Press, 1990).

Katouzian, H., *Musaddeq and the Struggle for Power in Iran*, (London and New York: I.B. Tauris, 1990).

Katzman, K., "The Iran Sanctions Act", *CRS Report for Congress*, (25 January 2007), http://www.fas.org/sgp/crs/row/RS20871.pdf.

Keddie, N. R., *Religion and Rebellion in Iran: The Tobacco Protest of 1881–1882*, (London: Frank Cass, 1966).

——, *Roots of Revolution: An Interpretative History of Modern Iran*, (London and New Haven: Yale University Press, 1981).

Kemp, R., and J. Rotmans, "Managing the transition to a sustainable mobility" in: *System Innovation and the Transition to Sustainability: Theory, Evidence and Policy*, ed. B. Elzen, F. Geels, and K. Green, (Cheltenham: Edward Elgar, 2004).

Kemp, R., and D. Loorbach, "Transition Management: a reflexive governance approach" in: *Reflexive Governance for Sustainable Development*, ed. J-P. Voss, D. Bauknecht, and R. Kemp, (Cheltenham: Edward Elgar, 2006).

Keohane, R. O., *After Hegemony: Cooperation and Discord in the World Political Economy*, (Princeton NJ: Princeton University Press, 1984).

Kern, F., and A. Smith, "Restructuring energy systems for sustainability? Energy transition policy in the Netherlands", *Energy Policy*, doi:10.1016/j.enpol.2008.06.018., (2008).

Klare, M., *Blood and Oil: How America's Thirst for Petrol is Killing Us*, (Harmondsworth: Penguin, 2004).

Klid, B., "Ukraine as a Transportation Corridor for Caspian Sea Oil to Europe", *Caspian Crossroads*, vol. 3, no. 1, (1997), 18–22.

Klusmann, B., Third International Conference of the Energy Program Asia, Beijing, *"Global Cooperation Towards Energy Efficiency: Barriers and Opportunities"*, (24–25 June 2006), http://www.iias.nl/epa/node/5.

Kok, W. C., "De nieuwe energietransitie in het licht van de vorige", *KEMA*, (6 October 2004).

Koppenjan J., and J. Groenewegen, "Institutional Design for Complex Technological Systems", *International Journal Technology, Policy and Management*, vol. 5, no. 3, (2005), 240–257.

Kudelia, S., "Revolutionary Bargain: The Unmaking of Ukraine's Autocracy through Pacting", *Journal of Communist Studies and Transition Politics*, vol. 23, no. 1, (2007), 77–100.

Kumaraswamy, P. R., "Delhi—Between Tehran and Washington", *Middle East Quarterly*, vol. 15, no. 1, (2008), 41–47.

Künneke, R., "Institutional Reform and Technological Practice: the Case of Electricity", *Industrial and Corporate Change*, vol. 17, (April 2008), 233–265.

——, A. F. Correljé, and J. P. M. Groenewegen, *Institutional Reform, Regulation and Privatization; Process and outcomes in Infrastructure Industries*, (Cheltenham: Edward Elgar, 2005).

——, J. Groenewegen, and C. Ménard, *Aligning institutions with technology: critical transactions in the reform of infrastructures*, discussion paper, (2008 Forthcoming).

Lai, H. H., "China's Oil Diplomacy: Is it a Global Security Threat?", *Third World Quarterly*, vol. 28, no. 3, (2007), 519–539.

Lan, X. Z., "Power to the People", *Beijing Review*, (20 March 2008).

Lee, B., Froggatt, A. et al., *Changing Climates: Interdependencies on Energy and Climate Security for China and Europe*, Chatham House Report, (London: Chatham House, November 2007).

Lenczowski, G., *Russia and the West in Iran (1918–1948): A Study in Big Power Rivalry*, (Ithaca, NY: Cornell University Press, 1978).

LeVine, S., *the Oil and the Glory*, (New York: Random House, 2007).

Lewis, R. A., ed., *Geographic Perspectives on Central Asia*, (London: Routledge, 1992).

Lewis, J., "China's Climate Change Strategy", *China Brief*, vol. 7, no. 13, (27 June 2007).

Liao, X., "The Petroleum Factor in Sino-Japanese Relations: Beyond Energy Cooperation", *International Relations of the Asia-Pacific*, vol. 7, (2007), 23–46.

Li, P. M. et al., "Zhongguo yingdui gaoyoujiade zhanlüe xuanze" (China's strategic options in dealing with high oil prices), *Hongguan jingji yanjiu (Macroeconomic research)*, vol. 12, (2005).

Li, Z., "China's Long-term Energy Outlook and the Implications for Global Governance", *Asia-Pacific Review*, vol. 14, no. 1, (2007), 13–27.

Ling, Z., "China's Approach Towards Addressing Global Warming", *Georgetown International Environmental Law Review*, vol. 12, (2000), 494–515.

Liu, S. J., "Challenges and Transformation of Development Mode Facing China in the New Development Stage", *China Development Review*, vol. 1, (2008), 1–14.

Loorbach, D., R. van der Brugge, and M. Taanman, "Governance in the Energy Transition: Practice of Transition Management in the Netherlands", *International Journal of Environmental Technology and Management*, vol. 9, Nos. 2/3, (2008), 294–315.

Lorenzetti, M., "Oil Industry Sees Mixed Signals on US Iran Policy", *Oil and Gas Journal*, (1 April 2002), http://www.ogj.com.

Loskot-Strachota, A. Pelczynska-Nalecz, K., *Gasprom's Expansion in the EU—Cooperation or Domination?*, Warsaw: Centre for Eastern Studies, (April 2008).

Lovins, A. B., "Twenty Hydrogen Myths", *Rocky Mountain Institute*, E03-05, (2003).

Luisi, P. L., "Autopoiesis: A Review and a Reappraisal", *Naturwissenschaften*, vol. 90, no. 2, (2003), 49–59.

Luo, Z. X., Zhang, L. M., "Zhongwai shiyou gongsi shangyou chanye jingzhengli pingjia" (An evaluation of the competitiveness of the upstream sector of Chinese and foreign oil companies), *Guoji shiyou jingji (International Oil Economy)*, vol. 9, (2007), 62–66.

Ma, M. J., "Gao youjiage xiadi meizhiyou zhengce quxiang" (Policy tendencies of oil from coal under high oil prices), *Zhongguo fazhan pinglun (Discussions of China's development)*, vol. 3, (2008), 45–51.

Mackinder, H. J., "The Geographical Pivot of History", *Geographical Journal*, vol. 23, (1904), 421–42.

——, *Democratic Ideals and Reality: A Study in the Politics of Reconstruction*, (London: Constable, 1919).

Madan, T., *The Brookings Foreign Policy Studies Energy Policy Series: India*, (Washington: Brookings Institution, 2006).

Maeda, S., "China's Environmental and Energy Problems and the Possibility of Japan-China Technical Cooperation", *Science & Technology Trends*, vol. 22, (2007), 76–85.

Malik, M., "China Responds to the US-India Nuclear Deal", *Jamestown Foundation China Brief*, vol. 6, no. 7, (2006), 3–5, http://www.jamestown.org/china_brief/article.php?articleid=2373178.

Manji, F., Marks, S., (Eds.), *African Perspectives on China in Africa*, (Oxford and Cape Town: Fahamu, 2007).

Manning, R., *The Asian Energy Factor: Myths and Dilemmas of Energy, Security and the Pacific Future*, (New York: Palgrave, 2000).

Marschall, C., *Iran's Persian Gulf Policy-From Khomeini to Khatami*, (London and New York: Routledge Curzon, 2003).

Martin, V., *Islam and Modernism: The Iranian Revolution of 1906*, (Syracuse: Syracuse University Press, 1989).

Masood, T., "*Gwadar Deep Water Port Project: Implications for Pakistan and China*", International Assessment and Strategy Center, (2004), http://www.strategycenter.net/research/pubID.4/pub_detail.asp.

McDowall W., and M. Eames, "Forecasts, Scenarios, Visions, Backcasts and Roadmaps to the Hydrogen Economy: a Review of the Hydrogen Futures Literature", *Energy Policy*, vol. 34. Elsevier, doi: 10.1016/j.enpol.2005.12.006, (2005).

McDowall W., and M. Eames, "Towards a sustainable hydrogen economy: A multi-criteria sustainability appraisal of competing hydrogen futures", *International Journal of Hydrogen Energy*, doi: 10.1016/j.ijhydene.2007.06.020, (2007).

McHaffie, P., "Decoding the Globe: Globalism, Advertising and Corporate Practice", *Environment and Planning D: Society and Space*, vol. 15, no. 1, (1997), 73–86.

Meidan, M., "China in a Post-Kyoto Architecture", *China Perspectives*, vol. 1, (2007), 65–71.

Ménard, C., and M. M. Shirley, *Handbook of New Institutional Economics,* (New York: Springer, 2005).
Mertha, A. C., *China's Water Warriors: Citizen Action and Policy Change,* (Ithaca, NY: Cornell University Press, 2008).
Mesbahi, M., "Tajikistan, Iran, and the International Politics of the 'Islamic Factor' ", *Central Asian Survey,* vol. 16, vol. 20, (1997), 141–158.
Millison, D., "*Turkmenistan-Afghanistan-Pakistan(-India) Natural Gas Pipeline Project*", Manila: Asian Development Bank, South Asia Energy Division, (2006).
Mingers, J., *Self-Producing Systems,* (London: Kluwer Academic/Plenum, 1994).
Ministry of Economic Affairs, *Innovatie in het energiebeleid; Energietransitie: stand van zaken en het vervolg,* Publication number 04ME06, (2004).
——, *Nu voor later*: Energierapport 2005, Publication number 05E01, (2005).
Mojtahed-Zadeh, P. (ed.), *Boundary Politics and International Boundaries of Iran,* (Boca Raton: Universal Publishers, 2006).
Moradi, M., "Caspian Pipeline Politics and Iran-EU Relations", *UNISCI Discussion Papers,* (10 January2006), 173–184.
Moshaver, Z., "Revolution, Theocratic Leadership and Iran's Foreign Policy: Implications for Iran-EU Relations", *The Review of International Affairs,* vol. 3, no. 2, (Winter 2003), 283–305.
Moslem, M., *Factional Politics in post-Khomeini Iran,* (Syracuse: Syracuse University Press, 2002).
Mudiam, P. R., "Indian Power Projection in the Greater Middle East: Tools and Objectives" in: M. P. Amineh (ed.) *The Greater Middle East in Global Politics: Social Science Perspectives on the Changing Geography of the World Politics,* (Leiden and Boston: Brill Academic Publishers, 2007), 405–427.
Mulder, F., "Does Transport Offer a Basis for Developing the Hydrogen Economy?", *Senternovem presentation,* (25 February 2005).
Murmann, J. P., *Knowledge and Competitive Advantage: the Coevolution of Firms, Technology, and National Institutions,* (Cambridge: Cambridge University Press, 2003).
Nahavandi, F., "Russia, Iran and Azerbaijan-The Historic Origins of Iranian Foreign Policy" in: B. Coppetiers (ed.) *Contested Borders in the Caucasus,* (Brussels: VUB Press, 1996).
Naipaul, V. S., *India: A Million Mutinies Now,* (New York: Viking, 1990).
National Development and Reform Commission, *Outlines of New Energy and Renewable Energy Development,* (1995).
——, *10th Five-Year Plan of New Energy and Renewable Energy Development,* (2001).
——, *Medium and Long-term Plan of Renewable Energy Development,* (2003).
——, *11th Five-Year Plan of Renewable Energy Development,* (2006).
——, *China's National Climate Change Programme,* Beijing: People's Republic of China, (June 2007).
National Environmental Policy Plan (NMP4), *Een wereld en een wil. Werken aan duurzaamheid,* VROM, The Hague, (2000).
Narain, U., and Krupnick, A., "The Impact of Delhi's CNG Program on Air Quality", *Resources for the Future,* (February 2007).
Nel, W. P., and Cooper, C. J., "A Critical Review of IEA's Oil Demand Forecast for China", *Energy Policy,* vol. 1, (2008), 1096–1106.
Nelson, R., "The Co-evolution of Technology, Industrial Structure, and Supporting Institutions", *Industrial and Corporate Change,* vol. 3, no. 1, (1994), 47–62.
Nelson, R. R., "Recent Evolutionary Theorizing About Economic Change", *Economic Journal XXXIII,* (1995), 48–90.
——, and S. G. Winter, *An Evolutionary Theory of Economic Change,* (Cambridge: Harvard University Press, 1982).
Netherlands Environmental Assessment Agency, *Global Fossil CO_2 Emissions for 2006,* The Hague, (21 June 2007).

Ni, J. M. et al. (Eds.), *Guojia nengyuan anquan baogao (Report on national energy security)*, Beijing: Renmin Pub, (2005).

Niazi, T., "Gwadar: China"s Naval Outpost on the Indian Ocean", *Jamestown Foundation*, (16 February 2005), http://www.jamestown.org/news_details.php?news_id=93.

Nili, M., "Picture of Iran's Oil-Dependent Economy", *Iran International Magazine*, vol. 40, (May 2006), http://www.iraninternationalmagazine.com.

Norgaard, R. B., *Development Betrayed: the End of Progress and a Coevolutionary Revisioning of the Future*, (London: Routledge, 1994).

——, "Bubbles in a back eddy: a commentary on 'the origin, diagnostic attributes and practical application of coevolutionary theory', "*Ecological Economics*, vol. 54, (2005), 362–365.

Nuclear Safety Commission, *Regulatory Guide for Reviewing Seismic Design of Nuclear Power Reactor Facilities*, (19 September 2006), http://www.nsc.go.jp/english/taishin.pdf.

Ogen, O., "China-Nigerian Economic Relations: Chinese Imperialism or South-South Partnership?", *China Aktuell*, vol. 3, (2008), 78–102.

Olcott, M. B., "International Gas Trade in Central Asia: Turkmenistan, Iran, Russia and Afghanistan", in: *Natural Gas and Geopolitics: From 1970 to 2040*, eds. D.G. Victor, A. M. Jaffe and M. H. Hayes, (Cambridge: Cambridge University Press, 2006), 202–233.

OPEC, Oil Outlook to 2025, *OPEC Review Paper*, Blackwell Publishing, UK, (2004).

——, *Oil outlook to 2030: OPEC Secretariat Background Paper*, 11th International Energy Forum, Rome, (20–22 April 2008).

Organisation for Economic Cooperation and Development, *OECD Environmental Performance Reviews: China*, Paris: OECD, (2007).

Ó'Tuathail, G., "Geopolitical Structure and Cultures: Towards Conceptual Clarity in the Critical Study of Geopolitics", in: *Geopolitics: Global Problems and Regional Concerns*, the Center for Defense and Security Studies, Bison Paper 4, (January 1996).

——, *Critical Geopolitics: The Politics of Writing Global Space*, (Minneapolis: University of Minnesota Press, 1996).

—— and S. Dalby, *Rethinking Geopolitics*, (London: Routledge, 1998).

Pandian, S. G., "Energy Trade as a Confidence-Building Measure Between India and Pakistan: A Study of the Indo-Iran Trans-Pakistan Pipeline Project", *Contemporary South Asia*, vol. 14, no. 3, (2005a), 307–320.

——, "The Political Economy of Trans-Pakistan Gas Pipeline Project: Assessing the Political and Economic Risks for India", *Energy Policy*, vol. 33, (2005b), 659–670.

Pant, H. V., "Saudi Arabia Woos China and India", *Middle East Quarterly*, vol. 13, no. 4, (2006), 45–52.

Peimani, H., "Trans-Asia Rail Network A Potential Benefit to Central Asian and Caucasian States", *Central Asia-Caucasus Analyst*, (15 November 2006), http://www.cacianalyst.org.

Perez, C., "Technological change and opportunities for development as a moving target", *CEPAL Review*, vol. 75, (2001), 109–130.

Pickett, S., "Japan's nuclear energy policy", *Energy Policy*, vol. 30, (2002), 1337–1355.

Pipeline and Gas Journal, "Iran-Pakistan-India Ironing Out Pipeline Details", *Pipeline and Gas Journal*, vol. 233, no. 2, (2006), 14.

PSA Press Release, "*PSAI Signs Agreement to Operate Gwadar Port*", PSA International, (6 February 2007).

Raczka, W., "A Sea or a Lake? The Caspian's Long Odyssey", *Central Asian Survey*, vol. 19, no. 2, (2000), 189–221.

Rakel, E. P., *Power, Islam, and Political Elite in Iran—A Study on the Iranian Political Elite from Khomeini to Ahmadinejad*, (Leiden, Boston, and London: Brill Academic Publishers, 2009).

Ramesh, J., *Making Sense of Chindia: Reflections on China and India*, (New Delhi: India Research Press, 2005).

Ramo, J. C., "The Beijing Consensus", *The Foreign Policy Centre*, (May 2004).

Rashid, A., "The New Great Game: The Battle for Central Asia's Oil", *Far Eastern Economic Review*, (10 April 1997).

Rip, A., "A co-evolutionary approach to reflexive governance and its ironies", in: *Reflexive Governance for Sustainable Development*, ed. J-P. Voss, D. Bauknecht, and R. Kemp, (Cheltenham: Edward Elgar, 2006).

Roads2HyCom, *Hydrogen and Fuel Cells—A Handbook for Communities*, European Commission, Hydrogen and Fuel Cell Platform, (2007).

Rotmans, J., R. Kemp, M. Van Asselt, F. Geels, G. Verbong, and K. Molendijk, "Transities en Transitiemanagement. De casus van een emissiearme energievoorziening", final report of study *Transitions and Transition Management* for the 4th National Environmental Policy Plan (NMP-4) of the Netherlands, Maastricht: ICIS and MERIT, (October 2000).

Rotmans, J., R. Kemp, and M. van Asselt, "More Evolution than Revolution; Transition Management in Public Policy", *Foresight*, vol. 3, no. 1, (2001), 15–31.

Roy, O., *The Failure of Political Islam*, (Cambridge: Harvard University Press, 1996/97).

SABB Notes, *Saudi-China Trade Relations*, (10 February 2009).

Saikal, A., *The Rise and Fall of the Shah*, (Princeton: Princeton University Press, 1980).

Saivetz, C., "Perspectives on the Caspian Sea Dilemma: Russian Policies Since the Soviet Demise", *Eurasian Geography and Economics*, vol. 44, no. 8, (2003), 588–606.

Saviotti, P., "Systems Theory and Technological Change", *Futures*, Butterworth & Co (Publishers) Ltd., (December 1986).

—— and J. S. Metcalfe, *Evolutionary Theories of Economic and Technical Change*, (Chur. Harwood Academic Publishers, 1991).

Saunders, S., *Jane's Fighting Ships 2006–2007*, (Coulsdon: Jane's Information Group, 2006).

Saurbek, Z., "Kazakh-Chinese Energy Relations: Economic Pragmatism or Political Cooperation?", *China and Eurasia Forum Quarterly*, vol. 6, no. 1, (2008), 79–94.

Seifzadeh, H. S., "The Landscape of Factional Politics and Its Future in Iran", *Middle East Journal*, vol. 57, (Winter 2001), 57–75.

SenterNovem. 2004. *Waterstofinnovatie in Nederland: een overzicht*. Ministerie van Economische Zaken.

——, *Hydrogen and Fuel Cells in The Netherlands: Who's who?*, SenterNovem Publication Centre: 02DEN04–18, (2004).

——, *Waterstof in Perspectief*, SenterNovem, EV /04, (2005).

——, "EnergieTransitie", *Senternovem*, (6 February 2009), main pages: (http://www.senternovem.nl/energietransitie/),(www.senternovem.nl/energietransitiedm/index.asp), (www.senternovem.nl/energietransitieng/index.asp), (www.senternovem.nl/eos/) and (www.senternovem.nl/eos/infotheek/projectencatalogus/Projecten/unieke_kansen_regeling/Index.asp).

Shackley, S., and K. Green, "A conceptual framework for exploring transitions to decarbonised energy systems in the United Kingdom", *Energy*, vol. 32, (2007), 221–236.

Shaoul, R., "An evaluation of Japan's current energy policy in the context of the Azadegan Oil Field Agreement signed in 2004", *Japanese Journal of Political Science*, vol. 6, no. 3, (2005), 1–27.

Shengliang, D./Xiaojie, X., "The Oil Dragon's Move: What It Means for China and the World", in: *The Dragon Millennium. Chinese Business in the Coming World Economy*, edited by Frank-Jürgen Richter, Westport, (Co.-London: Quorum Books, 2000), 79–89.

Shichor, Y., "Blocking the Hormuz Strait: China's Energy Dilemma", *China-Brief*, vol. 8, no. 18, (23 September 2008).

Shu, X. L., "Shiyou huoban shengyu jingzheng duishou", (Better oil partners than competing opponents), *Zhongguo Shiyou Qiye (China Petroleum Enterprise)*, vol. 1, (2007), 120–124.

SIPRI, *Importer/Exporter TIV Tables database*, (2007), http://armstrade.sipri.org/arms_trade/values.php.

Smith, A., *The Failures of Japan's Nuclear Fuel Cycle Program 1956–2007*, Green Action, Kyoto, (2007).

—— and A. Stirling, "Social-ecological Resilience and Socio-technical Transitions: Critical Issues for Sustainability Governance", *STEPS Working Paper 8*, (Brighton: STEPS Centre, 2008).

Smith, C. E., "2007 Construction Lags 2006, But More Projects Lie Ahead", *Oil and Gas Journal*, (19 February 2007), 48.

Smit, R., M. Weeda, and A. de Groot, "Hydrogen infrastructure development in The Netherlands", *International Journal of Hydrogen Energy*, doi: 10.1016/j.ijhydene.2006.10.044, 1387–1395, (2006).

Socor, V., "South Stream: Gazprom's New Mega Project", *Eurasia Daily Monitor*, (25 June2007), http://www.jamestown.org/edm/article.php?article_id=2372249.

SSGC, "*Karakoram Oil Pipeline: Gwadar to China*", SSGC company presentation, Islamabad, (25–27April 2006).

State Council Information Office of the People's Republic of China, white paper of *China's Energy Conditions and Policies*, (2007), http://www.scio.gov.cn.

State Council of the People's Republic of China, *The Medium- and Long-Term Nuclear Power Development Plan 2005–2020*, (31 March 2006), http://www.cnnc.com.cn/2006-03-31/000225425.html.

——, *White Paper; China's Policies and Actions for Addressing Climate Change*, (29 October 2008), http://english.gov.cn/official/2005-08/17/content_24165.htm#2008.

State Power Regulatory Committee, *Inspection Report on the Implementation Situation of the Purchase of Electricity from Renewable Energy and the Electricity Price Policy* (in Chinese), (21 January 2009), www.china5e.com.

Statistical Communiqué of the PRC on the 2008 Economic and Social Development, (26 February 2009).

Steinfeld, E. et al., "Greener Plants, Grayer Skies? A Report from the Front Lines of China's Energy Sector", *MIT China Energy Group*, (2008), http://web.mit.edu/ipc.

Stern, J., *The Future of Russian Gas and Gazprom*, (Oxford: Oxford University Press, 2005).

Stern, R., "The Iranian Petroleum Crisis and United States National Security", *Journal of National Academy of Science*, vol. 104, no. 1, (2 January2007), 377–382.

Storey, I., "China Seeks to Reduce its Dependence on Strait of Malacca", *Jane's Intelligence Review*, (May 2005), 36–39.

——, "China's 'Malacca Dilemma'", *Jamestown Foundation China Brief*, vol. 6, no. 8, (2006), 4.

Stulberg, A., "Moving Beyond the Great Game: The Geoeconomics of Russia's Influence in the Caspian Energy Bonanza", *Geopolitics*, vol. 10, no. 1, (2005), 1–25.

Taheri, A., *Holy Terror: Inside the World of Islamic Terrorism*, (Bethesda, MD: Adler and Adler, 1987).

Tarock, A., "Iran-Western Europe Relations on the Mend", *British Journal of Middle Eastern Studies*, vol. 26, no. 1, (1999), 41–61.

——, "Iran's Nuclear Program and the West", *Third World Quarterly*, vol. 27, no. 4, (2006), 645–664.

Taylor, P. J., "Geopolitical World Orders", in: P. J. Taylor (ed.) *the Political Geography of the Twentieth-Century: A Global Analysis*, (London, UK: Belhaven Press, 1993), 31–62.

Tellenbach, S., "Zur Änderung der Verfassung der Islamischen Republik vom 28. Juli 1989", *Orient*, vol. 31, no. 1, (1990), 45–66.
Thompson, D., "China's Global Strategy for Energy, Security, and Diplomacy", *China Brief*, vol. 5, no. 7, (29 March 2007).
The Institute of Energy Economic, Japan, *Asia/World Energy Outlook: Focusing on China and India Energy Outlook*, (15 October 2007).
Tian, C. R., "2006—nian Zhongguo shiyou jinchukou zhuangkuang fenxi" (Analysis of China's 2006 oil import and export situation), *Guoji shiyou jingji (International Oil Economy)*, vol. 3, (2007), 14–21.
Tian, H., "Zhongguo meicengqi ziyuan fazhan" (Development of China's Coal Bed Methane Resources), *Zhongguo nengyuan (Energy of China)*, vol. 4, (2007), 30–33.
Tokyo Electric Power Company (TEPCO), *Recent Challenges and Solutions of Energy and Electric Utility in Japan*, (2007).
Tu, J., "Energy Implications of the 2008 Sichuan Earthquake", *China Brief*, vol. 8, no. 14, (3 July 2008a).
——, "Smoke and Mirrors in China's Oil Statistics", *China Brief*, vol. 8, no. 11, (21 May 2008b).
——, "China's New National Energy Commission and Energy Policy", *China-Brief*, vol. 8, no. 7, (28 March 2008c).
——, "Coal Mining Safety: China's Achilles Heel", *China Security*, vol. 3, no. 2, (2007), 36–55.
Umbach, F., *Globale Energiesicherheit. Herausforderungen für die europäische und deutsche Außenpolitik*, (Munich: Oldenbourg Verlag, 2003).
——, "Future Impacts of Chinese and Asian Dependency on Energy Imports from the Middle East and Central Asia", in: *The Impact of Asian Powers on Global Developments*, edited by Erich Reiter and Peter Hazdra, (Heidelberg-New York, 2004), 143–163.
——, "Europe's Next Cold War: The European Union Needs a Plan to Secure Its Energy Security", *Internationale Politik* (Transatlantic Edition), (Summer 2006a, 64–71.
——, "Asiens Energieunsicherheit und Ressourcennationalismus: Kooperation oder Konfliktpotenzial für die Region?", in: *Regionale politische und wirtschaftliche Kooperation in Asien*, edited by Christine Berg and Günter Schucher, (Hamburg: Mitteilungen des IFA Hamburg, 2006b), 33–57.
——, "Energiesicherheit und Energiestrategie Japans", in: *Energie und Sicherheit in Ostasien - globale Dimensionen der Wachstumsregion*, edited by Xuewu Gu and Kirsten Kupfer, (Frankfurt/M.-New York: Campus-Verlag, 2006c), 41–60.
——, "Die Neuen Herren der Welt. Öl gleich Macht: Energie-Verbraucherländer müssen umdenken", *Internationale Politik* (September 2006), 52–59.
——, *Zielkonflikte der europäischen Energiesicherheit. Dilemmata zwischen Russland und Zentralasien* (DGAPanalyse, 28 pp.; No. 3), (Berlin: November 2007a).
——,"The Legs of the Triangle—The EU-China Relations", in: *The New Strategic Triangle: China, Europe and the United States in a Changing International System*, edited by Wilfried Jung, Beijing: Konrad-Adenauer-Stiftung, KAS-Schriftenreihe No. 76, Beijing: KAS, (2007b), 36–45.
——/Skiba, A, *Licht und Schatten auf dem EU-Frühjahrsgipfel 2007—Gemeinsame Energie- und Energieaußenpolitik oder nationale Sonderbeziehungen mit Russland?*, DGAP-Standpunkt (3), Berlin, (24 April 2007).
——/Riley, A., "Out of Gas. Looming Russian Gas Deficits Demand Readjustment of European Energy Policy", *Internationale Politik* (Global Edition), vol. 1, (2007), 83–90.
——, "Diversifizierung statt Protektorat. Energiepartnerschaft zwischen Russland und der EU", *Die politische Meinung* (September2008a), 25–30.

——, *Europe's Energy Dependence in Mid-term Perspective*, Washington D.C.: American Institute for Contemporary German Studies (AICGS), (February 2008b), http://www.aicgs.org/documents/advisor/umbach.gmf.pdf.

——, "German Debates on Energy Security and Impacts on Germany's 2007 EU Presidency" in: *Energy Security. Visions from Asia and Europe*, edited by A. Marquina, (Hampshire-New York: Palgrave Macmillan, 2008c), 1–23.

——, "Memorandum: The European Union and Russia—Perspectives of the Common „Strategic Energy Partnership", Personal Analysis for the Sub-Committee on Foreign Affairs, Defence and Development Policy, House of Lords/Great Britain, in: *The European Union and Russia*, edited by the House of Lords and its European Union Committee, Report with Evidence, 14th Report of Session 2007–08, HL Paper 98, (Norwich-London: The Stationary Office, 2008d), 185–188, http://www.publications.parliament.uk/pa/ld/ldeucom.htm.

——, "Chinas Energy Insecurity in the Context of Growing Geopolitical Competition: Implications for the Future EU-China Relations", in: *European-Chinese Security Cooperation: Possibilities and Limits* (Konferenzband), ed. Klaus Lange, Munich: Akademie für Akademie für Politik und Zeitgeschehen der Hanns-Seidel-Stiftung (HSS), Studies & Comments 6, (2008e), 43–59.

——, "Germany's Energy Insecurity", *Journal of Energy Security*, (October 2008f), http://www.ensec.org/index.php?option=com_content&view=article&id=153:germanysenergyinsecurity&catid=81:europe&Itemid=324.

UNDP, *World Energy Assessment: Overview, 2004 Update*, New York: UNDP, (2004).

United States Department of Defense, *Proliferation: Threat and Response*, Washington: US Department of Defense, (2001), http://www.fas.org/irp/threat/prolif00.pdf.

US Department of Defense, Joint Chiefs of Staff, *The National Military Strategy of the United States of America: A Strategy for Today; A Vision for Tomorrow*, Washington: US Department of Defense, (2004)

United States Energy Information Administration, *International Total Primary Energy Consumption and Intensity, (2005),* http://www.eia.doe.gov/emeu/international/energyproduction.html.

——, *Japan Country Analysis Brief*, (2008a), http://www.eia.doe.gov/emeu/cabs/Japan/Oil.html.

——, *World Energy Outlook 2008, (2008b),* http://www.eia.doe.gov/oiaf/ieo/.

——, *Future of the Chinese Nuclear Industry,* (2008), http://www.eia.doe.gov/cneaf/nuclear/page/nuc_reactors/china/outlook.html.

Uyterlinde, M. A., J. R. Ybema, and R. W. van den Brink, "De belofte van een duurzame Europese energiehuishouding", Energievisie van ECN en NRG, *Energy research Centre of the Netherlands / NRG*, ECN-E-07-061, (2007).

Vakil, S., "Iran: Balancing East Against West", *The Washington Quarterly*, vol. 29, no. 4, (2006), 51–60.

Varela, F., H. Maturana and R. Uribe, R., "Autopoiesis: The Organization of Living Systems, Its Characterization and a Model," *BioSystems*, vol. 5, (1974), 87–196.

Verbruggen, A., "Renewable and nuclear power: A common future?", *Energy Policy*, vol. 36, no. 11, (2008), 4036–47.

Verma, S. K., "Energy Politics and Iran-Pakistan-India Gas Pipeline", *Energy Policy*, vol. 35, (2007), 3280–3301.

Vicziany, M., "The Indian Economy in the Twenty-First Century: The Tough Questions That Just Won't Go Away", *South Asia: Journal of South Asian Studies*, vol. 28, no. 2, (2005), 221–222.

Vikas, S. and C. L. Ellsworth, "Special Report: India's Changing Energy Balance", *Oil and Gas Journal*, (23 January 2006).

Von Tunzelmann, N., "Historical coevolution of governance and technology in the industrial revolutions", *Structural Change and Economic Dynamics,* vol. 14, (2003), 365–384.

Wallerstein, I., *The Modern World-System: Capitalist Agriculture and the Origins of the European World-Economy in the Sixteenth Century,* (New York: Academic Press, 1974).

Waltz, K., *Man State and War,* (New York: Columbia University Press, 1959).

——, *Theory of World Politics,* (Reading, Mass.: Addison Wesley, 1979).

Wang, F-L., "Preservation, Prosperity and Power: What Motivates China's Foreign Policy?", *Journal of Contemporary China,* vol. 14, no. 45, (2005), 676–677.

Wang Zhongying, Li Junfeng, Annual Report on China's Renewable Energy Industry, *Chemical Industry Press,* (2007), 79.

Weitz, R., "China's Nuclear Gambit", *China-Brief,* vol. 8, no. 6, (14 March 2008).

Wietschel, M., U. Hasenauer, and A. de Groot, "Development of European hydrogen infrastructure scenarios—CO2 reduction potential and infrastructure investment", *Energy Policy,* vol. 34, (2006), 1284–1298.

Winder, N., B. S. McIntosh, and P. Jeffrey, "The origin, diagnostic attributes and practical application of co-evolutionary theory", *Ecological Economics,* vol. 54, vol. 4, (2005), 347–361.

Winrow, G., "Turkey and the East-West Gas Transportation Corridor", *Turkish Studies,* vol. 5, no. 2, (2004), 23–42.

World Bank, *East Asia and Pacific Updat.,* Washington D.C.: World Bank, (2007) http://www.worldbank.org.

——, *China Quick Facts,* Washington D.C.: World Bank, (2007), http://web.worldbank.org/WBSITE/EXTERNAL/COUNTRIES/EASTASIAPACIFICEXT/CHINAEXTN/0,,contentMDK:20680895~pagePK:1497618~piPK:217854~theSitePK:318950,00.html

World Nuclear Association, *Japanese Waste and MOX Shipments from Europe,* (2007), http://www.world-nuclear.org/info/inf39.html.

——, *Nuclear Power for Electricity—Table of the World's Nuclear Power Reactors,* (2008a), http://www.world-nuclear.org/info/reactors.html.

——, *Nuclear Power in Japan,* (2008b), http://www.world-nuclear.org/info/inf79.html.

——, *Nuclear Power in China,* (2008c), http://www.world-nuclear.org/info/inf63.html.

Wu, J., Garnett, S., Barnes, T., "Beyond an energy deal: Impacts of the Sino-Australia uranium agreement", *Energy Policy,* vol. 36, (2008), 413–422.

Xu, C.,. "Nuclear Energy in China: Contested regimes". *Energy* vol. 33, no. 8, (2008), 1197–205.

Yang, Z. H., "Zai baohu huanjing yu shengtaidi jichushang dali fazhan shuidian" (Greatly develop hydro-power on the basis of protection of the environment and the ecology), in: *'Shuidian kaifa yu lüse weilai' Xinhua luntan wenji* (Collection of articles of the Xinhua Hydro-power Investment Company Conference "Development of Hydro-power and a Green Future"), Beijing: China Waterpub, (2007), 5–7.

Yanjiu ketizu (Research Theme Group), *Zhongguo nengyuan zhanlüe yanjiu (2000–2050), (Study of China's Energy Strategy (2000–2050),* China Power Publishing House, (1997)

Yenikeyeff, S. M., *Kazakhstan's Gas: Export Markets and Export Routes,* (Oxford: Oxford Institute for Energy Studies, 2008).

Yu, S. X. et al., "Contrary thoughts about the effect of oil prices on the economy and society", *Zhongguo nengyuan (Energy of China),* vol. 4, (2007), 34–38.

Yun, W., Zhang, Z., "Electric power grid interconnection in Northeast Asia", *Energy Policy,* vol. 34, no. 15, (2006), 2298–2309.

Zha, D. J., *Zhongguo shiyou anquan guoji zhengzhi jingji fenxi (China's Oil Security: International Political and Economic Analysis),* (Beijing: Contemporary World Press, 2005).

Zha, D. J., Hu, W. X., "Promoting Energy Partnership in Beijing and Washington", *The Washington Quarterly,* vol. 4, (2007), 105–115.

Zhang, J. R., "Zhong'E nengyuan hezuozhongdi wentiji dui Zhongguodi yingxiang", (Problems in Sino-Russian energy cooperation and their effect on China), *Shehui kexue (Social Sciences)*, vol. 1, (2006), 59–69.
Zhang, L. B., Lee, J., "Untangling China's Energy Policy", *China Security*, vol. 4, no. 3, (2008), 58–62.
Zhang, Z., "China is moving away the pattern of 'develop first and then treat pollution", *Energy Policy*, vol. 35, no. 7, (2007), 3547–9.
Zhang, Z. M. et al., *Renewable Energy Development in China: The Potential and the Challenges*, Center for Resource Solutions, (2000).
Zhao, H., "China's oil venture in Africa", *East Asia*, vol. 24, (2007), 399–415.
Zhongguo huanjing chanye xiehui (China Environmental Industry Association), "Woguo huodianchang tuoliu hangye 2006-nian fazhan baogao" (Report on the development of the desulphurization sector for China's thermo-power stations), *Zhongguo huanjing chanye (China Environmental Protection Industry)*, vol. 10, (2007), 20–25.
Zhongguo nengyuan fazhan baogao (China Energy Development Report 2007), Beijing: China Waterpub, (2007).
Zweig, D., Jianhai, B., "China's Global Hunt for Energy", *Foreign Affairs*, (September–October 2005), 25–38.
—— and Ye, S. L., "A Crisis is Looming: China's Energy Challenge in the Eyes of University Students", *Journal of Contemporary China*, vol. 55, (2008), 273–296.

"ADB: TAP Line Feasible; Field Study Due", *Oil and Gas Journal*, (7 February 2005).
"EU auf der Suche nach neuen Transportwegen für Erdgas", *EurActiv*, (19 January 2006), http://www.euractiv.com.
"Interview with William C. Ramsey, Deputy Executive Director of the IEA", *Energies*, vol. 9, (March 2006).
"Memo Signed for Turkmen Gas Sales to Pakistan", *Oil and Gas Journal*, (6 March 2006).
"Qatar-Pakistan Gas Pipeline", *Economic Review*, vol. 5, no. 6, (2004), 26.
"Report on China's Photovoltaic Development", Chinese Renewable Energy Industry Association under *China Association of Resource Comprehensive Utilization, Greenpeace, European Photovoltaic Industry Association and World Wide Fund 2007*, (2008).

Press

Anwar, M., "Real Significance of Gwadar Port", *The News*, (20 March 2007). http://www.thenews.com.pk/daily_detail.asp?id=47664.
Auyezov, O., "China, Turkmenistan Agree to Boost Planned Gas Sales", *Reuters* (29 August 2008), http://www.reuters.com.
Aziz, F., "Musharraf Opens Pakistani Port, Promises Another", *Reuters India*, (20 March 2007). http://in.today.reuters.com/news/newsArticle.aspx?type=topNews&storyID=2007-03-20T202436Z_01_NOOTR_RTRJONC_0_India-291696-1.xml.
Bajpaee, C., "India, China Locked in Energy Game", *Asia Times*, (17 March 2005), (http://www.atimes.com/atimes/Asian_Economy/GC17Dk01.html).
Bhadrakumar, M. K., "India Finds a $40bn Friend in Iran", *Asia Times*, (11 January 2005), http://www.atimes.com.
——, "The Great Game Comes to South Asia", *Asia Times*, (24 May 2006), http://www.atimes.com/atimes/South_Asia/HE24Df04.html.
Blank, S., "The Shanghai Cooperation Organization: Cracks behind the Façade", *Eurasia Net.org*, (2006), http://www.eurasianet.org/departments/insight/articles/eav 062106.shtml.

Calamur, K., "Analysis: IPI Faces Dangers and Hurdles", *Alexander's Gas and Oil Connection*, (31 January 2008), http://gasandoil.com.
Dutta, S., "No US Pressure on Iran Pipeline: Deora to Aziz", *Times of India*, (7 April 2007), http://timesofindia.indiatimes.com/NEWS/India_Business/No_US_pressure_on_Iran_pipeline_Deora_to_Aziz/articleshow/1867983.cms.
Fazl-e-Haider, S., "Pakistan Port Opens New Possibilities", *Asia Times*, (22 March 2007), http://www.atimes.com/atimes/South_Asia/IC22Df02.html.
Johnston, E., "Nuclear plants rural Japan's economic fix", *Japan Times*, (4 September 2007), http://search.japantimes.co.jp/cgi-bin/nn20070904f1.html.
Koshoji, H., "Japan to boost its nuclear capabilities", *UPI Asia.com*, (17 April 2008), http://www.upiasia.com/Security/2008/04/18/japan_to_boost_its_nuclear_capabilities/6270/.
Leverett, F., "Race for Iran", *The New York Times*, (20 June2006), http://www.nytimes.com/2006/06/20/opinion/20leverett.html.
McGregor, R., Dickie, M., "China's Soaring Power Output Stuns Even Beijing", *Financial Times*, (7 February 2007).
Nair, A., "Oil Diplomacy: A Few Lessons from China", *Economic Times*, (11 May 2007), 17.
Ni Y., "Corridor of Cooperation", *Beijing Review*, (30 March 2006), http://www.bjreview.com.cn/quotes/txt/2007-01/09/content_52531_2.htm.
Ögütçü, M., "Kazakhstan's Expanding Cross-border Gas Links," *CEPMLP Internet Journal*, (2006), 17, http://www.dundee.ac.uk/cepmlp/journal/html/Vol17/Vol17_8.pdf.
Pradhan, B., "Hamid Karzai Says Afghan Stability Key to Regional Prosperity", *Bloomberg*, (20 November 2006), http://www.bloomberg.com/apps/news?pid=newsarchive&sid=aL_I521MFl4Y.
Rahman, M., "US Hits Proposed Pipeline for India", *Washington Times*, (15 March 2005).
Raman, S., "Cheers as Indian government wins vote", *BBC News Online*, (22 July 2008), http://news.bbc.co.uk/2/hi/south_asia/7520487.stm.
Siddiqui, J. A., "Upgrading National Trade Corridor", *Dawn*, (26 February 2007), http://www.dawn.com/2007/02/26/ebr14.htm.
Stakelbeck, F., China embraces the Atom, *Asia Times*, (4 March 2006), http://www.atimes.com/atimes/China_Business/HC04Cb05.html.
Su, S., "China's Unrefined Oil Data", *The Wall Street Journal*, (13 February 2007), 32.
Wang Y. and S. Engle., "Pakistan Considers Building Gas Pipeline to China", *Bloomberg.com*, (23 April 2007), http://quote.bloomberg.com/apps/news?pid=20601087&sid=agFWKJ36isgg.

"500b-Yuan Loss from Sulfur Cloud", *South China Morning Post*, (4 August 2006).
"ABB delivers longest underwater power link", *ABB Corporate*, (12 September 2008), http://www.abb.com/cawp/seitp202/9448d40b7659aeecc12574c5003805cc.aspx.
"Ausstoß von Treibhausgasen steigt weiter an", *Die Welt*, vol. 18, (November 2008), 9. China Customs Statistics Yearbooks (1991–2001, 2006–2008)
"A survey of oil", *The Economist*, vol. 375, no. 8424, (2005).
"China, Iran Sign Valuable Oil & Gas Contract", *Energy Bulletin*, (30 October 2004), http://www.energybulletin.net/2917.html.
"China, Iran sign biggest Oil and gas Deal", *China Daily*, (31 October 2004).
"China Lauds India-Pakistan Peace", *BBC News Online*, (22 November 2006), http://news.bbc.co.uk/2/hi/south_asia/6171616.stm.
"China becomes net coal importer", *Xinhua News Network*, (29 May 2007), http://www2.chinadaily.com.cn/bizchina/2007-05/29/content_882370.htm.
"China overtakes U.S. in greenhouse gas emissions", *Bloomberg News*, (20 June 2007), http://www.iht.com/articles/2007/06/20/business/emit.php.

"China's solar industry ranks world's No. 1", *Shanghai Securities News*, (12 December 2007).
"China may join gas pipeline project with Iran if India opts out", *Alexander's Gas and Oil Connection*, (11 February 2008), http://gasandoil.com.
"China ready to join gas pipeline project if India stays away", *Economic Times*, (11 February 2008), http://economictimes.indiatimes.com/News/International_Business/China_ready_to_join_gas_pipeline_project_if_India_stays_away_/articleshow/2773393.cms.
"China inflation hits fresh high", *BBC News Online*, (11 March 2008), http://news.bbc.co.uk/2/hi/business/7288940.stm.
"Darfur rebel leader plans attacks on Chinese oil firms", *Sudan Tribune*, (8 December 2007).
"Darfur rebel JEM wants Western oil Cos to replace Chinese", *Sudan Tribune*, (16 April 2008).
Dow Jones Newswire, (9 February 2009).
"Energy cooperation between China and Saudi Arabia", (January 2006), http://news.tom.com/2006/01/21.
"Franco-Russian Joint Statement on the occasion of the visit to Moscow of Mr. de Villepin, French Prime Minister", (14 February 2006), http://www.diplomatie.gouv.fr.
"Free Trading Neighbours", *Beijing Review*, (30 November 2006), http://www.bjreview.com.cn/qanda/txt/2006-12/10/content_50407.htm.
"Gwadar Port Master Plan—Implications for Energy Sector Development", Report presented by *Arthur D. Little Company*, (25–26 April 2006),
Hamshari (Daily Journal in Persian)
"India strike over fuel price rise", *BBC News Online*, (5 June 2008), http://news.bbc.co.uk/2/hi/south_asia/7437071.stm.
"Inflation surpasses 12% in India", *BBC News Online*, (8 August 2008), http://news.bbc.co.uk/2/hi/business/7548694.stm.
"Iran's Chinese Air Force", *Middle East Defense News*, vol. 2, no. 4, (21 November 1988), 1–2.
"Iran-EU link two agreements in energy sector", *Pyvand's Iran New*, (19 October 2002), http://www.netnative.com.
IRIB (Islamic Republic of Iran Broadcasting)
"Japan must lead way on nuclear energy, say advisors", *Agence France Presse*, (21 March 2008), http://afp.google.com/article/ALeqM5gEHaZZKFDHEoZA8hyy2uA7z8qFyw
Jomhouri Eslami (Iranian Newspaper)
"Minister: China job picture 'grim', unrest 'top concern'", *Xinhua News Network*, (20 November 2008), http://news.xinhuanet.com/english/2008-11/20/content_10386287.htm.
"Mitsubishi Heavy wins China order for nuclear power plant equipment", *The Nikkei Weekly*,(9 October 2007).
"Nabucco may not be possible without Russia", *Alexander's Gas and Oil Connection*, vol. 12, no. 12, (29 June 2007), http://www.gasandoil.com/goc/company/cne72650.htm.
"Now What? The Kyoto Protocol won't save the planet", *TIME*, vol. 48, (14 July 2008).
"Oil imports up 12.4% to 163.17 million in 2007", *China Briefing News*, (14 January 2008), http://www.china-briefing.com/news/2008/01/14/oil-imports-up-124-percent-to-16317-million-tons-in-2007.html.
"Oil Importer Lands LNG Contract with Iran", *China Daily*, (18 March 2004), http://china.org.cn/english/BAT/90612.htm.
"Pakistan demands US nuclear deal", *BBC News Online*, (2 October 2008), http://news.bbc.co.uk/2/hi/south_asia/7648435.stm.

"Polycrystalline silicon production: toxic pollutants and high energy consumption cannot be neglected", *China Quality Promotion Tour Magazine*, (8 May 2008), (www.china5e.com)
"Renewable Energy Law Creates New Opportunity", *People's Daily*, (1 March 2005), http://english.people.com.cn/200503/01/eng20050301_175162.html.
"Russia Clinches Pipeline Deal", *BBC News Online*, (12 May 2007), http://news.bbc.co.uk/2/hi/asia-pacific/6649169.stm.
"Serve agriculture, rural areas and farmers, balance urban and rural development ", *The Economic Daily*, (12 October 2007).
"Sinopec Group in SaudiArabia", (September 2005), http://sa.mofcom.gov.cn/aarticle/zxhz/tzwl/200509/20050900458773.html.
The Hindu
"The world's longest link", *ABB Corporate*, (18 December 2007), http://www.abb.com/cawp/seitp202/59E88322489BF9FBC1257402003B65AC.
aspx.www.china5e.com www.euractiv.com/

INDEX

INTERNATIONAL COMPARATIVE SOCIAL STUDIES

ISSN 1568-4474

In modern research, breaking boundaries between the different social sciences is becoming more and more popular. Discussions in which different disciplines are being invited to shed their light on such issues as migration, violence, urbanisation, trust and social capital are common in current academic discourse. Brill's *International Comparative Social Studies* focuses on presenting the results of comparative research by anthropologists, sociologists, political scientists and other social scientists.

1. Wilson, H.T. *Bureaucratic Representation.* Civil Servants and the Future of Capitalist Democracies. 2001. ISBN 90 04 12194 3
2. Rath, J. *Western Europe and its Islam.* 2001. ISBN 90 04 12192 7
3. Inayatullah, S. *Understanding Sarkar.* The Indian Episteme, Macrohistory and Transformative Knowledge. 2002. ISBN 90 04 12193 5 (hardcover)
   ISBN 90 04 12842 5 (paperback)
4. Gelissen, J. *Worlds of Welfare, Worlds of Consent?* Public Opinion on the Welfare State. 2002. ISBN 9004 12457 8
5. Wilson, H.T. *Capitalism after Postmodernism.* Neo-Conservatism, Legitimacy, and the Theory of Public Capital. 2002. ISBN 9004 12458 6
6. Roulleau-Berger, L. *Youth and Work in the Post-Industrial City of North America and Europe.* With an Epilogue by Saskia Sassen. 2003.
   ISBN 9004 12533 7
7. Aalberg, T. *Achieving Justice.* Comparative Public Opinion on Income Distribution. 2003. ISBN 9004 12990 1
8. Arnason, J.P. *Civilizations in Dispute.* Historical Questions and Theoretical Traditions. 2003. ISBN 9004 13282 1
9. Falzon, M.-A. *Cosmopolitan Connections.* The Sindhi diaspora, 1860-2000. 2004. ISBN 9004 14008 5
10. Ben-Rafael, E. and Y. Sternberg (eds.), *Comparing Modernities Pluralism Versus Homogenity.* Essays in Homage to Shmuel N. Eisenstadt. 2005. ISBN 90 04 14407 2
11. Douw, L. and K-b. Chan (eds.), *Conflict and Innovation. Joint Ventures in China.* 2006. ISBN 90 04 15188 5
12. Smith, J. With an Introduction by S.N. Eisenstadt. *Europe and the Americas.* State Formation, Capitalism and Civilizations in Atlantic Modernity. 2006.
    ISBN 13: 978 90 04 15229 8. ISBN-10: 90 04 15229 6.

13. Ben-Rafael, E., M. Lyubansky, O. Glöckner, P. Harris, Y. Israel, W. Jasper and J. Schoeps. *Building a Diaspora.* Russian Jews in Israel, Germany and the USA. 2006. ISBN-13: 978 90 04 15332 5. ISBN-10: 90 04 15332 2.
14. Arjomand, S.A. (ed.),*Constitutionalism and Political Reconstruction.* 2007. ISBN-13: 978 90 04 15174 1. ISBN-10: 90 04 15174 5.
15. Kwok-bun, C., J.W. Walls and D. Hayward (eds.), *East-West Identities.* Globalization, Localization, and Hybridization. 2007. ISBN-13: 978 90 04 15169 7. ISBN-10: 90 04 15169 9.
16. Meulemann, H. (ed.), *Social Capital in Europe: Similarity of Countries and Diversity of People?* Multi-level Analyses of the European Social Survey 2002. 2008. ISBN 978 90 04 16362 1.
17. Roberts, C.W. *"The" Fifth Modality: On Languages that Shape our Motivations and Cultures.* 2008. ISBN 978 90 04 16235 8.
18. Rakel, E.P. *Power, Islam, and Political Elite in Iran.* A Study on the Iranian Political Elite from Khomeini to Ahmadinejad. 2009. ISBN 978 90 04 17176 3.
19. Ben-Rafael, E. and Y. Sternberg (eds.), with Judit Bokser Liwerant and Yosef Gorny. *Transnationalism.* Diasporas and the advent of a new (dis)order. 2009. ISBN 978 90 04 17470 2
20. Stefan, A.M. *Democratization and Securitization.* The Case of Romania. 2009. ISBN 978 90 04 17739 0
21. Amineh, M.P. and Y. Guang (eds.), *The Globalization of Energy.* China and the European Union. 2010. ISBN 978 90 04 18112 0.
22. Smith, K.E. *Meaning, Subjectivity, Society.* Making Sense of Modernity. 2010. ISBN 978 90 04 18172 4